T0281535

Solution Techniques for Elementary Partial Differential Equations

"In my opinion, this is quite simply the best book of its kind that I have seen thus far."
—Professor Peter Schiavone, University of Alberta, from the Foreword to the Fourth Edition

Praise for the previous editions

"An ideal tool for students taking a first course in PDEs, as well as for the lecturers who teach such courses."
—Marian Aron, Plymouth University, UK

"This is one of the best books on elementary PDEs this reviewer has read so far. Highly recommended."
—CHOICE

Solution Techniques for Elementary Partial Differential Equations, Fourth Edition remains a top choice for a standard, undergraduate-level course on partial differential equations (PDEs). It provides a streamlined, direct approach to developing students' competence in solving PDEs, and offers concise, easily understood explanations and worked examples that enable students to see the techniques in action.

New to the Fourth Edition

- Two additional sections
- A larger number and variety of worked examples and exercises
- A companion pdf file containing more detailed worked examples to supplement those in the book, which can be used in the classroom and as an aid to online teaching

Christian Constanda, MS, PhD, DSc, is the Charles W. Oliphant Endowed Chair in Mathematical Sciences and director of the Center for Boundary Integral Methods at the University of Tulsa. He is also an emeritus professor at the University of Strathclyde and chairman of the International Consortium on Integral Methods in Science and Engineering. He is the author/editor of more than 30 books and more than 150 journal papers. His research interests include boundary value problems for elastic plates with transverse shear deformation, direct and indirect integral equation methods for elliptic problems and time-dependent problems, and variational methods in elasticity.

Solution Techniques for Elementary Partial Differential Equations

Fourth Edition

Christian Constanda
The University of Tulsa, USA

CRC Press
Taylor & Francis Group
Boca Raton London New York

CRC Press is an imprint of the
Taylor & Francis Group, an **informa** business

A CHAPMAN & HALL BOOK

Fourth edition published 2023

by CRC Press
6000 Broken Sound Parkway NW, Suite 300, Boca Raton, FL 33487-2742

and by CRC Press
4 Park Square, Milton Park, Abingdon, Oxon, OX14 4RN

© 2023 Christian Constanda

First edition published by CRC Press 2002
Second edition published by CRC Press 2010
Third edition published by CRC Press 2016

CRC Press is an imprint of Taylor & Francis Group, LLC

LCCN no. 2022007080

ISBN: 978-1-032-00166-1 (hbk)
ISBN: 978-1-032-00031-2 (pbk)
ISBN: 978-1-003-17304-5 (ebk)

DOI: 10.1201/9781003173045

Typeset in CMR10
by KnowledgeWorks Global Ltd.

Publisher's note: This book has been prepared from camera-ready copy provided by the authors.

Access the Support Material: www.Routledge.com/9781032001661

For Lia.
Also for Jack, Emily, and Archie.

Contents

Foreword

It is often difficult to persuade undergraduate students of the importance of mathematics. Engineering students in particular, geared towards the practical side of learning, often have little time for theoretical arguments and abstract thinking. In fact, mathematics is the language of engineering and applied science. It is the vehicle by which ideas are analyzed, developed, and communicated. It is no accident, therefore, that any undergraduate engineering curriculum requires several mathematics courses, each one designed to provide the necessary analytic tools to deal with questions raised by engineering problems of increasing complexity, for example in the modeling of physical processes and phenomena. The most effective way to teach students how to use these mathematical tools is by example. The more worked examples and practice exercises a textbook contains, the more effective it will be in the classroom.

Such is the case with *Solution Techniques for Elementary Partial Differential Equations* by Christian Constanda. The author, a skilled classroom performer with considerable experience, understands exactly what students want and has given them just that: a textbook that explains the essence of the method briefly and then proceeds to show it in action. The book contains a wealth of worked examples and exercises (half of them with answers) and is accompanied by an Instructor's Manual with full solutions to each problem, as well as a pdf file for use on a computer-linked projector. A new and very helpful element added to the third edition is the insertion of *Mathematica*® software code at the end of each worked example, which shows how the computed solution to a problem can be verified by computer. The distribution of the exercises to each relevant section and subsection instead of having them bunched up together at the end of the chapter is also a welcome development.

In my opinion, this is quite simply the best book of its kind that I have seen thus far. The book not only contains solution methods for some very important classes of PDEs, in easy-to-read format, but is also student-friendly and teacher-friendly at the same time. It is definitely a textbook for adoption.

Professor Peter Schiavone
Department of Mechanical Engineering
University of Alberta
Edmonton, AB, Canada

Preface to the Fourth Edition

The changes implemented in this edition include misprint repairs, some text improvements, the addition of two new sections, and an increase in the number and variety of worked examples and exercises. These two categories have been augmented by 70 and 130 items, respectively, with the latter including many more representative ones as replacements for the older, less user-friendly, versions. The exercises that require computational handling because they are not solvable by elementary means are identified by italicized numerical labels.

A new feature is a companion pdf file containing worked examples that supplement those in the book. Bearing the same numbers as their counterparts in the book and marked by one or several asterisks, they solve the same type of problem but with different data and much greater detail in the solution. I created these additional examples as a tool for teaching online during the fall of 2020 and spring of 2021. They can be used with a projector or solved by hand in the classroom, and can also prove very helpful in distance learning. This file will be made available to all instructors who have adopted the book.

I would like to thank Callum Fraser at Taylor & Francis for his assistance with this project, and Mansi Kabra, Michele Dimont and the other members of the production team, who handled the text with competence and care, observing all my special instructions.

My students, in particular Andrew Gray and his classroom colleagues, have been very helpful chasing misprints and making comments on the structure and readability of the material.

Special thanks are due to my wife, whose unflinching support has been truly commendable. Also, I wish to acknowledge the interest in science and learning shown by my very young grandchildren, of whom much is expected in years to come, for they are the ones who will take mathematical modelling one step closer to our complete understanding of the universe.

Christian Constanda
The University of Tulsa
January 2022

Preface to the Third Edition

Apart from some cosmetic adjustments, the changes in the third edition are aimed at making the book more user-friendly.

- Sections 2.4 and 11.5, and Subsections 5.2.3 and 8.1.3, are new.

- Many other sections (2.2, 3.1, 5.1, 5.2, 5.3, 5.5, 6.1, 6.2, 8.1, 8.2, 9.2, 10.1, 11.1, 12.3, and 12.4) have been split into several subsections, to make it easier for the reader to navigate the contents.

- The exercises have been distributed to each section/subsection, as appropriate, instead of being listed at the end of the chapters.

- The lists of exercises have been revamped, with some replacements and some new ones added in to enhance variety. The same treatment has been applied to the worked examples. The number of both has increased slightly (exercises from 604 to 615 and worked examples from 143 to 159).

- Almost all of the worked examples end with a brief *Mathematica*® program that shows how the results can be verified by computer.

The last item is meant to offer the students a way of checking the accuracy of their work without interfering with the solution procedure, which in this book, with a few exceptions, still has to be carried out by hand. Students have to come to terms with the fact that a computer is only a tool doing what its human handler instructs it to do. To write a program that solves a given analytic problem requires solid theoretical expertise and practical experience with the solution method, both acquired only by first energizing the time-tested combination of brain, hand, pencil, and paper.

The exceptions referred to in the preceding paragraph are some of the worked examples and the exercises with an italicized number. These need computer help for their solution, and in all of them, the adopted convention is that numerical results are rounded to the fourth decimal place. Also, to keep the notation simple, and without danger of ambiguity, the equality sign is used where such numbers occur, instead of the approximate equality symbol.

I would like to express my thanks to Sunil Nair and Alex Edwards at Taylor & Francis for their assistance with various issues during the completion of this project, and to Shashi Kumar and Marcus Fontaine for helping me tweak the LaTeX class file until it bowed to my aesthetic demands.

Classroom users of the book also deserve acknowledgement for their comments and constructive suggestions. I would be very pleased to continue receiving feedback from them and, in particular, to hear if anyone has discovered how to make students stop misusing the word 'like'.

I don't have enough words of praise and gratitude for my wife, who stoically endured my numerous episodes of computer time and showed no ill feelings toward my laptop.

Finally, special thanks are due to my grandson Jack, who, with his cheerful nature and nascent curiosity, provided many a welcome moment of distraction and entertainment during the tedious proofreading process. I hope that, in the fullness of time, he will be able to understand the content of this book and make good use of it.

Christian Constanda
The University of Tulsa
October 2015

Preface to the Second Edition

Following constructive suggestions received from some of the users of the book, I have made a number of changes in the second edition.

- Section 1.4 (Cauchy–Euler Equations) has been added to Chapter 1.

- Chapter 3 contains three new sections: 3.3 (Bessel Functions), 3.4 (Legendre Polynomials), and 3.5 (Spherical Harmonics).

- The new Section 4.4 in Chapter 4 lists additional mathematical models based on partial differential equations.

- Sections 5.4 and 7.4 have been added to Chapters 5 and 7, respectively, to show—by means of examples—how the methods of separation of variables and eigenfunction expansion work on equations other than heat, wave, and Laplace.

- Supplementary applications of the Fourier transformations are now shown in Section 8.3.

- The method of characteristics is applied to more general hyperbolic equations in the additional Section 12.4.

- Chapter 14 (Complex Variable Methods) is entirely new.

- The number of worked examples has increased from 110 to 143, and that of the exercises has almost quadrupled—from 165 to 604.

- The tables of Fourier and Laplace transforms in the Appendix have been considerably enhanced.

- The first coefficient of the Fourier series is now $\frac{1}{2} a_0$ instead of the previous a_0. Similarly, the direct and inverse full Fourier transformations are now defined with the normalizing factor $1/\sqrt{2\pi}$ in front of the integral; the Fourier sine and cosine transformations are defined with the factor $\sqrt{2/\pi}$.

While I still believe that students should be encouraged not to use electronic computing devices in their learning of the fundamentals of partial differential equations, I have made a concession when it comes to examples and exercises that involve special functions or exceedingly lengthy integration. The (new) exercises that require computational help because they are not solvable by elementary means have been given italicized numerical labels.

I wish to thank all the readers who sent me their comments and urge them to continue to do so in the future. It is only with their help that this book may undergo further improvement.

I would also like to thank Sunil Nair and Sarah Morris at Taylor & Francis for their professional and expeditious handling of this project.

Christian Constanda
The University of Tulsa
January 2010

Preface to the First Edition

There are many textbooks on partial differential equations on the market. The great majority of them are well written and very rigorous, with full background explanations, detailed proofs, and lots of comments. But they also tend to be rather voluminous and daunting for the average student. When I ask the undergraduates what they want from a book, their most common answers are (i) to understand without excessive effort most of what is being said; (ii) to be given full yet concise explanations of the essence of the topics discussed, in simple words; (iii) to have many worked examples, preferably of the type found in test papers, so they could learn the various techniques by seeing them in action and thus improve their chances of passing examinations; and (iv) to pay as little as possible for it in the bookstore. I do not wish to comment on the validity of these answers, but I am prepared to accept that even in higher education the customer may sometimes be right.

This book is an attempt to meet all the above requirements. It is designed as a no-frills text that explains a number of major methods completely but succinctly, in everyday classroom language. It does not indulge in multi-page, multicolored spiels. It includes many practical applications with solutions, and exercises with selected answers. It has a reasonable number of pages and is produced in a format that facilitates digital reproduction, thus helping keep costs down.

Teachers have their own individual notions regarding what makes a book ideal for use in coursework. They say—with good reason—that the perfect text is the one they themselves sketched in their classroom notes but never had the time or inclination to polish up and publish. We each choose our own material, the order in which the topics are presented, and how long we spend on them. This book is no exception. It is based on my experience of the subject for many years and the feedback received from my teaching's beneficiaries. The "use in combat" of an earlier version seems to indicate that average students can work from it independently, with some occasional instructor guidance, while the high flyers get a basic and rapid grounding in the fundamentals of the subject before progressing to more advanced texts (if they are interested in further details and want to get a truly sophisticated picture of the field). A list, by no means exhaustive, of such texts can be found in the Bibliography.

This book contains no example or exercise that needs a calculating device in its solution. Computing machines are now part of everyday life and we all use them routinely and extensively. However, I believe that if you really want to learn what mathematical analysis is all about, then you should exercise your mind and hand the long way, without any electronic help. (In fact, it seems that quite a few of my students are convinced that computers are better used for surfing the Internet than for solving homework problems.) The only prerequisites for reading this book are a first course in calculus and some basic knowledge of certain types of ordinary differential equations.

The topics are arranged in the order I have found to be the most convenient. After some essential but elementary ordinary differential equations, Fourier series, and Sturm-Liouville problems are discussed briefly, the heat, Laplace, and wave equations are introduced in quick succession as mathematical models of physical phenomena, and then a number of methods (separation of variables, eigenfunction expansion, Fourier and Laplace transformations, and Green's functions) are applied in turn to specific initial/boundary value problems for each of these equations. There follows a brief discussion of the general second-order linear equation with two independent variables. Finally, the method of characteristics and perturbation (asymptotic expansion) methods are presented. A number of useful tables and formulas are listed in the Appendix.

The style of the text is terse and utilitarian. In my experience, the teacher's classroom performance does more to generate undergraduate enthusiasm and excitement for a topic than the cold words in a book, however skillfully crafted. Since the aim here is to get the students well drilled in the main solution techniques and not in the physical interpretation of the results, the latter hardly gets a mention. The examples and exercises are formal, and in many of them the chosen data may not reflect plausible real-life situations. Due to space pressure, some intermediate steps—particularly the solutions of simple ODEs—are given without full working. It is assumed that the readers know how to derive them, or that they can refer without difficulty to the summary provided in Chapter 1. Personally, in class I always go through the full solution regardless, which appears to meet with the approval of the audience. Details of a highly mathematical nature, including formal proofs, are kept to a minimum, and when they are given, an assumption is made that any conditions required by the context (for example, the smoothness and behavior of functions) are satisfied.

The textbook will be accompanied by an Instructor's Manual containing the complete solutions of all the exercises. Also, on adoption of the book, a pdf file of the text (in suitably large sans-serif fonts) will be made available to instructors for use on classroom projectors.

My own lecturing routine consists in (i) using a projector to present a skeleton of the theory, so the students do not need to take notes and can follow the live explanations, and (ii) doing a selection of examples on the board with full details, which the students take down by hand. I found that this sequence of 'talking periods' and 'writing periods' helps the audience to maintain concentration and makes the lecture more enjoyable (if what the end-of-semester evaluations say is true).

Wanting to offer students complete, rigorous, and erudite expositions is highly laudable, but the market priorities appear to have shifted of late. With the current standards of secondary education manifestly lower than in the past, students come to us less and less equipped to tackle the learning of mathematics from a fundamental point of view. When this becomes unavoidable, they seem to prefer a concise text that shows them the method and then, without fuss and niceties of form, goes into as many worked examples as possible. Whether we like it or not, it seems that we have entered the era of the digest. It is to this uncomfortable reality that the present book seeks to offer a solution.

The last stages of preparation of this book were completed while I was a Visiting Professor in the Department of Mathematical and Computer Sciences at the University of Tulsa. I wish to thank the authorities of this institution and the faculty in the department for providing me with the atmosphere, conditions, and necessary facilities to finish the work on time. Particular thanks go to the following: Bill Coberly, the head of the department, who helped me engineer several summer visits and a couple of successful sabbatical years in Tulsa; Pete Cook, who heard my daily moans and groans from across the corridor and did not complain about it; Dale Doty, the resident Mathematica wizard who drew some of the figures and showed me how to do the others; and the *sui generis* company at the lunch table in the Faculty Club for whom, in time-honored academic fashion, no discussion topic was too trivial or taboo and no explanation too implausible.

I also wish to thank Sunil Nair, Helena Redshaw, Andrea Demby, and Jasmin Naim from Chapman & Hall/CRC for their help with technical advice and flexibility over deadlines.

Finally, I would like to state for the record that this book project would not have come to fruition had I not had the full support of my wife, who, not for the first time, showed a degree of patience and understanding far beyond the most reasonable expectations.

Christian Constanda

CHAPTER 1

Ordinary Differential Equations: Brief Review

In the process of solving partial differential equations (PDEs), we usually reduce the problem to the solution of certain classes of ordinary differential equations (ODEs). Here, we mention without proof some basic methods for integrating simple ODEs of the types encountered later in the text. We restrict our attention to *real* solutions of ODEs with *real* coefficients. In what follows, the set of real numbers is denoted by $\mathbb{R}$.

1.1 First-Order Equations

Separable equations. The general form of this type of ODE is

$$y' = \frac{dy}{dx} = f(x)g(y).$$

Taking standard precautions, we can rewrite the equation as

$$\frac{dy}{g(y)} = f(x)dx$$

and then integrate each side with respect to its corresponding variable.

1.1 Example. For the equation

$$y^2 y' - 2x = 0,$$

the above procedure leads to

$$\int y^2 \, dy = 2 \int x \, dx,$$

which yields

$$\tfrac{1}{3} y^3 = x^2 + c, \quad c = \text{const},$$

or

$$y(x) = (3x^2 + C)^{1/3}, \quad C = \text{const}.$$

Linear equations. Their general (normal) form is

$$y' + p(x)y = q(x),$$

where p and q are given functions. Computing an integrating factor μ by means of the formula

$$\mu(x) = \exp\left\{\int p(x)dx\right\},$$

we obtain the general solution

$$y(x) = \frac{1}{\mu(x)} \int \mu(x)q(x)\, dx.$$

An equivalent formula for the general solution is

$$y(x) = \frac{1}{\mu(x)}\left[\int_a^x \mu(t)q(t)\, dt + C\right], \quad C = \text{const},$$

where a is any point in the interval where the ODE is satisfied.

1.2 Example. The normal form of the equation

$$xy' + 2y - x^2 = 0, \quad x \neq 0,$$

is

$$y' + \frac{2}{x} y = x$$

with

$$p(x) = \frac{2}{x}, \quad q(x) = x,$$

so an integrating factor is

$$\mu(x) = \exp\left\{2\int \frac{dx}{x}\right\} = e^{\ln x^2} = x^2.$$

Then the general solution of the equation is

$$y(x) = \frac{1}{x^2} \int x^3\, dx = \tfrac{1}{4} x^2 + \frac{C}{x^2}, \quad C = \text{const}.$$

Exercises

Find the general solution of the given equation.

1 $(x^2 + 1)y' = 2xy$.

2 $y' - 3x^2(y + 1) = 0$.

3 $(x - 1)y' + 2y = x$, $x \neq 1$.

4 $x^2y' - 2xy = x^5 e^x$.

Answers to Odd-Numbered Exercises

1 $y(x) = C(x^2 + 1)$.

3 $y(x) = (x - 1)^{-2}(x^3/3 - x^2/2 + C)$.

1.2 Homogeneous Linear Equations with Constant Coefficients

First-order equations. These are equations of the form

$$y' + ay = 0, \quad a = \text{const.}$$

Such equations can be solved by means of an integrating factor or separation of variables, or by means of the characteristic equation

$$s + a = 0,$$

whose root $s = -a$ yields the general solution

$$y(x) = Ce^{-ax}, \quad C = \text{const.}$$

1.3 Example. The characteristic equation for the ODE

$$y' - 3y = 0$$

is

$$s - 3 = 0;$$

hence, the general solution of the equation is

$$y(x) = Ce^{3x}, \quad C = \text{const.}$$

Second-order equations. Their general form is

$$y'' + ay' + by = 0, \quad a, b = \text{const.}$$

If the characteristic equation

$$s^2 + as + b = 0$$

has two distinct real roots s_1 and s_2, then the general solution of the given ODE is

$$y(x) = C_1 e^{s_1 x} + C_2 e^{s_2 x}, \quad C_1, C_2 = \text{const.}$$

If $s_1 = s_2 = s_0$, then

$$y(x) = (C_1 + C_2 x)e^{s_0 x}, \quad C_1, C_2 = \text{const.}$$

Finally, if s_1 and s_2 are complex conjugate—that is, $s_1 = \alpha + i\beta$, $s_2 = \alpha - i\beta$, where α and β are real numbers, then the general solution is

$$y(x) = e^{\alpha x}[C_1 \cos(\beta x) + C_2 \sin(\beta x)], \quad C_1, C_2 = \text{const.}$$

1.4 Remarks. (i) When $s_1 = -s_2 = s_0$, s_0 real, the general solution of the equation can also be written as

$$y(x) = C_1 y_1(x) + C_2 y_2(x), \quad C_1, C_2 = \text{const},$$

where $y_1(x)$ and $y_2(x)$ are any two of the functions

$$\cosh\big(s_0(x - \alpha)\big), \quad \sinh\big(s_0(x - \alpha)\big), \quad \cosh\big(s_0(x - \beta)\big), \quad \sinh\big(s_0(x - \beta)\big)$$

and α and β are any distinct real numbers. Normally, α and β are chosen as the points where boundary conditions are given.

(ii) This can be generalized to the case of any two distinct real roots s_1 and s_2, where direct computation shows that $y_1(x)$ and $y_2(x)$ can be any two of the functions

$$e^{(s_1+s_2)(x-\alpha)/2} \cosh\left[\tfrac{1}{2}(s_1 - s_2)(x - \alpha)\right],$$

$$e^{(s_1+s_2)(x-\alpha)/2} \sinh\left[\tfrac{1}{2}(s_1 - s_2)(x - \alpha)\right],$$

$$e^{(s_1+s_2)(x-\beta)/2} \cosh\left[\tfrac{1}{2}(s_1 - s_2)(x - \beta)\right],$$

$$e^{(s_1+s_2)(x-\beta)/2} \sinh\left[\tfrac{1}{2}(s_1 - s_2)(x - \beta)\right].$$

1.5 Example. The characteristic equation for the ODE

$$y'' - 3y' + 2y = 0$$

is

$$s^2 - 3s + 2 = 0,$$

with roots $s_1 = 1$ and $s_2 = 2$, so the general solution of the ODE is

$$y(x) = C_1 e^x + C_2 e^{2x}, \quad C_1, C_2 = \text{const.}$$

1.6 Example. The general solution of the equation

$$y'' - 4y = 0$$

is

$$y(x) = C_1 e^{2x} + C_2 e^{-2x}, \quad C_1, C_2 = \text{const},$$

since the roots of the characteristic equation are $s_1 = -s_2 = 2$. According to Remark 1.4(i), we have alternative expressions for this in terms of hyperbolic functions. Thus, if the ODE is accompanied by, say, the boundary conditions

$$y(0) = -1, \quad y'(1) = 6,$$

then, with $\alpha = 0$ and $\beta = 1$, a useful way to write the general solution is

$$y(x) = C_1 \sinh(2x) + C_2 \cosh(2(x-1)), \quad C_1, C_2 = \text{const},$$

because $\sinh(2x)$ is equal to zero at $x = 0$ and the derivative of $\cosh(2(x-1))$ is equal to zero at $x = 1$, which helps us determine C_1 and C_2 very easily as

$$C_1 = 3\operatorname{sech} 2, \quad C_2 = -\operatorname{sech} 2.$$

1.7 Example. The roots of the characteristic equation for the ODE

$$y'' - 4y' + y = 0$$

are $s_1 = 2 + \sqrt{3}$ and $s_2 = 2 - \sqrt{3}$. Then

$$s_1 + s_2 = 4, \quad s_1 - s_2 = 2\sqrt{3},$$

so, if $y(0)$ and $y(1)$ are prescribed, we may prefer to use the statement in Remark 1.4(ii) with $\alpha = 0$ and $\beta = 1$ and write the general solution as

$$y(x) = C_1 e^{2x} \sinh\left(\sqrt{3}\,x\right) + C_2 e^{2(x-1)} \sinh\left(\sqrt{3}\,(x-1)\right), \quad C_1, C_2 = \text{const}.$$

1.8 Example. Since the roots of the characteristic equation for the ODE

$$y'' + 4y' + 4y = 0$$

are $s_1 = s_2 = -2$, the general solution is

$$y(x) = (C_1 + C_2 x)e^{-2x}, \quad C_1, C_2 = \text{const}.$$

1.9 Example. The characteristic equation for the ODE

$$y'' + 2y' + 10y = 0$$

has the roots $s_1 = -1 + 3i$ and $s_2 = -1 - 3i$, so the general solution of the ODE is

$$y(x) = e^{-x}\left[C_1 \cos(3x) + C_2 \sin(3x)\right], \quad C_1, C_2 = \text{const}.$$

1.10 Remark. The characteristic equation method can also be used to find the general solution of homogeneous linear ODEs of higher order.

1.11 Example. The characteristic equation for the ODE

$$y''' - 3y'' + 2y = 0$$

is

$$r^3 - 3r + 2 = 0,$$

with roots $r_1 = -2$ and $r_2 = r_3 = 1$, so its general solution is

$$y(x) = C_1 e^{-2x} + (C_2 + C_3 x)e^x.$$

Exercises

Find the general solution of the given equation.

1 $2y' + 5y = 0$.   **2** $3y' - 2y = 0$.

3 $y'' - 4y' + 3y = 0$.   **4** $2y'' - 5y' + 2y = 0$.

5 $4y'' + 4y' + y = 0$.   **6** $y'' - 6y' + 9y = 0$.

7 $y'' + 2y' + 5y = 0$.   **8** $y'' - 6y' + 13y = 0$.

9 $y'' - 16y = 0$.   **10** $4y'' - 9y = 0$.

Answers to Odd-Numbered Exercises

1 $y(x) = Ce^{-5x/2}$.   **3** $y(x) = C_1 e^x + C_2 e^{3x}$.

5 $y(x) = (C_1 + C_2 x)e^{-x/2}$.   **7** $y(x) = e^{-x}[C_1 \cos(2x) + C_2 \sin(2x)]$.

9 $y(x) = C_1 e^{4x} + C_2 e^{-4x}$, or

$y(x) = C_1 \cosh\left(4(x - \alpha)\right) + C_2 \cosh\left(4(x - \beta)\right)$, or

$y(x) = C_1 \cosh\left(4(x - \alpha)\right) + C_2 \sinh\left(4(x - \beta)\right)$, or

$y(x) = C_1 \sinh\left(4(x - \alpha)\right) + C_2 \cosh\left(4(x - \beta)\right)$, or

$y(x) = C_1 \sinh\left(4(x - \alpha)\right) + C_2 \sinh\left(4(x - \beta)\right)$,

where α and β are any distinct real numbers.

1.3 Nonhomogeneous Linear Equations with Constant Coefficients

The first-order equations in this category are of the form

$$y' + ay = f, \quad a = \text{const},$$

and the second-order equations can be written as

$$y'' + ay' + by = f, \quad a, b = \text{const},$$

where f is a given function. The general solution of such equations is the sum of the complementary function (the general solution of the corresponding homogeneous equation) and a particular integral (a particular solution of the nonhomogeneous equation). The latter is usually guessed from the structure of the function f, or may be found by some other method, such as variation of parameters.

1.12 Example. The complementary function for the ODE

$$y' - 3y = e^{-x}$$

is

$$y_{\text{CF}}(x) = Ce^{3x}, \quad C = \text{const}.$$

Guided by the nonhomogeneous term, we seek a particular integral of the form

$$y_{\text{PI}}(x) = ae^{-x}, \quad a = \text{const}.$$

Replacing it in the equation and collecting the like terms, we have

$$-4ae^{-x} = e^{-x},$$

so $a = -1/4$. Consequently, the general solution of the given ODE is

$$y(x) = Ce^{3x} - \tfrac{1}{4}e^{-x}, \quad C = \text{const}.$$

1.13 Example. If the function on the right-hand side in Example 1.12 is replaced by e^{3x}, then we cannot find a particular integral of the form ae^{3x}, $a = \text{const}$, since this is a solution of the corresponding homogeneous equation. Instead, we try

$$y_{\text{PI}}(x) = axe^{3x}, \quad a = \text{const},$$

and deduce, by replacing in the ODE, that $a = 1$; consequently, the general solution is

$$y(x) = Ce^{3x} + xe^{3x} = (C + x)e^{3x}, \quad C = \text{const}.$$

1.14 Example. For the equation

$$y'' - 4y' + 13y = 28 - 5x - 13x^2,$$

we seek a particular integral of the form

$$y_{\text{PI}}(x) = ax^2 + bx + c, \quad a, b, c = \text{const}.$$

Proceeding as in Example 1.12, we arrive at the equality

$$2a - 4(2ax + b) + 13(ax^2 + bx + c) = 28 - 5x - 13x^2,$$

which yields $a = -1$, $b = -1$, and $c = 2$. Since the complementary function is

$$y_{CF}(x) = e^{2x}\left[C_1 \cos(3x) + C_2 \sin(3x)\right],$$

it follows that the general solution of the given ODE is

$$y(x) = e^{2x}\left[C_1 \cos(3x) + C_2 \sin(3x)\right] - x^2 - x + 2, \quad C_1, C_2 = \text{const.}$$

1.15 Example. If

$$y'' + 2y' - 3y = 10 \sin x,$$

then

$$y_{CF}(x) = C_1 e^x + C_2 e^{-3x}.$$

Here, the method of undetermined coefficients suggests that we look for a particular integral of the form

$$y_{PI}(x) = a \cos x + b \sin x, \quad a, b = \text{const.}$$

Replaced in the equation, this leads to

$$(-4a + 2b) \cos x + (-2a - 4b) \sin x = 10 \sin x,$$

from which $a = -1$ and $b = -2$. Hence,

$$y_{PI}(x) = -\cos x - 2 \sin x,$$

so the general solution of the equation is

$$y(x) = C_1 e^x + C_2 e^{-3x} - \cos x - 2 \sin x, \quad C_1, C_2 = \text{const.}$$

Exercises

Find the general solution of the given equation.

1 $y' + 2y = 2x + e^{4x}$.

2 $2y' - 3y = -3x - 4 + e^x$.

3 $2y' - y = e^{x/2}$.

4 $y' + y = -x + 2e^{-x}$.

5 $y'' - 2y' - 8y = 4 + 4x - 8x^2$.

6 $y'' - y = x^2 - x + 2$.

7 $y'' - 25y = 30e^{-5x}$.

8 $4y'' + y = 8 \cos(x/2)$.

Answers to Odd-Numbered Exercises

1 $y(x) = Ce^{-2x} + x - 1/2 + e^{4x}/6$.

3 $y(x) = (C + x/2)e^{x/2}$.

5 $y(x) = C_1 e^{-2x} + C_2 e^{4x} + x^2 - x$.

7 $y(x) = C_1 e^{5x} + (C_2 - 3x)e^{-5x}$, or

$y(x) = C_1 \cosh(5x) + C_2 \cosh\left(5(x - \alpha)\right) - 3xe^{-5x}$, or

$y(x) = C_1 \cosh(5x) + C_2 \sinh\left(5(x - \alpha)\right) - 3xe^{-5x}$, or

$y(x) = C_1 \sinh(5x) + C_2 \cosh\left(5(x - \alpha)\right) - 3xe^{-5x}$, or

$y(x) = C_1 \sinh(5x) + C_2 \sinh\left(5(x - \alpha)\right) - 3xe^{-5x}$,

where α is any real number.

1.4   Cauchy–Euler Equations

The general form of a second-order Cauchy–Euler equation is

$$x^2 y'' + \alpha xy' + \beta y = 0, \quad \alpha, \beta = \text{const},$$

where, for simplicity, we assume that $x > 0$. We try a solution of the form

$$y(x) = x^r, \quad r = \text{const}.$$

Substituting in the equation, we arrive at

$$r^2 + (\alpha - 1)r + \beta = 0.$$

If the roots r_1 and r_2 of this quadratic equation are real and distinct, which is the case of interest for us, then the general solution of the given ODE is

$$y(x) = C_1 x^{r_1} + C_2 x^{r_2}, \quad C_1, C_2 = \text{const}.$$

1.16 Example. Following the above procedure, we see that the differential equation

$$2x^2 y'' + xy' - y = 0$$

yields $r_1 = 1$ and $r_2 = -1/2$, so its general solution is

$$y(x) = C_1 x + C_2 x^{-1/2}, \quad C_1, C_2 = \text{const}.$$

Exercises

Find the general solution of the given equation.

1 $2x^2 y'' + xy' - 3y = 0$.   **2** $x^2 y'' + 2xy' - 6y = 0$.

Answers to Odd-Numbered Exercises

1 $y(x) = C_1 x^{3/2} + C_2 x^{-1}$.

1.5 Functions and Operators

Throughout this book, we refer to a function as either f or $f(x)$, although, strictly speaking, the latter denotes the value of f at x. To avoid complicated notation, we also write $f(x) = c$ to designate a function f that takes the same value $c = $ const at all points x in its domain. When $c = 0$, we sometimes simplify this further to $f = 0$.

In the preceding sections we mentioned *linear* equations. Here we clarify the meaning of this concept.

1.17 Definition. Let $\mathcal{X}$ be a space of functions, and let L be an operator acting on the functions in $\mathcal{X}$ according to some rule. The operator L is called *linear* if

$$L(c_1 f_1 + c_2 f_2) = c_1(Lf_1) + c_2(Lf_2)$$

for any functions f_1, f_2 in $\mathcal{X}$ and any numbers c_1, c_2. Otherwise, L is called *nonlinear*. The space $\mathcal{X}$ will not normally be specified explicitly, its nature being inferred from the context.

1.18 Example. The operators of differentiation and definite integration acting on suitable functions of one independent variable are linear, since

$$L(c_1 f_1 + c_2 f_2) = (c_1 f_1 + c_2 f_2)' = c_1 f_1' + c_2 f_2' = c_1(Lf_1) + c_2(Lf_2),$$

$$L(c_1 f_1 + c_2 f_2) = \int_a^b \left[c_1 f_1(x) + c_2 f_2(x) \right] dx$$

$$= c_1 \int_a^b f_1(x)\, dx + c_2 \int_a^b f_2(x)\, dx = c_1(Lf_1) + c_2(Lf_2).$$

1.19 Example. Let α, β, and γ be given functions. In view of the preceding example, it is easy to verify that the operator L defined by

$$Lf = \alpha f'' + \beta f' + \gamma f$$

is linear.

1.20 Example. Let L be the operator defined for all continuously differentiable functions on $\mathbb{R}$ by

$$Lf = ff'.$$

Since the particular choice $c_1 = 1$, $c_2 = -1$, $f_1(x) = x$, $f_2(x) = 1$ yields

$$L(c_1 f_1 + c_2 f_2) = L(f_1 - f_2) = (f_1 - f_2)(f_1 - f_2)' = x - 1,$$

$$c_1 Lf_1 + c_2 Lf_2 = Lf_1 - Lf_2 = f_1 f_1' - f_2 f_2' = x \neq L(c_1 f_1 + c_2 f_2),$$

we conclude that L is nonlinear.

1.21 Definition. A differential equation of the form

$$Lu = g,$$

where L is a linear differential operator and g is a prescribed function, is called a *linear* equation. If the operator L is nonlinear, then the equation is also called *nonlinear*.

The following almost obvious result forms the basis of what is known as the *principle of superposition*.

1.22 Theorem. *If $Lu = g$ is a linear equation and u_1 and u_2 are solutions of this equation with $g = g_1$ and $g = g_2$, respectively, then $u_1 + u_2$ is a solution of the equation with $g = g_1 + g_2$; in other words, if*

$$Lu_1 = g_1, \quad Lu_2 = g_2,$$

then

$$L(u_1 + u_2) = g_1 + g_2.$$

Proof. This follows immediately from the definition of a linear operator. □

Exercises

Verify whether the given ODE is linear or nonlinear.

1 $xy'' - y' \sin x = xe^x$.

2 $y' + 2x \sin y = 1$.

3 $y'y'' - xy = 2x$.

4 $y'' + \sqrt{x}\, y = \ln x$.

Answers to Odd-Numbered Exercises

1 Linear.

3 Nonlinear.

CHAPTER 2

Fourier Series

It is well known that an infinitely differentiable function f can be expanded in a Taylor series around a point x_0 in the interval where it is defined. This series has the form

$$f(x) \sim \sum_{n=0}^{\infty} c_n (x - x_0)^n, \qquad (2.1)$$

with the coefficients c_n given by

$$c_n = \frac{f^{(n)}(x_0)}{n!}, \quad f^{(n)} = \frac{d^n f}{dx^n}, \quad n = 1, 2, \ldots .$$

If certain conditions are satisfied, then the above series converges to f pointwise (that is, at every point x) in an open interval centered at x_0, and we can use the equality sign between the two sides in (2.1).

In this chapter we discuss a different class of expansions, which are particularly useful in the study and solution of PDEs and can be constructed for functions with a much lower degree of smoothness.

2.1   The Full Fourier Series

This is an expansion of the form

$$f(x) \sim \tfrac{1}{2} a_0 + \sum_{n=1}^{\infty} \left(a_n \cos \frac{n\pi x}{L} + b_n \sin \frac{n\pi x}{L} \right), \qquad (2.2)$$

where L is a positive number and a_0, a_n, and b_n are constant coefficients.

2.1 Definition. A function f defined on $\mathbb{R}$ is called *periodic* if there is a number $T > 0$ such that

$$f(x + T) = f(x) \quad \text{for all } x \text{ in } \mathbb{R}.$$

The smallest number T with this property is called the *fundamental period* (or, simply, the *period*) of f. It is obvious that a periodic function also satisfies

$$f(x + nT) = f(x) \quad \text{for any integer } n \text{ and all } x \text{ in } \mathbb{R}.$$

2.2 Example. As is well known, the functions $\sin x$ and $\cos x$ are periodic with period 2π since, for all $x \in \mathbb{R}$,

$$\sin(x + 2\pi) = \sin x,$$
$$\cos(x + 2\pi) = \cos x.$$

We see that for each positive integer $n = 1, 2, \ldots,$

$$\cos \frac{n\pi(x + 2L)}{L} = \cos\left(\frac{n\pi x}{L} + 2n\pi\right) = \cos \frac{n\pi x}{L},$$
$$\sin \frac{n\pi(x + 2L)}{L} = \sin\left(\frac{n\pi x}{L} + 2n\pi\right) = \sin \frac{n\pi x}{L},$$

so the right-hand side in (2.2) is periodic with period $2L$. This suggests the following method of construction for the full Fourier series of a given function.

Let f, defined on $[-L, L]$, be represented by the graph in Fig. 2.1.

Figure 2.1: $f(x)$, $-L \leq x \leq L$.

We construct the periodic extension of f from $(-L, L]$ to $\mathbb{R}$, of period $2L$ (see Fig. 2.2, which shows the three 'central' periods). The value of f at $x = -L$ is left out so that the extension is correctly defined as a function.

Figure 2.2: $f(x + 2L) = f(x)$ for all x in $\mathbb{R}$.

For the extended function f, it is now possible to seek an expansion of the form (2.2). All we need to do is compute the coefficients a_0, a_n, b_n and discuss the convergence of the series.

It is easy to check by direct calculation that

$$\int_{-L}^{L} \cos \frac{n\pi x}{L} \, dx = 0, \quad \int_{-L}^{L} \sin \frac{n\pi x}{L} \, dx = 0, \quad n = 1, 2, \ldots, \tag{2.3}$$

$$\int_{-L}^{L} \cos \frac{n\pi x}{L} \cos \frac{m\pi x}{L} \, dx = \begin{cases} 0, & n \neq m, \\ L, & n = m, \end{cases} \quad n, m = 1, 2, \ldots, \tag{2.4}$$

$$\int_{-L}^{L} \sin \frac{n\pi x}{L} \sin \frac{m\pi x}{L} \, dx = \begin{cases} 0, & n \neq m, \\ L, & n = m, \end{cases} \quad n, m = 1, 2, \ldots, \tag{2.5}$$

$$\int_{-L}^{L} \cos \frac{n\pi x}{L} \sin \frac{m\pi x}{L} \, dx = 0, \quad n, m = 1, 2, \ldots. \tag{2.6}$$

Regarding (2.2) formally as an equality, we integrate it term by term over $[-L, L]$ and use (2.3) to obtain

$$a_0 = \frac{1}{L} \int_{-L}^{L} f(x) \, dx. \tag{2.7}$$

If we multiply (2.2) by $\cos(m\pi x/L)$, integrate the new relation over $[-L, L]$, and take (2.3), (2.4), and (2.6) into account, we see that all the integrals on the right-hand side vanish except that for which the summation index n is equal to m, when the integral is equal to La_m. Replacing m by n, we find that

$$a_n = \frac{1}{L} \int_{-L}^{L} f(x) \cos \frac{n\pi x}{L} \, dx, \quad n = 1, 2, \ldots. \tag{2.8}$$

Clearly, (2.8) incorporates (2.7) if we allow n to take the value 0 as well; that is, $n = 0, 1, 2, \ldots$.

Finally, we multiply (2.2) by $\sin(m\pi x/L)$ and repeat the above procedure, where this time we use (2.3), (2.5), and (2.6). The result is

$$b_n = \frac{1}{L} \int_{-L}^{L} f(x) \sin \frac{n\pi x}{L} \, dx, \quad n = 1, 2, \ldots, \tag{2.9}$$

which completes the construction of the (full) Fourier series for f. The numbers a_0, a_n, and b_n given by (2.7)–(2.9) are called the *Fourier coefficients* of f.

The series on the right-hand side in (2.2) is of interest to us only on $[-L, L]$, where the original function f is defined, and may be divergent, or may have a different sum than $f(x)$.

2.3 Definition. A function f is said to be *piecewise continuous* on an interval $[a, b]$ if it is continuous at all but finitely many points in $[a, b]$, where it has jump discontinuities; that is, at any discontinuity point x, the function has distinct right-hand side and left-hand side (finite) limits $f(x+)$ and $f(x-)$.

If f is differentiable and both f and f' are continuous on $[a, b]$, then f is called *smooth* on $[a, b]$. If at least one of f, f' is piecewise continuous on $[a, b]$, then f is said to be *piecewise smooth* on $[a, b]$.

2.4 Remarks.   (i) The function shown on the left in Fig. 2.3 is piecewise smooth; the one on the right is not, since $f(0-)$ does not exist: the graph indicates that the function increases without bound as the variable approaches the origin from the left.

(ii) If f is piecewise continuous on $[-L, L]$, then the values of f at its points of discontinuity do not affect the construction of its Fourier series. More precisely, $\int_{-L}^{L} f(x)\, dx$ exists for such a function and is independent of the values assigned to it at its (finitely many) discontinuity points.

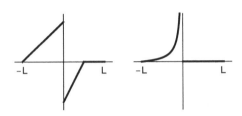

Figure 2.3: Left: both $f(0-)$ and $f(0+)$ exist. Right: $f(0-)$ does not exist.

2.5 Theorem. *If f is piecewise smooth on $[-L, L]$, then its (full) Fourier series converges pointwise to*

(i) *the periodic extension of f to $\mathbb{R}$ at all points x where this extension is continuous;*

(ii) $\frac{1}{2}[f(x-) + f(x+)]$ *at the points x where the periodic extension of f has a (finite) discontinuity jump.*

This means that at each x in $(-L, L)$ where f is continuous, the sum of series (2.2) is equal to $f(x)$. Also, if the function f is continuous on $[-L, L]$ and such that

$$f(-L) = f(L),$$

then the sum of (2.2) is equal to $f(x)$ at all points x in $[-L, L]$.

2.6 Example. Consider the function defined by

$$f(x) = \begin{cases} x + 2, & -2 \le x < 0, \\ 0, & 0 \le x \le 2, \end{cases}$$

whose graph is shown in Fig. 2.4. Clearly, here $L = 2$.

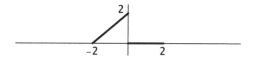

Figure 2.4: $f(x)$, $-2 \le x \le 2$.

The periodic extension of period $2L = 4$ of f, defined by $f(x + 4) = f(x)$ for all x in $\mathbb{R}$, is shown in Fig. 2.5.

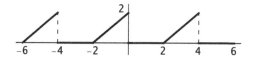

Figure 2.5: $f(x + 4) = f(x)$ for all x in $\mathbb{R}$.

Using (2.7)–(2.9) with $L = 2$ and integration by parts, we find that

$$a_0 = \frac{1}{2} \int_{-2}^{2} f(x)\,dx = \frac{1}{2} \int_{-2}^{0} (x + 2)\,dx = 1,$$

$$a_n = \frac{1}{2} \int_{-2}^{2} f(x) \cos \frac{n\pi x}{2}\,dx = \frac{1}{2} \int_{-2}^{0} (x + 2) \cos \frac{n\pi x}{2}\,dx = \left[1 - (-1)^n\right] \frac{2}{n^2 \pi^2},$$

$$b_n = \frac{1}{2} \int_{-2}^{2} f(x) \sin \frac{n\pi x}{2}\,dx = \frac{1}{2} \int_{-2}^{0} (x + 2) \sin \frac{n\pi x}{2}\,dx = -\frac{2}{n\pi}, \quad n = 1, 2, \ldots,$$

so (2.2) yields the Fourier series

$$f(x) \sim \frac{1}{2} + \sum_{n=1}^{\infty} \left\{ \left[1 - (-1)^n\right] \frac{2}{n^2 \pi^2} \cos \frac{n\pi x}{2} - \frac{2}{n\pi} \sin \frac{n\pi x}{2} \right\}. \qquad (2.10)$$

By Theorem 2.5, on $-2 \le x \le 2$

$$(\text{series}) = \begin{cases} x + 2, & -2 \le x < 0, \\ 1, & x = 0, \\ 0, & 0 < x \le 2; \end{cases} \qquad (2.11)$$

that is, the series converges to $f(x)$ for $-2 \le x < 0$ and $0 < x \le 2$, where the periodic extension of f (see Fig. 2.5) is continuous; at the point $x = 0$, where that extension has a jump discontinuity, the series converges to

$$\tfrac{1}{2}\left[f(0-) + f(0+)\right] = \tfrac{1}{2}(2 + 0) = 1.$$

The graph of the function defined by the sum of the above series on $\mathbb{R}$ is shown in Fig. 2.6.

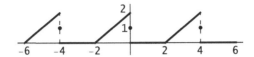

Figure 2.6: Graphical representation of the sum of the series.

VERIFICATION WITH MATHEMATICA®. We compare the graph of f with that of the function obtained by adding the first few terms of its series representation, to see if the latter is a reasonable approximation of the former. This procedure will be followed throughout the book without further explanation where infinite series are involved. Here, with truncation after $n = 5$, the input

```
GenTerm = ((1 - (-1)^n) * 2/(n^2 * Pi^2)) * Cos[n * Pi * x/2]
    - (2/(n * Pi)) * Sin[n * Pi * x/2];
f = Piecewise[{{x + 2, -2<x<0}, {0, 0<x<2}}];
Plot[{1/2 + Sum[GenTerm, {n,1,5}], f}, {x, -2,2},
    PlotRange -> {{-2.2,2.2}, {-0.5,2.4}}, PlotStyle -> Black,
    ImageSize -> Scaled[0.25], Ticks -> {{-2,2},{2}},
    AspectRatio -> 0.5]
```

generates the graphical output

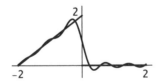

which confirms that our computation is correct.

Exercises

Construct the full Fourier series of the given function f. In each case, use appropriate sketches and discuss the convergence of the series on the interval $[-L, L]$ where f is defined.

1 $f(x) = \begin{cases} 1, & -1 \le x \le 0, \\ 0, & 0 < x \le 1. \end{cases}$

2 $f(x) = \begin{cases} 0, & -\pi \le x \le 0, \\ -3, & 0 < x \le \pi. \end{cases}$

3 $f(x) = \begin{cases} -2, & -1 \le x \le 0, \\ 3, & 0 < x \le 1. \end{cases}$

4 $f(x) = \begin{cases} 1, & -\pi/2 \le x \le 0, \\ 2, & 0 < x \le \pi/2. \end{cases}$

5 $f(x) = x + 1, \quad -1 \le x \le 1.$

6 $f(x) = 1 - 2x, \quad -2 \le x \le 2.$

7 $f(x) = \begin{cases} -1, & -2 \le x \le 0, \\ 2 - x, & 0 < x \le 2. \end{cases}$

8 $f(x) = \begin{cases} 1 + 2x, & -1/2 \le x \le 0, \\ 4, & 0 < x \le 1/2. \end{cases}$

9 $f(x) = \begin{cases} x, & -1 \le x \le 0, \\ 2x - 1, & 0 < x \le 1. \end{cases}$

10 $f(x) = \begin{cases} x - 1, & -\pi \le x \le 0, \\ 2x + 1, & 0 < x \le \pi. \end{cases}$

11 $f(x) = x^2 - 2x + 3, \quad -1 \le x \le 1.$

12 $f(x) = \begin{cases} x^2, & -1 \le x \le 0, \\ 1 + 2x, & 0 < x \le 1. \end{cases}$

13 $f(x) = e^x, \quad -2 \le x \le 2.$

14 $f(x) = \begin{cases} 1, & -\pi/2 \le x \le 0, \\ \sin x, & 0 < x \le \pi/2. \end{cases}$

15 $f(x) = \begin{cases} 0, & -2 \le x \le -1, \\ 1 + x, & -1 < x \le 2. \end{cases}$

16 $f(x) = \begin{cases} 3, & -\pi \le x \le \pi/2, \\ 1, & \pi/2 < x \le \pi. \end{cases}$

Answers to Odd-Numbered Exercises

1 $f(x) \sim 1/2 + \sum_{n=1}^{\infty} [(-1)^n - 1](1/(n\pi)) \sin(n\pi x).$

3 $f(x) \sim 1/2 + \sum_{n=1}^{\infty} [1 - (-1)^n](5/(n\pi)) \sin(n\pi x).$

5 $f(x) \sim 1 + \sum_{n=1}^{\infty} (-1)^{n+1}(2/(n\pi)) \sin(n\pi x).$

7 $f(x) \sim \sum_{n=1}^{\infty} \{[1 - (-1)^n](2/(n^2\pi^2)) \cos(n\pi x/2)$
$$+ [3 - (-1)^n](1/(n\pi)) \sin(n\pi x/2)\}.$$

9 $f(x) \sim -1/4 + \sum_{n=1}^{\infty} \{[(-1)^n - 1](1/(n^2\pi^2)) \cos(n\pi x)$
$$- [2(-1)^n + 1](1/(n\pi)) \sin(n\pi x)\}.$$

11 $f(x) \sim 10/3 + \sum_{n=1}^{\infty} (-1)^n (4/(n^2\pi^2))[\cos(n\pi x) + n\pi \sin(n\pi x)].$

13 $f(x) \sim (e^4 - 1)/(4e^2) + \sum_{n=1}^{\infty} (-1)^n [(e^4 - 1)/(e^2(4 + n^2\pi^2))]$
$$\times [2\cos(n\pi x/2) - n\pi \sin(n\pi x/2)].$$

15 $f(x) \sim 9/8 + \sum_{n=1}^{\infty} \{(2/(n^2\pi^2))[(-1)^n - \cos(n\pi/2)] \cos(n\pi x/2)$
$$+ [(-1)^{n+1}(3/(n\pi)) + (2/(n^2\pi^2)) \sin(n\pi/2)] \sin(n\pi x/2)\}.$$

2.2 Fourier Sine and Cosine Series

For some classes of functions, series (2.2) has a simpler form.

2.7 Definition. A function f defined on an interval symmetric with respect to the origin is called *odd* if $f(-x) = -f(x)$ for all x in the given interval; if, on the other hand, $f(-x) = f(x)$ for all x in that interval, then f is called an *even* function.

2.8 Examples.   (i) The functions $\sin(n\pi x/L)$, $n = 1, 2, \ldots$, are odd. The functions $\cos(n\pi x/L)$, $n = 0, 1, 2, \ldots$, are even.

(ii) The graph (symmetric with respect to the origin) shown on the left in Fig. 2.7 represents an odd function; the one (symmetric with respect to the y-axis) on the right represents an even function. The function graphed in Fig. 2.4 is neither odd nor even.

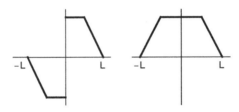

Figure 2.7: Left: graph of an odd function. Right: graph of an even function.

2.9 Remarks.   (i) It is easy to verify that the product of two odd functions is even, the product of two even functions is even, and the product of an odd function and an even function is odd.

(ii) If f is odd on $[-L, L]$, then

$$\int_{-L}^{0} f(x)\,dx = \int_{L}^{0} f(-x)\,d(-x) = -\int_{0}^{L} f(x)\,dx;$$

hence,

$$\int_{-L}^{L} f(x)\,dx = \int_{-L}^{0} f(x)\,dx + \int_{0}^{L} f(x)\,dx = 0.$$

If f is even on $[-L, L]$, then

$$\int_{-L}^{0} f(x)\,dx = \int_{L}^{0} f(-x)\,d(-x) = \int_{0}^{L} f(x)\,dx;$$

consequently,

$$\int_{-L}^{L} f(x)\,dx = \int_{-L}^{0} f(x)\,dx + \int_{0}^{L} f(x)\,dx = 2\int_{0}^{L} f(x)\,dx.$$

(iii) If f is odd, then so is $f(x)\cos(n\pi x/L)$ and, in view of (2.7), (2.8), and (ii) above, we have

$$a_n = 0, \quad n = 0, 1, 2, \ldots;$$

that is, the Fourier series of an odd function on $[-L, L]$ contains only sine terms. Similarly, if f is even, then, by Remark 2.9(i), $f(x)\sin(n\pi x/L)$ is odd, so (2.9) implies that

$$b_n = 0, \quad n = 1, 2, \ldots;$$

this means that the Fourier series of an even function on $[-L, L]$ has only cosine terms, including the constant term, which is obviously an even function.

2.2.1 Fourier Sine Series

Remark 2.9(iii) implies that if f is defined on $[0, L]$, then it can be expanded in a *Fourier sine series*. Let f be the function represented by the graph in Fig. 2.8.

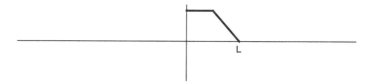

Figure 2.8: $f(x)$, $0 \leq x \leq L$.

We construct the odd extension of f from $(0, L]$ to $[-L, L]$ by setting $f(-x) = -f(x)$ for all x in $[-L, L]$, $x \neq 0$, and $f(0) = 0$ (see Fig. 2.9).

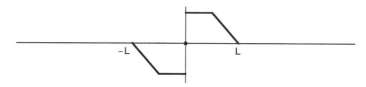

Figure 2.9: $f(-x) = -f(x)$, $-L \leq x \leq L$.

Next, we construct the periodic extension with period $2L$ of this odd function from $(-L, L]$ to $\mathbb{R}$, by requiring that $f(x + 2L) = f(x)$ for all x in $\mathbb{R}$ (see Fig. 2.10).

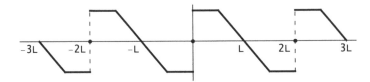

Figure 2.10: $f(x + 2L) = f(x)$ for all x in $\mathbb{R}$.

Finally, we construct the Fourier (sine) series of this last function, which is of the form

$$f(x) \sim \sum_{n=1}^{\infty} b_n \sin \frac{n\pi x}{L},$$

where, according to Remarks 2.9(ii),(iii) and formula (2.9), the coefficients are given by

$$b_n = \frac{2}{L} \int_0^L f(x) \sin \frac{n\pi x}{L} \, dx, \quad n = 1, 2, \ldots. \qquad (2.12)$$

2.10 Example. Consider the function

$$f(x) = 1 - x, \quad 0 \le x \le 1,$$

graphed in Fig. 2.11.

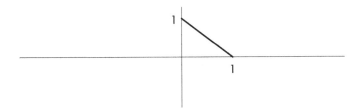

Figure 2.11: $f(x)$, $0 \le x \le 1$.

As explained above, the odd extension of this function to $[-1, 1]$ is defined by $f(-x) = -f(x)$ for all x in $[-1, 1]$, $x \ne 0$, and $f(0) = 0$, and is shown in Fig. 2.12.

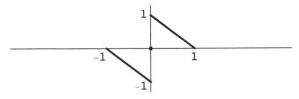

Figure 2.12: $f(-x) = -f(x)$, $-1 \le x \le 1$.

The periodic extension to $\mathbb{R}$, of period $2L = 2$, of the function in Fig. 2.12, defined by $f(x + 2) = f(x)$ for all x in $\mathbb{R}$, is shown in Fig. 2.13.

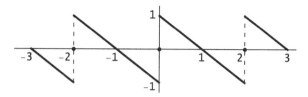

Figure 2.13: $f(x + 2) = f(x)$ for all x in $\mathbb{R}$.

Using (2.12) with $L = 1$ and integration by parts, we find that

$$b_n = 2 \int_0^1 (1 - x) \sin(n\pi x)\, dx = \frac{2}{n\pi}, \quad n = 1, 2, \ldots;$$

hence,

$$f(x) \sim \sum_{n=1}^{\infty} \frac{2}{n\pi} \sin(n\pi x). \tag{2.13}$$

By Theorem 2.5, for $0 \le x \le 1$,

$$(\text{series}) = \begin{cases} 0, & x = 0, \\ 1 - x, & 0 < x \le 1; \end{cases} \tag{2.14}$$

that is, the series converges to $f(x)$ for $0 < x \le 1$, where the periodic extension of f to $\mathbb{R}$ is continuous, and to

$$\tfrac{1}{2}\left[f(0-) + f(0+) \right] = \tfrac{1}{2}(-1 + 1) = 0$$

at $x = 0$, where the periodic extension has a jump discontinuity. The graph in Fig. 2.13 is also the representation of the function defined by the sum of the above series on $\mathbb{R}$.

VERIFICATION WITH MATHEMATICA®. The input

```
GenTerm = (2/(n * Pi)) * Sin[n * Pi * x];
f = 1 - x;
Plot[{Sum[GenTerm, {n,1,7}], f}, {x,0,1},
   PlotRange -> {{-0.1,1.1}, {-0.2,1.1}}, PlotStyle -> Black,
   ImageSize -> Scaled[0.2], Ticks -> {{1},{1}}, AspectRatio -> 0.9]
```

generates the graphical output

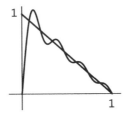

2.11 Example. The graph of the function

$$f(x) = \begin{cases} 1, & 0 \le x \le \pi/2, \\ 2, & \pi/2 < x \le \pi, \end{cases}$$

is shown in Fig. 2.14, where, for convenience, the values at the end points of the two branches are not specifically indicated.

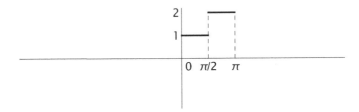

Figure 2.14: $f(x)$, $0 \le x \le \pi$.

To construct the Fourier sine series of f, we first extend it as an odd function to $[-\pi, \pi]$, as shown in Fig. 2.15.

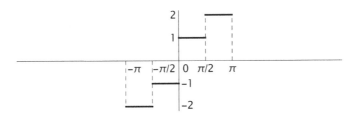

Figure 2.15: $f(-x) = -f(x)$, $-\pi \le x \le \pi$.

Finally, we extend the new odd function periodically to $\mathbb{R}$, with period $2L = 2\pi$ (see Fig. 2.16).

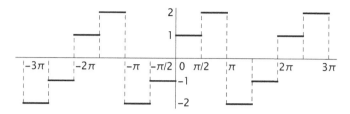

Figure 2.16: $f(x + 2\pi) = f(x)$ for all x in $\mathbb{R}$.

By (2.12) with $L = \pi$, we find that

$$b_n = \frac{2}{\pi} \int_0^{\pi} f(x)\, dx = \frac{2}{\pi} \left\{ \int_0^{\pi/2} \sin(nx)\, dx + \int_{\pi/2}^{\pi} 2\sin(nx)\, dx \right\}$$

$$= \frac{2}{n\pi} \left\{ \cos \frac{n\pi}{2} + 2(-1)^{n+1} + 1 \right\}, \quad n = 1, 2, \ldots,$$

so

$$f(x) \sim \sum_{n=1}^{\infty} \frac{2}{n\pi} \left\{ \cos \frac{n\pi}{2} + 2(-1)^{n+1} + 1 \right\} \sin(nx).$$

According to Theorem 2.5, for $0 \le x \le \pi$ we have

$$(\text{series}) = \begin{cases} 1, & 0 < x < \pi/2, \\ 2, & \pi/2 < x < \pi, \\ 0, & x = 0, \pi, \\ 3/2, & x = \pi/2. \end{cases}$$

This is illustrated by the graph of the approximation obtained from the truncation of the series after the first 7 terms, shown below.

VERIFICATION WITH MATHEMATICA®. The input

```
GenTerm = (2/(n*Pi)) * (Cos[n*Pi/2]
  + 2*(-1)^(n+1) + 1) * Sin[n*x];
f = Piecewise[{{1,0<x<Pi/2},{2,Pi/2<x<Pi}}];
Plot[{Sum[GenTerm, {n,1,10}], f}, {x,0,Pi},
  PlotRange -> {{0,Pi+0.2}, {-0.2,2.4}},
  PlotStyle -> {Black, Thickness[0.011]}
  ImageSize -> Scaled[0.2], Ticks -> {{Pi},{0,1,2}},
  AspectRatio -> 0.7]
```

generates the graphical output

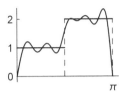

2.2.2 Fourier Cosine Series

By Remark 2.9(iii), a function defined on $[0, L]$ can also be expanded in a *Fourier cosine series*. The construction is similar to that of a Fourier sine series.

Let f be a function defined on $[0, L]$, as shown graphically in Fig. 2.17.

Figure 2.17: $f(x)$, $0 \leq x \leq L$.

We extend f to an even function on $[-L, L]$ by setting $f(-x) = f(x)$ for all x in $[-L, L]$ (see Fig. 2.18).

Figure 2.18: $f(-x) = f(x)$, $-L \leq x \leq L$.

Next, we construct the periodic extension of this even function from $(-L, L]$ to $\mathbb{R}$, of period $2L$, which is defined by $f(x + 2L) = f(x)$ for all x in $\mathbb{R}$ (see Fig. 2.19).

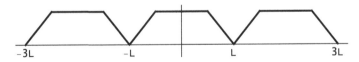

Figure 2.19: $f(x + 2L) = f(x)$ for all x in $\mathbb{R}$.

Finally, we write the Fourier (cosine) series of this last function, which is of the form

$$f(x) \sim \frac{1}{2} a_0 + \sum_{n=1}^{\infty} a_n \cos \frac{n\pi x}{L},$$

where, according to (2.7), (2.8), and Remark 2.9(ii),

$$a_0 = \frac{2}{L} \int_0^L f(x)\,dx,$$

$$a_n = \frac{2}{L} \int_0^L f(x) \cos \frac{n\pi x}{L}\,dx, \quad n = 1, 2, \ldots. \tag{2.15}$$

As noticed earlier, the first equality (2.15) is covered by the second one if we let $n = 0, 1, 2, \ldots$.

2.12 Example. Consider the function

$$f(x) = 1 - x, \quad 0 \le x \le 1,$$

graphed in Fig. 2.20.

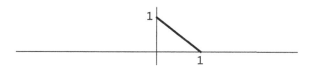

Figure 2.20: $f(x)$, $0 \le x \le 1$.

The even extension of f to $[-1, 1]$, defined by $f(-x) = f(x)$ for all x in $[-1, 1]$, is shown in Fig. 2.21.

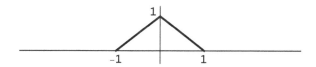

Figure 2.21: $f(-x) = f(x)$, $-1 \le x \le 1$.

The periodic extension of period $2L = 2$ of this even function to $\mathbb{R}$, which is defined by $f(x + 2) = f(x)$ for all x in $\mathbb{R}$, is shown in Fig. 2.22.

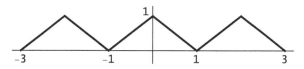

Figure 2.22: $f(x + 2) = f(x)$ for all x in $\mathbb{R}$.

By (2.15) with $L = 1$ and integration by parts,

$$a_0 = 2 \int_0^1 (1 - x)\,dx = 1,$$

$$a_n = 2 \int_0^1 (1-x)\cos(n\pi x)\, dx = \left[1 - (-1)^n\right]\frac{2}{n^2\pi^2}, \quad n = 1, 2, \ldots,$$

so the Fourier cosine series of f is

$$f(x) \sim \frac{1}{2} + \sum_{n=1}^{\infty} \left[1 - (-1)^n\right]\frac{2}{n^2\pi^2} \cos(n\pi x). \tag{2.16}$$

By Theorem 2.5,

$$\text{(series)} = 1 - x \quad \text{for } 0 \leq x \leq 1, \tag{2.17}$$

because the periodic extension of f to $\mathbb{R}$ is continuous everywhere. The graph in Fig. 2.22 is also the representation of the function defined by the sum of the above series on $\mathbb{R}$.

VERIFICATION WITH MATHEMATICA®. The input

```
GenTerm = ((1 - (-1)^n) * 2/(n^2 * Pi^2)) * Cos[n * Pi * x];
f = 1 - x;
Plot[{1/2 + Sum[GenTerm, {n,1,1}], f}, {x,0,1},
    PlotRange -> {{-0.2,1.1}, {-0.2,1.1}}, PlotStyle -> Black,
    ImageSize -> Scaled[0.2], Ticks -> {{1},{1}}, AspectRatio -> 0.9]
```

generates the graphical output

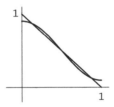

2.13 Example. We start again with the function

$$f(x) = \begin{cases} 1, & 0 \leq x \leq \pi/2, \\ 2, & \pi/2 < x \leq \pi, \end{cases}$$

graphed in Fig. 2.23, which is the same as that in Example 2.11.

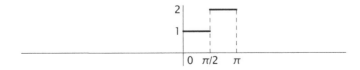

Figure 2.23: $f(x)$, $0 \leq x \leq \pi$.

This time, however, we want to construct its Fourier cosine series representation, so we begin by extending f as an even function to $[-\pi, \pi]$ (see Fig. 2.24).

Figure 2.24: $f(-x) = f(x)$, $-\pi \leq x \leq \pi$.

According to the general procedure, we now extend this new even function periodically to $\mathbb{R}$, with period $2L = 2\pi$, as shown in Fig. 2.25.

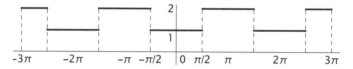

Figure 2.25: $f(x + 2\pi) = f(x)$ for all x in $\mathbb{R}$.

Then, by (2.15) with $L = \pi$,

$$a_0 = \frac{2}{\pi} \int_0^\pi f(x\,dx = \frac{2}{\pi}\left\{ \int_0^{\pi/2} dx + \int_{\pi/2}^\pi 2\,dx \right\} = 3,$$

$$a_n = \frac{2}{\pi} \int_0^{2\pi} f(x)\cos(nx)\,dx$$

$$= \frac{2}{\pi}\left\{ \int_0^{\pi/2} \cos(nx)\,dx + \int_{\pi/2}^\pi 2\cos(nx)\,dx \right\} = -\frac{2}{n\pi}\sin\frac{n\pi}{2};$$

therefore,

$$f(x) \sim \frac{3}{2} + \sum_{n=1}^\infty -\frac{2}{n\pi}\sin\frac{n\pi}{2}\cos(nx).$$

By Theorem 2.5,

$$(\text{series}) = \begin{cases} 1, & 0 \leq x < \pi/2, \\ 2, & \pi/2 < x \leq \pi, \\ 3/2, & x = \pi/2. \end{cases}$$

The figure below illustrates the graph of the approximation provided by the first 6 terms of the series.

VERIFICATION WITH MATHEMATICA®. The input

```
GenTerm = - (2/(n*Pi)) * Sin[n*Pi/2] * Cos[n*x];
f = Piecewise[{{1,0<x<Pi/2},{2,Pi/2<x<Pi}}];
Plot[{3/2+Sum[GenTerm, {n,1,5}], f}, {x,0,Pi},
   PlotRange -> {{0,Pi+0.2}, {-0.2,2.4}},
   PlotStyle -> {Black, Thickness[0.011],
   ImageSize -> Scaled[0.2], Ticks -> {{Pi},{0,1,2}},
   AspectRatio -> 0.7]
```

generates the graphical output

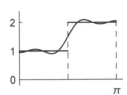

Exercises

Construct the Fourier sine series and the Fourier cosine series of the given function f. In each case, use appropriate sketches and discuss the convergence of the series on the interval $[0, L]$ where f is defined.

1 $f(x) = \begin{cases} 0, & 0 \le x \le 1, \\ 1, & 1 < x \le 2. \end{cases}$

2 $f(x) = \begin{cases} 2, & 0 \le x \le \pi, \\ 0, & \pi < x \le 2\pi. \end{cases}$

3 $f(x) = \begin{cases} 1, & 0 \le x \le 1, \\ -1, & 1 < x \le 2. \end{cases}$

4 $f(x) = \begin{cases} -2, & 0 \le x \le \pi/2, \\ 3, & \pi/2 < x \le \pi. \end{cases}$

5 $f(x) = 2 - x, \quad 0 \le x \le 1.$

6 $f(x) = 3x + 1, \quad 0 \le x \le 2\pi.$

7 $f(x) = \begin{cases} x, & 0 \le x \le 1, \\ -2, & 1 < x \le 2. \end{cases}$

8 $f(x) = \begin{cases} 2x - 1, & 0 \le x \le 1, \\ 2x + 1, & 1 < x \le 2. \end{cases}$

9 $f(x) = \begin{cases} 2 + x, & 0 \le x \le 1, \\ 1 - x, & 1 < x \le 2. \end{cases}$

10 $f(x) = x^2 + x - 1, \quad 0 \le x \le 2.$

11 $f(x) = \begin{cases} x + 1, & 0 \le x \le 1, \\ x^2 - 2x, & 1 < x \le 2. \end{cases}$

12 $f(x) = 1 + e^x, \quad 0 \le x \le 1.$

13 $f(x) = x + \sin x, \quad 0 \le x \le \pi.$

14 $f(x) = \begin{cases} \cos x, & 0 \le x \le \pi/2, \\ -1, & \pi/2 < x \le \pi. \end{cases}$

Answers to Odd-Numbered Exercises

1 $f(x) \sim \sum\limits_{n=1}^{\infty} (2/(n\pi))[\cos(n\pi/2) - (-1)^n] \sin(n\pi x/2);$

$f(x) \sim 1/2 - \sum\limits_{n=1}^{\infty} (2/(n\pi)) \sin(n\pi/2) \cos(n\pi x/2).$

3 $f(x) \sim \sum\limits_{n=1}^{\infty} (2/(n\pi))[1 + (-1)^n - 2\cos(n\pi/2)] \sin(n\pi x/2);$

$f(x) \sim \sum\limits_{n=1}^{\infty} (4/(n\pi)) \sin(n\pi/2) \cos(n\pi x/2).$

5 $f(x) \sim \sum\limits_{n=1}^{\infty} [2 - (-1)^n](2/(n\pi)) \sin(n\pi x);$

$f(x) \sim 3/2 + \sum\limits_{n=1}^{\infty} [1 - (-1)^n](2/(n^2\pi^2)) \cos(n\pi x).$

7 $f(x) \sim \sum\limits_{n=1}^{\infty} (2/(n\pi))[-3\cos(n\pi/2) + (2/(n\pi))\sin(n\pi/2) + (-1)^n 2]$
$$\times \sin(n\pi x/2);$$

$f(x) \sim -3/4 + \sum\limits_{n=1}^{\infty} (2/(n\pi))[3\sin(n\pi/2) + (2/(n\pi))\cos(n\pi/2) - 2/(n\pi)]$
$$\times \cos(n\pi x/2).$$

9 $f(x) \sim \sum\limits_{n=1}^{\infty} (2/(n\pi))[2 + (-1)^n - 3\cos(n\pi/2) + (4/(n\pi))\sin(n\pi/2)]$
$$\times \sin(n\pi x/2);$$

$f(x) \sim 1 + \sum\limits_{n=1}^{\infty} \{(6/(n\pi))\sin(n\pi/2) + (8/(n^2\pi^2))\cos(n\pi/2)$
$$+ [(-1)^{n+1} - 1](4/(n^2\pi^2))\} \cos(n\pi x/2).$$

11 $f(x) \sim \sum\limits_{n=1}^{\infty} (2/(n^3\pi^3))[8(-1)^n + n^2\pi^2 - (8 + 3n^2\pi^2)\cos(n\pi/2)$
$$+ 2n\pi \sin(n\pi/2)] \sin(n\pi x/2);$$

$f(x) \sim 5/12 + \sum\limits_{n=1}^{\infty} (2/(n^3\pi^3))\{[2(-1)^n - 1]2n\pi + 2n\pi \cos(n\pi/2)$
$$+ (8 + 3n^2\pi^2) \sin(n\pi/2)\} \cos(n\pi x/2).$$

13 $f(x) \sim 3\sin x + \sum\limits_{n=2}^{\infty} (-1)^{n+1}(2/n) \sin(nx);$

$f(x) \sim 2/\pi + \pi/2 - (4/\pi)\cos x$
$$+ \sum\limits_{n=2}^{\infty} [2(1 - (-1)^n - 2n^2)/((n^2 - 1)n^2\pi)] \cos(nx).$$

2.3   Convergence and Differentiation

There are several important points that need to be made at this stage.

2.14 Remarks.   (i) It is obvious that a function f defined on $[0, L]$ can be expanded in a full Fourier series, or in a sine series, or in a cosine series. While the last two are unique to the function, the first one is not: since the full Fourier series requires no symmetry, we may extend f from $[0, L]$ to $[-L, L]$ in infinitely many ways. We usually choose the type of series that seems to be the most appropriate for the problem. A function defined on $[-L, L]$ can be expanded, in general, only in a full Fourier series.

(ii) Earlier, we saw that we cannot automatically use the equality sign between a function f and its Fourier series. According to Theorem 2.5, for a full series we can do so at all the points in $[-L, L]$ where f is continuous; we can also do it at $x = -L$ and $x = L$ provided that the periodic extension of f is continuous at these points from the right and from the left, respectively, and $f(-L) = f(L)$. For a sine series, this can be done at $x = L$ if f is continuous there from the left and $f(L) = 0$. For a cosine series, this is always possible if f is continuous from the left at $x = L$. Similar arguments can be used for the sine and cosine series at $x = 0$. In what follows, we work exclusively with piecewise smooth functions; therefore, to avoid cumbersome notation and possible confusion, we will use the equality sign in all circumstances, assuming tacitly that equality is understood in the sense of Theorem 2.5, Remark 2.4(ii), and the above comments.

In practical applications, we often need to differentiate Fourier series term by term. Consequently, it is important to know when this operation may be performed.

2.15 Theorem.   (i) *If f is continuous on $[-L, L]$,*

$$f(-L) = f(L),$$

and f' is piecewise smooth on $[-L, L]$, then the full Fourier series of f can be differentiated term by term, and the result is the full Fourier series of f', which converges to $f'(x)$ at all x where $f''(x)$ exists.

(ii) *If f is continuous on $[0, L]$,*

$$f(0) = f(L) = 0,$$

and f' is piecewise smooth on $[0, L]$, then the Fourier sine series of f can be differentiated term by term, and the result is the Fourier cosine series of f', which converges to $f'(x)$ at all x where $f''(x)$ exists.

(iii) *If f is continuous on $[0, L]$ and f' is piecewise smooth on $[0, L]$, then the Fourier cosine series of f can be differentiated term by term, and the result is the Fourier sine series of f', which converges to $f'(x)$ at all x where $f''(x)$ exists.*

2.16 Remark. In the PDE problems that we discuss in the rest of the book, the conditions laid down in Theorem 2.15 will always be satisfied and we will formally differentiate the corresponding Fourier series term by term without explicit mention of these conditions.

2.4   Series Expansion of More General Functions

It is important to know how functions defined on various types of intervals and having the necessary degree of smoothness can be expanded in an infinite series by means of the procedures set out in Sections 2.1 and 2.2.

Function defined on $[-L,L]$. If a function like this is neither odd nor even, then it can be expanded only in a full Fourier series.

Function defined on $[0,L]$. We can expand such a function in a Fourier sine series or a Fourier cosine series. The latter requires the computation of an additional term (a_0), but, because the extension of the function to $\mathbb{R}$ is even and, therefore, symmetric with respect to the y-axis, the cosine series generally converges to the function on a larger interval than the sine series. We can also expand the function in a full Fourier series if we first extend it by, say, 0 to $[-L, L]$.

Function defined on $[a,b]$. A function of this type can be expanded by means of the full Fourier series technique if we make the substitution

$$x = \frac{1}{2L}(b-a)X + \tfrac{1}{2}(b+a), \tag{2.18}$$

which maps the interval $[a, b]$ for x onto the interval $[-L, L]$ for X. We can also expand the function using a Fourier sine or cosine series construction if we substitute

$$x = \frac{1}{L}(b-a)X + a, \tag{2.19}$$

which maps the interval $[a, b]$ for x onto the interval $[0, L]$ for X.

2.17 Example. In the case of the function

$$f(x) = \begin{cases} \frac{4}{3}(x-2), & 2 \le x \le 7/2, \\ 0, & 7/2 < x \le 5 \end{cases}$$

we have $a = 2$ and $b = 5$, so, with the choice $L = 2$, substitution (2.18) is

$$x = \tfrac{3}{4}X + \tfrac{7}{2},$$

or

$$X = \tfrac{2}{3}(2x - 7),$$

which transforms $f(x)$ to

$$F(X) = f\left(\tfrac{3}{4}X + \tfrac{7}{2}\right) = \begin{cases} X + 2, & -2 \le X \le 0, \\ 0, & 0 < X \le 2. \end{cases}$$

This is the function expanded in Example 2.6, therefore, by (2.10),

$$F(X) \sim \frac{1}{2} + \sum_{n=1}^{\infty} \left\{ \left[1 - (-1)^n \right] \frac{2}{n^2 \pi^2} \cos \frac{n\pi X}{2} - \frac{2}{n\pi} \sin \frac{n\pi X}{2} \right\}.$$

According to our substitution, this leads to

$$f(x) \sim \frac{1}{2} + \sum_{n=1}^{\infty} \left\{ \left[1 - (-1)^n \right] \frac{2}{n^2 \pi^2} \cos \frac{n\pi(2x-7)}{3} - \frac{2}{n\pi} \sin \frac{n\pi(2x-7)}{3} \right\},$$

with convergence expressed by (2.11) as

$$(\text{series}) = \begin{cases} \frac{4}{3}(x-2), & 2 \le x < 7/2, \\ 1, & x = 7/2, \\ 0, & 7/2 < x \le 5. \end{cases}$$

VERIFICATION WITH MATHEMATICA®. The input

```
GenTerm = ((1-(-1)^n)*2/(n^2*Pi^2))*Cos[n*Pi*(2*x-7)/3]
  - (2/(n*Pi))*Sin[n*Pi*(2*x-7)/3];
f = Piecewise[{{(4/3)*(x-2), 2<x<7/2}, {0, 7/2<x<5}}];
Plot[{1/2+Sum[GenTerm, {n,1,5}], f}], {x,2,5},
  PlotRange->{{1.5,5.5}, {-0.5,2.4}}, PlotStyle->Black,
  ImageSize->Scaled[0.25], Ticks->{{},{1,2}},
  AspectRatio->0.5]
```

generates the graphical output

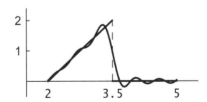

2.18 Example. Let

$$f(x) = \frac{1}{5}(2-x), \quad -3 \le x \le 2.$$

With $a = -3$, $b = 2$, and, say, $L = 1$, formula (2.19) yields

$$x = 5X - 3,$$

or

$$X = \frac{1}{5}(x+3).$$

The given function f now becomes

$$F(X) = f(5X - 3) = 1 - X, \quad 0 \le X \le 1,$$

which was expanded in a Fourier sine series and a Fourier cosine series in Examples 2.10 and 2.12, respectively:

$$F(X) \sim \sum_{n=1}^{\infty} \frac{2}{n\pi} \sin(n\pi X),$$

$$F(X) \sim \frac{1}{2} + \sum_{n=1}^{\infty} \left[1 - (-1)^n\right] \frac{2}{n^2\pi^2} \cos(n\pi X);$$

hence,

$$f(x) \sim \sum_{n=1}^{\infty} \frac{2}{n\pi} \sin \frac{n\pi(x+3)}{5},$$

$$f(x) \sim \frac{1}{2} + \sum_{n=1}^{\infty} \left[1 - (-1)^n\right] \frac{2}{n^2\pi^2} \cos \frac{n\pi(x+3)}{5}.$$

From (2.14) and (2.17), we deduce that the convergence of the first of these expansions is summed up by

$$(\text{series}) = \begin{cases} 0, & x = -3, \\ \frac{1}{5}(2 - x), & -3 < x \le 2, \end{cases}$$

and that of the second one by

$$(\text{series}) = \frac{1}{5}(2 - x), \quad -3 \le x \le 2.$$

VERIFICATION WITH MATHEMATICA®. The two sets of input

```
GenTerm1 = (2/(n*Pi)) * Sin[n*Pi*(x+3)/5];
f = (2-x)/5;
Plot[{Sum[GenTerm1, {n,1,7}], f}, {x,-3,2},
   PlotRange -> {{-3.5,2.5}, {-0.2,1.1}}, PlotStyle -> Black,
   ImageSize -> Scaled[0.2], Ticks -> {{},{1}}, AspectRatio -> 0.9]
```

```
GenTerm = ((1-(-1)^n) * 2/(n^2*Pi^2)) * Cos[n*Pi*(x+3)/5];
f = (2-x)/5;
Plot[{1/2 + Sum[GenTerm, {n,1,1}], f}, {x,-3,2},
   PlotRange -> {{-3.5,2.5}, {-0.2,1.1}}, PlotStyle -> Black,
   ImageSize -> Scaled[0.2], Ticks -> {{-3},{1}}, AspectRatio -> 0.9]
```

generate the graphical output

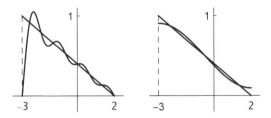

Function defined on an infinite interval. Obviously, this type of function cannot be expanded by means of the Fourier series method. Its proper handling is explained in Chapter 8.

CHAPTER 3

Sturm-Liouville Problems

A particular class of problems involving second-order ODEs plays an essential role in the solution of partial differential equations. Below, we present the main results characterizing such problems, which will be used extensively in the methods of separation of variables and eigenfunction expansion developed in Chapters 5 and 7.

3.1 Regular Sturm–Liouville Problems

To avoid complicated notation, in what follows we denote intervals generically by I, regardless of whether they are closed, open, or half-open, finite or infinite. Specific intervals are described either in terms of their endpoints or by means of inequalities. The functions defined on such intervals are assumed to be integrable on them.

3.1 Definition. Let $\mathcal{X}$ be a space of functions defined on the same interval I. A linear differential operator L acting on $\mathcal{X}$ is called *symmetric* on $\mathcal{X}$ if

$$\int_I \left[f_1(x)(Lf_2)(x) - f_2(x)(Lf_1)(x) \right] dx = 0$$

for any functions f_1 and f_2 in $\mathcal{X}$.

3.2 Example. Consider the space $\mathcal{X}$ of functions that are twice continuously differentiable on $[0, 1]$ and equal to zero at $x = 0$ and $x = 1$. If L is the linear operator defined by

$$L = \frac{d^2}{dx^2},$$

then, using integration by parts, we find that for any f_1 and f_2 in $\mathcal{X}$,

$$\int_0^1 \left[f_1(Lf_2) - f_2(Lf_1) \right] dx = \int_0^1 (f_1 f_2'' - f_2 f_1'') \, dx$$

$$= \left[f_1 f_2' \right]_0^1 - \int_0^1 f_1' f_2' \, dx - \left[f_2 f_1' \right]_0^1 + \int_0^1 f_1' f_2' \, dx = 0.$$

Hence, L is symmetric on $\mathcal{X}$.

3.3 Remark. An operator may be symmetric on a space of functions but not on another. If no restriction is imposed on the values at $x = 0$ and $x = 1$ of the functions in Example 3.2, then L is not symmetric on the new space $\mathcal{X}$.

3.4 Definition. Let σ be a function defined on I, with the property that $\sigma(x) > 0$ for all x in I. Two functions f_1 and f_2, also defined on I, are called *orthogonal with weight σ on I* if

$$\int_I f_1(x) f_2(x) \sigma(x) \, dx = 0.$$

If $\sigma(x) = 1$, then f_1 and f_2 are simply called *orthogonal*. A set of functions that are pairwise orthogonal on I is called an *orthogonal set*.

3.5 Example. The functions $f_1(x) = 1$ and $f_2(x) = 9x - 5$ are orthogonal with weight $\sigma(x) = x + 1$ on $[0, 1]$ since

$$\int_0^1 (9x - 5)(x + 1) \, dx = \int_0^1 (9x^2 + 4x - 5) \, dx = 0.$$

Alternatively, we can say that the functions $f_1(x) = x + 1$ and $f_2(x) = 9x - 5$ are orthogonal on $[0, 1]$.

3.6 Example. The functions $\sin(3x)$ and $\cos(2x)$ are orthogonal on the interval $[-\pi, \pi]$ because

$$\int_{-\pi}^\pi \sin(3x) \cos(2x) \, dx = \int_{-\pi}^\pi \tfrac{1}{2} \left[\sin(5x) + \sin x \right] dx = 0.$$

3.7 Definition. Let L be a linear differential operator acting on a space $\mathcal{X}$ of functions defined on an interval (a, b). An equation of the form

$$(Lf)(x) + \lambda \sigma(x) f(x) = 0, \quad a < x < b, \tag{3.1}$$

where λ is a parameter and σ is a given function such that $\sigma(x) > 0$ for all x in (a, b), is called an *eigenvalue problem*. The numbers λ for which (3.1) has nonzero solutions in $\mathcal{X}$ are called *eigenvalues*; the corresponding nonzero solutions are called *eigenfunctions*.

3.8 Theorem. *If the operator L of the eigenvalue problem (3.1) is symmetric, then*

(i) *all the eigenvalues λ are real;*

(ii) *the eigenvalues form an infinite sequence $\lambda_1 < \lambda_2 < \cdots < \lambda_n < \cdots$ such that $\lambda_n \to \infty$ as $n \to \infty$;*

(iii) *eigenfunctions associated with distinct eigenvalues are orthogonal with weight σ on (a, b).*

3.9 Definition. Let $[a, b]$ be a finite interval, let p, q, and σ be real-valued functions, and let κ_1, κ_2, κ_3, and κ_4 be real numbers such that

(i) p is continuously differentiable on $[a, b]$ and $p(x) > 0$ for all x in $[a, b]$;

(ii) q and σ are continuous on $[a, b]$ and $\sigma(x) > 0$ for all x in $[a, b]$;

(iii) κ_1, κ_2 are not both zero and κ_3, κ_4 are not both zero.

An eigenvalue problem of the form

$$[p(x)f'(x)]' + q(x)f(x) + \lambda\sigma(x)f(x) = 0, \quad a < x < b, \qquad (3.2)$$

with the boundary conditions (BCs)

$$\kappa_1 f(a) + \kappa_2 f'(a) = 0, \qquad (3.3)$$
$$\kappa_3 f(b) + \kappa_4 f'(b) = 0, \qquad (3.4)$$

is called a *regular Sturm–Liouville* (S–L) *eigenvalue problem.*

3.10 Example. The choice

$$p(x) = 1, \quad q(x) = 0, \quad \sigma(x) = 1,$$
$$\kappa_1 = 1, \quad \kappa_2 = 0, \quad \kappa_3 = 1, \quad \kappa_4 = 0, \quad a = 0, \quad b = L$$

generates the regular S–L problem

$$f''(x) + \lambda f(x) = 0, \quad 0 < x < L,$$
$$f(0) = 0, \quad f(L) = 0.$$

3.11 Example. If p, q, σ, a, and b are as in Example 3.10 but $\kappa_1 = 0$, $\kappa_2 = 1$, $\kappa_3 = 0$, and $\kappa_4 = 1$, then we have the regular S–L problem

$$f''(x) + \lambda f(x) = 0, \quad 0 < x < L,$$
$$f'(0) = 0, \quad f'(L) = 0.$$

3.12 Example. Taking p, q, σ, a, and b as in Example 3.10 and the coefficients

$$\kappa_1 = 1, \quad \kappa_2 = 0, \quad \kappa_3 = h = \text{const}, \quad \kappa_4 = 1,$$

we obtain the regular S–L problem

$$f''(x) + \lambda f(x) = 0, \quad 0 < x < L,$$
$$f(0) = 0, \quad f'(L) + hf(L) = 0.$$

3.13 Example. The eigenvalue problem

$$f''(x) + 2f'(x) + \lambda f(x) = 0, \quad 0 < x < 1,$$
$$f(0) = 0, \quad f(1) = 0$$

may, at first glance, not seem to be a regular S–L problem. However, if we use the coefficient of f' to construct the function e^{2x} and multiply every term in the equation by this function, we bring the equation to the form

$$\left(e^{2x} f'(x)\right)' + \lambda e^{2x} f(x) = 0, \quad 0 < x < 1,$$

which indicates that we have, in fact, a regular S–L problem with coefficients

$$p(x) = \sigma(x) = \exp\left\{ \int 2\, dx \right\} = e^{2x}, \quad q(x) = 0$$

satisfying the requirements in Definition 3.9.

3.14 Theorem. *The operator*

$$Lf = (pf')' + qf \tag{3.5}$$

defined by the left-hand side in (3.2) *and acting on a space $\mathcal{X}$ of functions satisfying* (3.3) *and* (3.4) *is symmetric on $\mathcal{X}$.*

Proof. Suppose, for definiteness, that $\kappa_1 \neq 0$ and $\kappa_4 \neq 0$. Then, by (3.3) and (3.4), any function f in $\mathcal{X}$ satisfies

$$f(a) = -\frac{\kappa_2}{\kappa_1} f'(a), \quad f'(b) = -\frac{\kappa_3}{\kappa_4} f(b).$$

Consequently, using (3.5) and integration by parts, we find that for any f_1 and f_2 in $\mathcal{X}$,

$$\int_a^b \left[f_1(Lf_2) - f_2(Lf_1) \right] dx$$

$$= \int_a^b \left\{ f_1\left[(pf_2')' + qf_2\right] - f_2\left[(pf_1')' + qf_1\right] \right\} dx$$

$$= \left[f_1(pf_2') \right]_a^b - \int_a^b pf_2' f_1'\, dx - \left[f_2(pf_1') \right]_a^b + \int_a^b pf_1' f_2'\, dx$$

$$= p(b)\left[f_1(b)f_2'(b) - f_2(b)f_1'(b) \right] - p(a)\left[f_1(a)f_2'(a) - f_2(a)f_1'(a) \right]$$

$$= p(b)\left[-\frac{\kappa_3}{\kappa_4} f_1(b)f_2(b) + \frac{\kappa_3}{\kappa_4} f_2(b)f_1(b) \right]$$

$$\qquad\qquad - p(a)\left[-\frac{\kappa_2}{\kappa_1} f_1'(a)f_2'(a) + \frac{\kappa_2}{\kappa_1} f_2'(a)f_1'(a) \right] = 0,$$

as required. All other combinations of nonzero constants κ_1, κ_2, κ_3, and κ_4 are treated similarly. □

3.15 Corollary. *The eigenvalues and eigenfunctions of a regular Sturm–Liouville problem have all the properties listed in Theorem 3.8.*

Exercises

Verify if the given problem is a regular S–L eigenvalue problem.

1 $f''(x) + \lambda f(x) = 0, \quad 0 < x < 1, \quad f(0) + 2f'(0) = 0, \quad f'(1) = 0.$

2 $f''(x) - xf(x) + \lambda(x^2 + 1)f(x) = 0, \quad 0 < x < 1, \quad f(0) = 0, \quad f'(1) = 0.$

3 $(x - 2)f''(x) + f'(x) + (1 + \lambda)f(x) = 0, \quad 0 < x < 2, \quad f'(0) = 0,$

   $f(x), f'(x)$ bounded as $x \to 2-$.

4 $f''(x) - f'(x) + \lambda f(x) = 0, \quad 0 < x < 2, \quad f'(0) = 0, \quad f(2) - f'(2) = 0.$

5 $2f''(x) + f'(x) + (\lambda + x)f(x) = 0, \quad 0 < x < \infty, \quad f(0) = 0,$

   $f(x), f'(x)$ bounded as $x \to \infty$.

6 $x^2 f''(x) + xf'(x) + (2\lambda x - 1)f(x) = 0, \quad 0 < x < 1,$

   $f(x), f'(x)$ bounded as $x \to 0+, \quad f(1) - f'(1) = 0.$

Answers to Odd-Numbered Exercises

1 Regular.   **3** Not regular.   **5** Not regular.

3.1.1 Eigenvalues and Eigenfunctions of Some Basic Problems

Before actually computing the eigenvalues and eigenfunctions of specific regular Sturm–Liouville problems, we derive a very useful formula relating these quantities. Multiplying equation (3.2) by f and integrating over (a, b), we easily see that

$$0 = \int_a^b \left[f(pf')' + qf^2 \right] dx + \lambda \int_a^b \sigma f^2 \, dx$$

$$= \int_a^b \left[(pff')' - p(f')^2 + qf^2 \right] dx + \lambda \int_a^b \sigma f^2 \, dx$$

$$= \left[pff' \right]_a^b - \int_a^b \left[p(f')^2 - qf^2 \right] dx + \lambda \int_a^b \sigma f^2 \, dx.$$

Since f is an eigenfunction (nonzero solution of (3.2)) and $\sigma(x) > 0$ for all x in (a, b), the coefficient of λ in the last term is strictly positive. Hence, we can solve for λ and obtain

$$\lambda = \frac{\int\limits_a^b [p(f')^2 - qf^2]\, dx - [pff']_a^b}{\int\limits_a^b \sigma f^2\, dx}. \tag{3.6}$$

This expression is known as the *Rayleigh quotient*.

3.16 Example. For the regular S–L problem considered in Example 3.10—that is,

$$f''(x) + \lambda f(x) = 0, \quad 0 < x < L,$$
$$f(0) = 0, \quad f(L) = 0,$$

we have

$$\int\limits_a^b [p(x)(f')^2(x) - q(x)f^2(x)]\, dx = \int\limits_0^L (f')^2(x)\, dx \geq 0,$$

$$[p(x)f(x)f'(x)]_a^b = [f(x)f'(x)]_0^L = 0,$$

$$\int\limits_a^b \sigma(x)f^2(x)\, dx = \int\limits_0^L f^2(x)\, dx > 0,$$

so (3.6) yields

$$\lambda = \frac{\int\limits_0^L (f')^2(x)\, dx}{\int\limits_0^L f^2(x)\, dx} \geq 0. \tag{3.7}$$

It is clear that $\lambda = 0$ if and only if $f'(x) = 0$ on $[0, L]$, which means that $f(x) = \text{const}$. But this is unacceptable because, according to the BCs, the only constant solution is the zero solution. Consequently, $\lambda > 0$, and the general solution of the equation in this problem is

$$f(x) = C_1 \cos\left(\sqrt{\lambda}\, x\right) + C_2 \sin\left(\sqrt{\lambda}\, x\right), \quad C_1, C_2 = \text{const}. \tag{3.8}$$

Using the condition $f(0) = 0$, we find that $C_1 = 0$. Then the condition $f(L) = 0$ leads to

$$C_2 \sin\left(\sqrt{\lambda}\, L\right) = 0.$$

We cannot have $C_2 = 0$ because this would imply that $f = 0$, and we are seeking nonzero solutions; therefore, we must conclude that $\sin\left(\sqrt{\lambda}\, L\right) = 0$, from which we obtain $\sqrt{\lambda}\, L = n\pi$, $n = 1, 2, \ldots$, or

$$\lambda_n = \left(\frac{n\pi}{L}\right)^2, \quad n = 1, 2, \ldots.$$

These are the eigenvalues of the problem. The corresponding eigenfunctions (nonzero solutions) are

$$f_n(x) = \sin \frac{n\pi x}{L}, \quad n = 1, 2, \ldots.$$

For convenience, we have taken $C_2 = 1$; any other nonzero value of C_2 would simply produce a multiple of $\sin(n\pi x/L)$. In other words, the space of eigenfunctions corresponding to each eigenvalue is one-dimensional.

It is easy to see that the properties in Theorem 3.8 are satisfied. The orthogonality of the eigenfunctions on $[0, L]$ follows immediately from (2.5) since the integrand is an even function.

3.17 Example. The same technique is used to compute the eigenvalues and eigenfunctions of the S–L problem in Example 3.11, namely

$$f''(x) + \lambda f(x) = 0, \quad 0 < x < L,$$
$$f'(0) = 0, \quad f'(0) = 0.$$

Inequality (3.7) remains valid, but now we cannot reject the case $\lambda = 0$ since the functions $f(x) = $ const corresponding to it satisfy both BCs. Taking the constant to be, say, $1/2$, we conclude that the problem has the eigenvalue–eigenfunction pair

$$\lambda_0 = 0, \quad f_0(x) = \tfrac{1}{2}.$$

As in Example 3.16, for $\lambda > 0$ the general solution of the equation is (3.8), so

$$f'(x) = -C_1\sqrt{\lambda}\sin\left(\sqrt{\lambda}\,x\right) + C_2\sqrt{\lambda}\cos\left(\sqrt{\lambda}\,x\right).$$

The condition $f'(0) = 0$ immediately yields $C_2 = 0$; therefore, $f'(L) = 0$ implies that

$$C_1\sqrt{\lambda}\sin\left(\sqrt{\lambda}\,L\right) = 0.$$

Since we want nonzero solutions, it follows that $\sin\left(\sqrt{\lambda}\,L\right) = 0$. This means that $\sqrt{\lambda}L = n\pi$, $n = 1, 2, \ldots$, so, by (3.8) with $C_2 = 0$ and $C_1 = 1$, the eigenvalue–eigenfunction pairs are

$$\lambda_n = \left(\frac{n\pi}{L}\right)^2, \quad f_n(x) = \cos\frac{n\pi x}{L}, \quad n = 1, 2, \ldots.$$

3.18 Remarks. (i) The full set of eigenvalues and eigenfunctions (with $f_0(x) = 1$ instead of $1/2$) of the problem can be written as above, but with $n = 0, 1, 2, \ldots$. For convenience, in Chapter 7 we will indeed make the choice $f_0(x) = 1$.

(ii) Once again, we note that the eigenvalues and eigenfunctions satisfy the properties in Theorem 3.8. The orthogonality of the latter on $[0, L]$ follows from the first formula (2.3) and (2.4).

The eigenvalues and eigenfunctions of regular S–L problems with mixed BCs are generated in the same way.

3.19 Example. For the S–L problem

$$f''(x) + \lambda f(x) = 0, \quad 0 < x < L,$$
$$f(0) = 0, \quad f'(L) = 0,$$

we have $\left[f(x) f'(x) \right]_0^L = 0$, so the Rayleigh quotient shows that $\lambda \geq 0$. The BC at $x = 0$, however, eliminates the possibility $\lambda = 0$. This means that $\lambda > 0$, which leads to

$$f(x) = C_1 \cos\left(\sqrt{\lambda}\,x\right) + C_2 \sin\left(\sqrt{\lambda}\,x\right),$$
$$f'(x) = -C_1 \sqrt{\lambda} \sin\left(\sqrt{\lambda}\,x\right) + C_2 \sqrt{\lambda} \sin\left(\sqrt{\lambda}\,x\right).$$

Using the BCs and the fact that C_1 and C_2 cannot both be equal to 0, we then find that

$$C_1 = 0, \quad \cos\left(\sqrt{\lambda}\,L\right) = 0.$$

The latter equality implies that

$$\sqrt{\lambda}\,L = \frac{(2n-1)\pi}{2}, \quad n = 1, 2, \ldots,$$

which means that the eigenvalues and eigenfunctions of this problem are, respectively,

$$\lambda_n = \frac{(2n-1)^2 \pi^2}{4L^2}, \quad f_n(x) = \sin\frac{(2n-1)\pi x}{2L}, \quad n = 1, 2, \ldots.$$

3.20 Example. A similar handling of the regular S–L problem

$$f''(x) + \lambda f(x) = 0, \quad 0 < x < L,$$
$$f'(0) = 0, \quad f(L) = 0$$

produce the equalities

$$C_2 = 0, \quad \cos\left(\sqrt{\lambda}\,L\right) = 0,$$

which lead to the eigenvalue–eigenfunction pairs

$$\lambda_n = \frac{(2n-1)^2 \pi^2}{4L^2}, \quad f_n(x) = \cos\frac{(2n-1)\pi x}{2L}, \quad n = 1, 2, \ldots.$$

3.21 Example. Suppose that $h > 0$ in the regular S–L problem considered in Example 3.12. Then, writing the second BC of that problem in the form

$$f'(L) = -h f(L),$$

we see that

$$[p(x)f(x)f'(x)]_a^b = [f(x)f'(x)]_0^L$$
$$= f(L)f'(L) - f(0)f'(0)$$
$$= -hf^2(L).$$

Consequently, (3.6) yields

$$\lambda = \frac{\int\limits_0^L (f')^2(x)\,dx + hf^2(L)}{\int\limits_0^L f^2(x)\,dx} \geq 0,$$

with $\lambda = 0$ if and only if

$$f'(x) = 0 \quad \text{on } [0, L], \quad f(L) = 0.$$

But this implies that $f = 0$, which is not acceptable. Hence, the eigenvalues of this S–L problem are positive. To find them, we note that the general solution of the equation is again given by (3.8), and that the condition $f(0) = 0$ leads to $C_1 = 0$. Using the second BC, namely

$$f'(L) + hf(L) = 0,$$

we arrive at

$$C_2\left[\sqrt{\lambda}\cos\left(\sqrt{\lambda}\,L\right) + h\sin\left(\sqrt{\lambda}\,L\right)\right] = 0.$$

For nonzero solutions, we must have $C_2 \neq 0$; in other words,

$$\sqrt{\lambda}\cos\left(\sqrt{\lambda}\,L\right) + h\sin\left(\sqrt{\lambda}\,L\right) = 0.$$

Since $\cos\left(\sqrt{\lambda}\,L\right) \neq 0$ (otherwise $\sin\left(\sqrt{\lambda}\,L\right)$ would be zero as well, which is impossible), we can divide the above equality by $\cos\left(\sqrt{\lambda}\,L\right)$ and conclude that λ must be a solution of the transcendental equation

$$\tan\left(\sqrt{\lambda}\,L\right) = -\frac{\sqrt{\lambda}}{h} = -\frac{1}{hL}\left(\sqrt{\lambda}\,L\right).$$

Setting

$$\sqrt{\lambda}\,L = \zeta,$$

we see that the values of ζ can be computed from the points of intersection of the graphs of the functions

$$y = \tan\zeta, \quad y = -\frac{\zeta}{hL}.$$

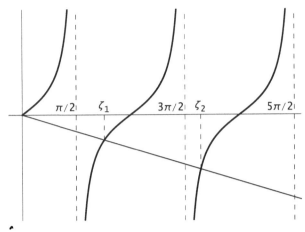

Figure 3.1: The graphs of $y = \tan \zeta$ (heavy line) and $y = -\zeta/(hL)$ (light line).

As Fig. 3.1 shows, there are countably many such points ζ_n, so this S–L problem has an infinite sequence of positive eigenvalues

$$\lambda_n = \left(\frac{\zeta_n}{L}\right)^2, \quad n = 1, 2, \ldots, \quad \lambda_n \to \infty \text{ as } n \to \infty,$$

with corresponding eigenfunctions (given by (3.8) with $C_1 = 0$ and $C_2 = 1$)

$$f_n(x) = \sin \left(\sqrt{\lambda_n}\, x\right) = \sin \frac{\zeta_n x}{L}, \quad n = 1, 2, \ldots.$$

By Theorem 3.8(iii), the set $\{f_n\}_{n=1}^{\infty}$ is orthogonal on $[0, L]$.

3.22 Example. For the S–L problem introduced in Example 3.13 we use a slightly modified approach. Here, the roots of the characteristic equation $s^2 + 2s + \lambda = 0$ are

$$s_1 = -1 + \sqrt{1 - \lambda}, \quad s_2 = -1 - \sqrt{1 - \lambda}.$$

Since the nature of the roots changes according as $\lambda < 1$, $\lambda = 1$, or $\lambda > 1$, we consider these cases separately.

(i) If $\lambda < 1$, then $1 - \lambda > 0$; hence, s_1 and s_2 are real and distinct, and the general solution of the equation is

$$f(x) = C_1 e^{s_1 x} + C_2 e^{s_2 x}, \quad C_1, C_2 = \text{const.}$$

Using the conditions $f(0) = 0$ and $f(1) = 0$, we arrive at the pair of equations

$$C_1 + C_2 = 0, \quad C_1 e^{s_1} + C_2 e^{s_2} = 0.$$

Since $s_1 \neq s_2$, this system has only the solution $C_1 = C_2 = 0$; consequently, $f = 0$, and we conclude that there are no eigenvalues $\lambda < 1$.

(ii) If $\lambda = 1$, then

$$f(x) = (C_1 + C_2 x)e^{-x}, \quad C_1, C_2 = \text{const.}$$

The condition $f(0) = 0$ yields $C_1 = 0$, while $f(1) = 0$ gives $C_2 e^{-1} = 0$, which means that $C_2 = 0$. This generates $f = 0$, so $\lambda = 1$ is not an eigenvalue.

(iii) If $\lambda > 1$, then $\lambda - 1 > 0$ and the general solution of the equation is

$$f(x) = e^{-x}\left[C_1 \cos\left(\sqrt{\lambda - 1}\, x\right) + C_2 \sin\left(\sqrt{\lambda - 1}\, x\right)\right], \quad C_1, C_2 = \text{const.}$$

Since $f(0) = 0$, we find that $C_1 = 0$, and from the condition $f(1) = 0$ it follows that

$$C_2 e^{-1} \sin \sqrt{\lambda - 1} = 0.$$

For nonzero solutions we must have

$$\sin \sqrt{\lambda - 1} = 0;$$

in other words,

$$\sqrt{\lambda - 1} = n\pi, \quad n = 1, 2, \ldots.$$

This implies that the eigenvalues of the problem are

$$\lambda_n = n^2 \pi^2 + 1, \quad n = 1, 2, \ldots,$$

with corresponding eigenfunctions

$$f_n(x) = e^{-x} \sin\left(\sqrt{\lambda_n - 1}\, x\right) = e^{-x} \sin(n\pi x), \quad n = 1, 2, \ldots.$$

VERIFICATION WITH MATHEMATICA®. The input

```
λn = n^2 * Pi^2 + 1;
fn = E^( - x) * Sin[n * Pi * x];
{D[fn,x,x] + 2 * D[fn,x] + λn * fn, fn /. x -> 0, fn /. x -> 1}
  // FullSimplify
```

generates the output $\{0, 0, 0\}$.

3.23 Remark. In Example 3.22, we may have been tempted to discard case (i) without detailed investigation because the restriction $\lambda < 1$ seems to contradict Corollary 3.15, which states that the eigenvalues of a regular S–L problem form a sequence whose terms increase without bound. In fact, for certain types of BCs and values of the constant coefficients a, b, and c, an equation of the form

$$f''(x) + af'(x) + bf(x) + c\lambda f(x) = 0$$

may produce an eigenvalue–eigenfunction pair even when the roots of the characteristic equation are real and distinct.

3.24 Example. The characteristic equation for the regular S–L problem

$$f''(x) + 2f'(x) - f(x) + \lambda f(x) = 0, \quad 0 < x < 1,$$
$$f(0) = 0, \quad 2f(1) - 3f'(1) = 0$$

is $s^2 + 2s - (1 - \lambda) = 0$, with roots

$$s_1 = -1 + \sqrt{2 - \lambda}, \quad s_2 = -1 - \sqrt{2 - \lambda}.$$

Suppose that $\lambda < 2$, which means that $2 - \lambda > 0$. Then s_1 and s_2 are real and distinct, therefore, the general solution of the equation can be written as $f(x) = C_1 e^{s_1 x} + C_2 e^{s_2 x}$. But, as it turns out, this leads to rather untidy computation, so, instead, we use Remark 1.4(ii) with

$$\alpha = \beta = 0, \quad s_1 + s_2 = -2, \quad s_1 - s_2 = 2\sqrt{2 - \lambda},$$

and work with the alternative form

$$f(x) = e^{-x} \left[C_1 \cosh\left(\sqrt{2 - \lambda}\, x\right) + C_2 \sinh\left(\sqrt{2 - \lambda}\, x\right) \right].$$

The BC at $x = 0$ yields $C_1 = 0$; hence, applying the BC at $x = 1$ (with, as before, a nonzero value for C_2), we arrive at the equality

$$5 \sinh \sqrt{2 - \lambda} - 3\sqrt{2 - \lambda} \cosh \sqrt{2 - \lambda} = 0,$$

or, since the hyperbolic cosine does not vanish for any value of its argument,

$$\tanh \sqrt{2 - \lambda} = \tfrac{3}{5} \sqrt{2 - \lambda}.$$

As the graphs of the functions defined by the two sides above clearly show (see Fig. 3.2), this equation has only one root satisfying $\lambda < 2$, which is approximately -0.2868. Consequently, we have the eigenvalue–eigenfunction pair

$$\lambda_0 \cong -0.2868, \quad f_0(x) = e^{-x} \sinh\left(\sqrt{2 - \lambda_0}\, x\right) \cong e^{-x} \sinh(1.5122\, x).$$

The cases $\lambda = 2$ and $\lambda > 2$ are treated as in Example 3.22.

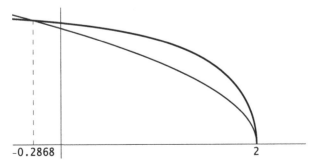

Figure 3.2: The graphs of $y = \tanh \sqrt{2 - \lambda}$ (heavy line) and $y = \tfrac{3}{5} \sqrt{2 - \lambda}$ (light line).

VERIFICATION WITH MATHEMATICA®. The input

```
λ0 = - 0.2868;
f0 = E^( - x) * Sinh[1.5122 * x];
{D[f0,x,x] + 2 * D[f0,x] - f0 + λ0 * f0, f0 /. x -> 0,
  (2 * f0 - 3 * D[f0,x]) /. x -> 1}
  //FullSimplify
```

generates, to four decimal places, the output $\{0, 0, 0\}$.

3.25 Remark. Using the same techniques as in Examples 3.13 and 3.22, we find that the eigenvalue problem

$$f''(x) + af'(x) + bf(x) + c\lambda f(x) = 0, \quad 0 < x < L,$$
$$f(0) = 0, \quad f(L) = 0,$$

where a, b, $c = $ const, $c > 0$, is a regular S–L problem with

$$p(x) = e^{ax}, \quad q(x) = be^{ax}, \quad \sigma(x) = ce^{ax},$$

eigenvalues

$$\lambda_n = \frac{1}{4c}\left(\frac{4n^2\pi^2}{L^2} + a^2 - 4b\right), \quad n = 1, 2, \ldots,$$

and eigenfunctions

$$f_n(x) = e^{-(a/2)x}\sin\frac{n\pi x}{L}, \quad n = 1, 2, \ldots.$$

The eigenvalue–eigenfunction pairs computed for some of the regular S–L problems in this chapter are summarized in Appendix A.2.

Exercises

In **1–10**, compute the eigenvalues and eigenfunctions of the given regular S–L problem.

1 $f''(x) + \lambda f(x) = 0, \quad 0 < x < \pi, \quad f(0) = 0, \quad f'(\pi) = 0.$

2 $f''(x) + \lambda f(x) = 0, \quad 0 < x < 1, \quad f'(0) = 0, \quad f(1) = 0.$

3 $f''(x) + \lambda f(x) = 0, \quad 0 < x < 1, \quad f(0) - f'(0) = 0, \quad f(1) = 0.$

4 $f''(x) + \lambda f(x) = 0, \quad 0 < x < 1, \quad f'(0) = 0, \quad f(1) + f'(1) = 0.$

5 $f''(x) + \lambda f(x) = 0, \quad 0 < x < 1, \quad f(0) - f'(0) = 0, \quad f'(1) = 0.$

6 $2f''(x) + (\lambda - 1)f(x) = 0, \quad 0 < x < 2, \quad f(0) = 0, \quad f(2) = 0.$

7 $f''(x) + 4f'(x) + 3\lambda f(x) = 0, \quad 0 < x < 1, \quad f(0) = 0, \quad f(1) = 0.$

8 $f''(x) - f'(x) + \lambda f(x) = 0$, $0 < x < 1$, $f'(0) = 0$, $f'(1) = 0$.

9 $2f''(x) + 3f'(x) + \lambda f(x) = 0$, $0 < x < 1$, $f(0) = 0$, $f'(1) = 0$.

10 $f''(x) - 5f'(x) + 2\lambda f(x) = 0$, $0 < x < \pi/2$, $f'(0) = 0$, $f(\pi/2) = 0$.

In **11–16**, show that the regular S–L problem

$$f''(x) + af'(x) + bf(x) + c\lambda f(x) = 0, \quad 0 < x < 1,$$
$$\kappa_1 f(0) + \kappa_2 f'(0) = 0, \quad \kappa_3 f(1) + \kappa_4 f'(1) = 0$$

with the constants a, b, c and κ_1, κ_2, κ_3, κ_4 as indicated, has an exponential-type eigenfunction, and compute the corresponding eigenvalue–eigenfunction pair, rounding all numbers to the fourth decimal place.

11 $a = 3$, $b = -1$, $c = 2$, $\kappa_1 = 1$, $\kappa_2 = 0$, $\kappa_3 = 1$, $\kappa_4 = -2$.

12 $a = 4$, $b = -2$, $c = 1$, $\kappa_1 = 1$, $\kappa_2 = 0$, $\kappa_3 = 2$, $\kappa_4 = -1$.

13 $a = -2$, $b = -2$, $c = 5$, $\kappa_1 = 1$, $\kappa_2 = 0$, $\kappa_3 = 4$, $\kappa_4 = -1$.

14 $a = -1$, $b = -2$, $c = 2$, $\kappa_1 = 1$, $\kappa_2 = 0$, $\kappa_3 = 5$, $\kappa_4 = -2$.

15 $a = 1$, $b = 2$, $c = 5$, $\kappa_1 = 2$, $\kappa_2 = 1$, $\kappa_3 = 1$, $\kappa_4 = 0$.

16 $a = -3$, $b = 1$, $c = 1$, $\kappa_1 = 3$, $\kappa_2 = 2$, $\kappa_3 = 1$, $\kappa_4 = 0$.

Answers to Odd-Numbered Exercises

1 $\lambda_n = (2n - 1)^2/4$, $f_n(x) = \sin\big((2n - 1)x/2\big)$, $n = 1, 2, \ldots$.

3 $\lambda_n = \zeta_n^2$, where ζ_n are the roots of the equation $\tan\zeta = -\zeta$,
   $f_n(x) = \zeta_n \cos(\zeta_n x) + \sin(\zeta_n x)$, $n = 1, 2, \ldots$.

5 $\lambda_n = \zeta_n^2$, where ζ_n are the roots of the equation $\cot\zeta = \zeta$,
   $f_n(x) = \zeta_n \cos(\zeta_n x) + \sin(\zeta_n x)$, $n = 1, 2, \ldots$.

7 $\lambda_n = (4 + n^2\pi^2)/3$, $f_n(x) = e^{-2x} \sin(n\pi x)$, $n = 1, 2, \ldots$.

9 $\lambda_n = 2\zeta_n^2 + 9/8$, where ζ_n are the roots of the equation $\tan\zeta = 4\zeta/3$,
   $f_n(x) = e^{-3x/4} \sin(\zeta_n x)$, $n = 1, 2, \ldots$.

11 $\lambda_0 = -0.2086$, $f_0(x) = e^{-3x/2} \sinh(1.9150\,x)$.

13 $\lambda_0 = -1.9817$, $f_0(x) = e^x \sinh(2.9847\,x)$.

15 $\lambda_0 = -0.6817$, $f_0(x) = e^{-x/2} \sinh\big(1.2878(x - 1)\big)$.

3.1.2 Generalized Fourier Series

Regular Sturm–Liouville problems have two important additional properties.

3.26 Theorem. (i) *Only one linearly independent eigenfunction f_n exists for each eigenvalue λ_n, $n = 1, 2, \ldots$.*

(ii) *The set $\{f_n\}_{n=1}^{\infty}$ of eigenfunctions is* complete; *that is, any piecewise smooth function u on $[a, b]$ has a* unique *generalized Fourier series (or eigenfunction) expansion of the form*

$$u(x) \sim \sum_{n=1}^{\infty} c_n f_n(x), \quad a \leq x \leq b, \tag{3.9}$$

which converges pointwise to

$$\tfrac{1}{2}\left[u(x-) + u(x+)\right], \quad a \leq x \leq b$$

(in particular, to $u(x)$ if u is continuous at x).

3.27 Remarks. (i) The coefficients c_n are computed by means of the properties mentioned in Theorem 3.26(i) and Theorem 3.8(iii). Thus, treating (3.9) formally as an equality, multiplying it by $f_m(x)\sigma(x)$, and integrating over $[a, b]$, we obtain

$$\int_a^b u(x) f_m(x) \sigma(x) \, dx = \sum_{n=1}^{\infty} c_n \int_a^b f_n(x) f_m(x) \sigma(x) \, dx$$

$$= c_m \int_a^b f_m^2(x) \sigma(x) \, dx.$$

Replacing m by n, we can now write

$$c_n = \frac{\int\limits_a^b u(x) f_n(x) \sigma(x) \, dx}{\int\limits_a^b f_n^2(x) \sigma(x) \, dx}, \quad n = 1, 2, \ldots. \tag{3.10}$$

The denominator in (3.10) cannot be zero for any n since the f_n are nonzero solutions of (3.2)–(3.4) and $\sigma(x) > 0$.

(ii) In the case of the S–L problems discussed in Examples 3.10 and 3.16, and in Examples 3.11 and 3.17, expansion (3.9) coincides, respectively, with the Fourier sine and cosine series of u, and (3.10) yields formulas (2.12) and (2.15). (For the problem discussed in Examples 3.11 and 3.17, we have $n = 0, 1, 2, \ldots$.)

3.28 Example. Suppose that we want to expand the function

$$u(x) = 1, \quad 0 \le x \le 2\pi,$$

in the (complete) set of the eigenfunctions

$$f_n = \sin\left((2n-1)x/4\right), \quad n = 1, 2, \ldots,$$

of the regular S–L problem discussed in Example 3.19:

$$u(x) \sim \sum_{n=1}^{\infty} c_n \sin \frac{(2n-1)x}{4}.$$

Direct computation shows that

$$\int_0^{2\pi} \sin \frac{(2n-1)x}{4} \, dx = \frac{4}{2n-1},$$

$$\int_0^{2\pi} \sin^2 \frac{(2n-1)x}{4} \, dx = \pi,$$

so, by (3.10) with $\sigma(x) = 1$,

$$c_n = \frac{4}{(2n-1)\pi}, \quad n = 1, 2, \ldots;$$

hence,

$$u(x) \sim \sum_{n=1}^{\infty} \frac{4}{(2n-1)\pi} \sin \frac{(2n-1)x}{4}.$$

VERIFICATION WITH MATHEMATICA®. The input

```
fn=Sin[(2*n-1)*x/4];
GenTerm = (4/((2*n-1)*Pi))*fn;
Plot[{Sum[GenTerm,n,10], 1}, {x,0,2*Pi},
   PlotRange ->{{-0.4, 2*Pi+0.2}, {-0.2, 1.3}},
   PlotStyle ->Black, ImageSize -> Scaled[0.3],
   Ticks ->{{2*Pi}, {1}}, AspectRatio ->0.4]
```

generates the graphical output

3.29 Example. Suppose that we want to expand the function

$$u(x) = x + 1, \quad 0 \le x \le 1,$$

in the eigenfunctions of the regular S–L problem discussed in Examples 3.12 and 3.21 with $L = h = 1$. The first five positive roots of the equation

$$\tan \zeta = -\zeta$$

mentioned in the latter, rounded to the fourth decimal place, are

$$\zeta_1 = 2.0288, \quad \zeta_2 = 4.9132, \quad \zeta_3 = 7.9787, \quad \zeta_4 = 11.0855, \quad \zeta_5 = 14.2074.$$

Then, using (3.10) with $\sigma(x) = 1$ and $f_n(x) = \sin(\zeta_n x)$, we find the approximate coefficients

$$c_1 = 1.9184, \quad c_2 = 0.1572, \quad c_3 = 0.3990, \quad c_4 = 0.1307, \quad c_5 = 0.1696,$$

so (3.9) takes the form

$$u(x) \sim 1.9184 \sin(2.0288\,x) + 0.1572 \sin(4.9132\,x)$$
$$+ 0.3390 \sin(7.9787\,x) + 0.1307 \sin(11.0855\,x)$$
$$+ 0.1696 \sin(14.2074\,x) + \cdots. \tag{3.11}$$

VERIFICATION WITH MATHEMATICA®. The input (based on the computed expansion (3.11))

```
Approx = 1.9184 * Sin[2.0288 * x] + 0.1572 * Sin[4.9132 * x]
  + 0.3390 * Sin[7.9787 * x] + 0.1307 * Sin[11.0855 * x]
  + 0.1696 * Sin[14.2074 * x];
Plot[{Approx, x+1}, {x,0,1}, PlotRange -> {{-0.1,1.1},
  {-0.2,2.2}}, PlotStyle -> Black, ImageSize -> Scaled[0.22],
  Ticks -> {{1}, {1,2}}, AspectRatio -> 0.9]
```

generates the graphical output

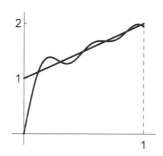

3.30 Example. Let us expand the function

$$u(x) = \begin{cases} 0, & 0 \le x \le 1/2, \\ 1, & 1/2 < x \le 1, \end{cases}$$

in the (complete) set of the eigenfunctions $f_n(x) = e^{-x}\sin(n\pi x)$, $n = 1, 2, \dots$, of the regular S–L problem discussed in Examples 3.13 and 3.22, where it was also specified that $\sigma(x) = e^{2x}$. Using a formula in Appendix A.2, for $n = 1, 2, \dots$ we have

$$\int_0^1 u(x) f_n(x) \sigma(x)\, dx = \int_{1/2}^1 e^x \sin(n\pi x)\, dx$$

$$= \frac{e^x}{1 + n^2\pi^2}\Big[-n\pi\cos(n\pi x) + \sin(n\pi x) \Big]_{1/2}^1$$

$$= \frac{1}{1 + n^2\pi^2}\Big\{ (-1)^{n+1} n\pi e$$

$$+ \Big(n\pi\cos\frac{n\pi}{2} - \sin\frac{n\pi}{2} \Big) e^{1/2} \Big\}$$

and

$$\int_0^1 f_n^2(x) \sigma(x)\, dx = \int_0^1 \sin^2(n\pi x)\, dx = \tfrac{1}{2},$$

which, replaced in (3.10), yield the coefficients

$$c_n = \frac{2}{1 + n^2\pi^2}\Big\{ (-1)^{n+1} n\pi e + \Big(n\pi\cos\frac{n\pi}{2} - \sin\frac{n\pi}{2} \Big) e^{1/2} \Big\},$$

and, thus, the expansion

$$u(x) \sim \sum_{n=1}^\infty c_n e^{-x} \sin(n\pi x). \tag{3.12}$$

VERIFICATION WITH MATHEMATICA®. The input (based on the computed expansion (3.12))

```
cn = 2 / (1 + m^2 * Pi^2) * ((-1)^(n+1) * m * Pi * E +
   (n * Pi * Cos[n * Pi /2] - Sin[n * Pi /2]) * E^(1/2));
GenTerm = cn * E^(-x) * Sin[n * Pi * x];
Plot[{Sum[GenTerm, {n,1,10}], 1}, {x,0,1},
   PlotRange -> {{-0.4, 1.2}, {-0.4, 1.3}}, PlotStyle -> Black,
   ImageSize -> Scaled[0.3], Ticks -> {{0.5,1},{1}},
   AspectRatio -> 0.4]
```

generates the graphical output

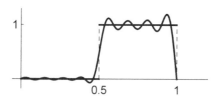

Exercises

In **1–8**, construct the generalized Fourier series expansion for the given function u in the eigenfunctions

$$\left\{\sin\left((2n-1)x/2\right)\right\}_{n=1}^{\infty}$$

of the S–L problem discussed in Example 3.19 with $L = \pi$.

1 $u(x) = 1, \quad 0 \le x \le \pi.$

2 $u(x) = 2x - 1, \quad 0 \le x \le \pi.$

3 $u(x) = 3x + 2, \quad 0 \le x \le \pi.$

4 $u(x) = \begin{cases} -1, & 0 \le x \le \pi/2, \\ 2, & \pi/2 < x \le \pi. \end{cases}$

5 $u(x) = \begin{cases} 2, & 0 \le x \le \pi/2, \\ -3, & \pi/2 < x \le \pi. \end{cases}$

6 $u(x) = \begin{cases} 2x - 1, & 0 \le x \le \pi/2, \\ -1, & \pi/2 < x \le \pi. \end{cases}$

7 $u(x) = \begin{cases} 2, & 0 \le x \le \pi/2, \\ x + 1, & \pi/2 < x \le \pi. \end{cases}$

8 $u(x) = \begin{cases} 2x + 1, & 0 \le x \le \pi/2, \\ 3 - 2x, & \pi/2 < x \le \pi. \end{cases}$

In **9–16**, construct the generalized Fourier series expansion for the given function u in the eigenfunctions

$$\left\{\cos\left((2n-1)\pi x/2\right)\right\}_{n=1}^{\infty}$$

of the S–L problem discussed in Example 3.20 with $L = 1$.

9 $u(x) = 1/2, \quad 0 \le x \le 1.$

10 $u(x) = x + 1, \quad 0 \le x \le 1.$

11 $u(x) = 2 - x, \quad 0 \le x \le 1.$

12 $u(x) = \begin{cases} 3, & 0 \le x \le 1/2, \\ 4, & 1/2 < x \le 1. \end{cases}$

13 $u(x) = \begin{cases} -2, & 0 \le x \le 1/2, \\ 1, & 1/2 < x \le 1. \end{cases}$

14 $u(x) = \begin{cases} 3 - 2x, & 0 \le x \le 1/2, \\ 2, & 1/2 < x \le 1. \end{cases}$

15 $u(x) = \begin{cases} 1, & 0 \le x \le 1/2, \\ 2x - 3, & 1/2 < x \le 1. \end{cases}$ **16** $u(x) = \begin{cases} 1 - x, & 0 \le x \le 1/2, \\ 2x + 1, & 1/2 < x \le 1. \end{cases}$

In **17–24**, compute the first five terms of the generalized Fourier series expansion for the given function u in the eigenfunctions of the S–L problem discussed in Example 3.21 with $h = 1$ and L as defined by the given interval for x, rounding all numbers to the fourth decimal place.

17 $u(x) = 1, \quad 0 \le x \le 1.$

18 $u(x) = 1 - 2x, \quad 0 \le x \le 1.$

19 $u(x) = 2 - x, \quad 0 \le x \le 2.$

20 $u(x) = \begin{cases} 1, & 0 \le x \le 1/2, \\ -3, & 1/2 < x \le 1. \end{cases}$

21 $u(x) = \begin{cases} -2, & 0 \le x \le 1/2, \\ 4, & 1/2 < x \le 1. \end{cases}$

22 $u(x) = \begin{cases} -1, & 0 \le x \le 1, \\ x + 2, & 1 < x \le 2. \end{cases}$

23 $u(x) = \begin{cases} 2x, & 0 \le x \le 1/2, \\ 1, & 1/2 < x \le 1. \end{cases}$

24 $u(x) = \begin{cases} 1 - x, & 0 \le x \le 1/2, \\ 1 + x, & 1/2 < x \le 1. \end{cases}$

In **25–32**, compute the first five terms of the generalized Fourier series expansion for the given function u in the eigenfunctions

$$\left\{ e^{-2x} \sin(n\pi x) \right\}_{n=1}^{\infty}$$

of the S–L problem discussed in Remark 3.25 with $a = 4$ and $L = 1$, rounding all numbers to the fourth decimal place.

25 $u(x) = 1, \quad 0 \le x \le 1.$

26 $u(x) = x, \quad 0 \le x \le 1.$

27 $u(x) = 2x + 1, \quad 0 \le x \le 1.$

28 $u(x) = \begin{cases} -2, & 0 \le x \le 1/2, \\ -1, & 1/2 < x \le 1. \end{cases}$

29 $u(x) = \begin{cases} 1, & 0 \le x \le 1/2, \\ -3, & 1/2 < x \le 1. \end{cases}$

30 $u(x) = \begin{cases} 2, & 0 \le x \le 1/2, \\ 2x + 1, & 1/2 < x \le 1. \end{cases}$

31 $u(x) = \begin{cases} x - 1, & 0 \le x \le 1/2, \\ -1, & 1/2 < x \le 1. \end{cases}$

32 $u(x) = \begin{cases} x + 1, & 0 \le x \le 1/2, \\ 3 - 2x, & 1/2 < x \le 1. \end{cases}$

Answers to Odd-Numbered Exercises

1 $u(x) \sim \sum_{n=1}^{\infty} [4/((2n - 1)\pi)] \sin((2n - 1)x/2).$

3 $u(x) \sim \sum_{n=1}^{\infty} [8(2n - 1 - 3(-1)^n)/((2n - 1)^2\pi)] \sin((2n - 1)x/2).$

5 $u(x) \sim \sum_{n=1}^{\infty} \{[8 - 20\sin((3 - 2n)\pi/4)]/((2n - 1)\pi)\} \sin((2n - 1)x/2).$

7 $u(x) \sim \sum\limits_{n=1}^{\infty} [2/((2n-1)^2\pi)]\{(2n-1)(\pi-2)\sin((3-2n)\pi/4)$

$$- 4[1+(-1)^n - 2n + \sin((2n-1)\pi/4)]\}\sin((2n-1)x/2).$$

9 $u(x) \sim \sum\limits_{n=1}^{\infty} [(-1)^{n+1}2/((2n-1)\pi)]\cos((2n-1)\pi x/2).$

11 $u(x) \sim \sum\limits_{n=1}^{\infty} [(8-4(-1)^n(2n-1)\pi)/((2n-1)^2\pi^2)]\cos((2n-1)\pi x/2).$

13 $u(x) \sim \sum\limits_{n=1}^{\infty} \{4[(-1)^{n+1} - 3\sin((2n-1)\pi/4)]/((2n-1)\pi)\}$

$$\times \cos((2n-1)\pi x/2).$$

15 $u(x) \sim \sum\limits_{n=1}^{\infty} [4/((2n-1)^2\pi^2)]\{4(2n-1)\pi\sin((2n-1)\pi/4)$

$$- 4\sin((3-2n)\pi/4) + (2n-1)\pi[(-1)^n - \sin((2n-1)\pi/4)]\}$$

$$\times \cos((2n-1)\pi x/2).$$

17 $u(x) \sim 1.1892\sin(2.0288\,x) + 0.3134\sin(4.9132\,x) + 0.2776\sin(7.9787\,x)$

$$+ 0.1629\sin(11.0855\,x) + 0.1499\sin(14.2074\,x) + \cdots.$$

19 $u(x) \sim 0.9639\sin(1.1444\,x) + 0.8718\sin(2.5435\,x) + 0.4227\sin(4.0481\,x)$

$$+ 0.3836\sin(5.5863\,x) + 0.2583\sin(7.1382\,x) + \cdots.$$

21 $u(x) \sim 2.4222\sin(2.0288\,x) - 2.9144\sin(4.9132\,x) - 1.3509\sin(7.9787\,x)$

$$+ 0.3704\sin(11.0855\,x) + 0.3322\sin(14.2074\,x) + \cdots.$$

23 $u(x) \sim 1.0549\sin(2.0288\,x) + 0.0227\sin(4.9132\,x) - 0.0157\sin(7.9787\,x)$

$$- 0.3785\sin(11.0855\,x) + 0.0242\sin(14.2074\,x) + \cdots.$$

25 $u(x) \sim e^{-2x}[3.8004\sin(\pi x) - 1.8466\sin(2\pi x) + 1.7035\sin(3\pi x)$

$$- 0.9917\sin(4\pi x) + 1.0511\sin(5\pi x) + \cdots].$$

27 $u(x) \sim e^{-2x}[8.3031\sin(\pi x) - 5.7781\sin(2\pi x) + 4.5576\sin(3\pi x)$

$$- 3.2366\sin(4\pi x) + 2.8691\sin(5\pi x) + \cdots].$$

29 $u(x) \sim e^{-2x}[-6.4533\sin(\pi x) + 9.8385\sin(2\pi x) - 4.7668\sin(3\pi x)$

$$+ 1.9083\sin(4\pi x) - 2.4786\sin(5\pi x) + \cdots].$$

31 $u(x) \sim e^{-2x}[-3.3732\sin(\pi x) + 2.1406\sin(2\pi x) - 1.8243\sin(3\pi x)$

$$+ 0.7873\sin(4\pi x) - 1.0104\sin(5\pi x) + \cdots].$$

3.2 Other Problems

Many important Sturm–Liouville problems are not regular.

3.31 Definition. With the notation in Definition 3.9, an eigenvalue problem that consists in solving equation (3.2) with the conditions

$$f(a) = f(b), \quad f'(a) = f'(b) \tag{3.13}$$

is called a *periodic Sturm–Liouville problem*.

3.32 Remark. It can be shown that the differential operator L introduced in (3.5) is also symmetric on the new space of functions defined by conditions (3.13), so Theorem 3.8 is again applicable. However, in contrast to regular S–L problems, here we may have two linearly independent eigenfunctions for the same eigenvalue. Nevertheless, we can always choose a pair of eigenfunctions that are orthogonal with weight σ on $[a, b]$. For such a choice, Theorem 3.26(ii) remains valid.

3.33 Example. Consider the periodic S–L problem where $p(x) = 1$, $q(x) = 0$, $\sigma(x) = 1$, $a = -L$, and $b = L$, $L > 0$; that is,

$$f''(x) + \lambda f(x) = 0, \quad -L < x < L,$$
$$f(-L) = f(L), \quad f'(-L) = f'(L).$$

Using the Rayleigh quotient argument and the above BCs, we arrive again at inequality (3.7). We accept the case $\lambda = 0$ since, as in Example 3.17, the corresponding constant solutions satisfy both BCs; therefore, the problem has the eigenvalue–eigenfunction pair

$$\lambda_0 = 0, \quad f_0(x) = \tfrac{1}{2}.$$

For $\lambda > 0$, the general solution of the equation is

$$f(x) = C_1 \cos \left(\sqrt{\lambda}\, x \right) + C_2 \sin \left(\sqrt{\lambda}\, x \right), \quad C_1, C_2 = \text{const},$$

with derivative

$$f'(x) = -C_1 \sqrt{\lambda} \sin \left(\sqrt{\lambda}\, x \right) + C_2 \sqrt{\lambda} \cos \left(\sqrt{\lambda}\, x \right).$$

From the BCs and the fact that $\cos(-\alpha) = \cos \alpha$ and $\sin(-\alpha) = -\sin \alpha$ it follows that C_1 and C_2 satisfy the system

$$C_1 \cos \left(\sqrt{\lambda}\, L \right) - C_2 \sin \left(\sqrt{\lambda}\, L \right) = C_1 \cos \left(\sqrt{\lambda}\, L \right) + C_2 \sin \left(\sqrt{\lambda}\, L \right),$$
$$C_1 \sin \left(\sqrt{\lambda}\, L \right) + C_2 \cos \left(\sqrt{\lambda}\, L \right) = -C_1 \sin \left(\sqrt{\lambda}\, L \right) + C_2 \cos \left(\sqrt{\lambda}\, L \right),$$

which reduces to

$$C_1 \sin \left(\sqrt{\lambda}\, L \right) = 0, \quad C_2 \sin \left(\sqrt{\lambda}\, L \right) = 0.$$

Since we want nonzero solutions, we cannot have $\sin\left(\sqrt{\lambda}\,L\right) \neq 0$ because this would imply $C_1 = C_2 = 0$—that is, $f = 0$; hence, $\sin\left(\sqrt{\lambda}\,L\right) = 0$. As we have seen in previous examples, this yields the eigenvalues

$$\lambda_n = \left(\frac{n\pi}{L}\right)^2, \quad n = 1, 2, \ldots.$$

Given that both C_1 and C_2 remain arbitrary, from the general solution we can extract (by taking $C_1 = 1$, $C_2 = 0$ and then $C_1 = 0$, $C_2 = 1$) for every λ_n the two linearly independent eigenfunctions

$$f_{1n}(x) = \cos\frac{n\pi x}{L}, \quad f_{2n}(x) = \sin\frac{n\pi x}{L}, \quad n = 1, 2, \ldots,$$

which, as noted in Chapter 2, are orthogonal on $[-L, L]$. Expansion (3.9) in this case becomes

$$u(x) \sim c_0 f_0(x) + \sum_{n=1}^{\infty}\left[c_{1n} f_{1n}(x) + c_{2n} f_{2n}(x)\right]$$

$$= \tfrac{1}{2}c_0 + \sum_{n=1}^{\infty}\left(c_{1n}\cos\frac{n\pi x}{L} + c_{2n}\sin\frac{n\pi x}{L}\right),$$

which we recognize as the full Fourier series of u. The coefficients c_0, c_{1n}, and c_{2n}, computed by means of (3.10) with $\sigma(x) = 1$, $a = -L$, $b = L$, and $n = 0, 1, 2, \ldots$ (to allow for the additional eigenfunction f_0), coincide with those given by (2.7)–(2.9).

3.34 Remark. It is easily shown that the eigenvalues and eigenfunctions computed in Example 3.33 remain the same if the interval $-L < x < L$ is replaced by $0 < x < 2L$ or any other interval of length $2L$.

3.35 Definition. An eigenvalue problem involving equation (3.2) is called a *singular Sturm–Liouville problem* if it exhibits one or more of the following features:

(i) $p(a) = 0$ or $p(b) = 0$ or both;

(ii) any of $p(x)$, $q(x)$, and $\sigma(x)$ becomes infinite as $x \to a+$ or as $x \to b-$ or both;

(iii) the interval (a, b) is infinite at a or at b or at both.

3.36 Remark. In the case of a singular Sturm–Liouville problem, some new boundary conditions need to be chosen to ensure that the operator L defined by (3.5) remains symmetric. For example, if $p(a) = 0$ but $p(b) \neq 0$, then we may require $f(x)$ and $f'(x)$ to remain bounded as $x \to a+$ and to satisfy (3.4). If the interval where the equation holds is of the form (a, ∞), then we may require $f(x)$ and $f'(x)$ to be bounded as $x \to \infty$ and to satisfy (3.3). Other types of boundary conditions, dictated by analytic and/or physical necessity, may also be considered.

3.37 Example. The eigenvalue problem

$$(x^2 + 1)f''(x) + 2xf'(x) + (\lambda - x)f(x) = 0, \quad x > 1,$$
$$f'(1) = 0, \quad f(x), \ f'(x) \text{ bounded as } x \to \infty$$

is a singular S–L problem: the ODE can be written in the form (3.2) with $p(x) = x^2 + 1$, $q(x) = -x$, $\sigma(x) = 1$, and $(a, b) = (1, \infty)$, and the boundary conditions are of the type indicated in the latter part of Remark 3.36.

A number of singular S–L problems occurring in the study of mathematical models give rise to important classes of functions. We discuss some of them explicitly.

3.3 Bessel Functions

Consider the singular S–L problem

$$x^2 f''(x) + xf'(x) + (\lambda x^2 - m^2)f(x) = 0, \quad 0 < x < a, \qquad (3.14)$$
$$f(x), \ f'(x) \text{ bounded as } x \to 0+, \quad f(a) = 0, \qquad (3.15)$$

where a is a given positive number and m is a nonnegative integer. Writing (3.14) in the form

$$\left[xf'(x)\right]' - \frac{m^2}{x} f(x) + \lambda x f(x) = 0,$$

we verify that this is (3.2) with $p(x) = x$, $q(x) = -m^2/x$, and $\sigma(x) = x$.

First, we consider the case $m = 0$, when $q = 0$ and (3.14) reduces to

$$xf''(x) + f'(x) + \lambda x f(x) = 0, \quad 0 < x < a.$$

Taking (3.15) into account, we see that

$$p(x)(f')^2(x) - q(x)f^2(x) = x(f')^2(x) \geq 0, \quad 0 < x < a,$$
$$\left[p(x)f(x)f'(x)\right]_0^a = \left[xf(x)f'(x)\right]_0^a$$
$$= af(a)f'(a) - \lim_{x \to 0+} xf(x)f'(x) = 0;$$

consequently, according to the Rayleigh quotient (3.6),

$$\lambda = \frac{\int\limits_0^a x(f')^2(x) \, dx}{\int\limits_0^a x f^2(x) \, dx} \geq 0.$$

In fact, as the above formula shows, we have $\lambda = 0$ if and only if $f'(x) = 0$, $0 < x \leq a$; that is, if and only if $f(x) = \text{const}$. But this is impossible, since the boundary condition $f(a) = 0$ would imply that f is the zero solution, which contradicts the definition of an eigenfunction. We must therefore conclude that $\lambda > 0$.

Now let $m \geq 1$. Multiplying (3.14) by $f(x)$ and integrating from 0 to a, we arrive at

$$\int_0^a \left[x^2 f''(x) f(x) + x f'(x) f(x) + \lambda x^2 f^2(x) - m^2 f^2(x) \right] dx = 0. \qquad (3.16)$$

Integration by parts, the interpretation of $f(x)$ and $f'(x)$ at $x = 0$ in the limiting sense as above, and (3.15) yield the equalities

$$\int_0^a x f'(x) f(x) \, dx = \int_0^a \tfrac{1}{2} x (f^2(x))' \, dx = \tfrac{1}{2} \left[x f^2(x) \right]_0^a - \tfrac{1}{2} \int_0^a f^2(x) \, dx$$

$$= -\tfrac{1}{2} \int_0^a f^2(x) \, dx,$$

$$\int_0^a x^2 f''(x) f(x) \, dx = \left[x^2 f'(x) f(x) \right]_0^a - \int_0^a \left[x^2 (f'(x))^2 + 2x f'(x) f(x) \right] dx$$

$$= -\int_0^a \left[x^2 (f'(x))^2 - f^2(x) \right] dx,$$

which, replaced in (3.16), lead to

$$\lambda = \frac{\int_0^a \left[x^2 (f'(x))^2 + \left(m^2 - \tfrac{1}{2} \right) f^2(x) \right] dx}{\int_0^a x^2 f^2(x) \, dx} \geq 0.$$

The same argument used for $m = 0$ rules out the value $\lambda = 0$; hence, $\lambda > 0$. For $m \geq 0$, let $\xi = \sqrt{\lambda}\, x$ and $f(x) = f(\xi/\sqrt{\lambda}) = g(\xi)$. By the chain rule,

$$f'(x) = \sqrt{\lambda}\, g'(\xi), \quad f''(x) = \lambda g''(\xi),$$

and (3.14) becomes

$$\xi^2 g''(\xi) + \xi g'(\xi) + (\xi^2 - m^2) g(\xi) = 0, \qquad (3.17)$$

which is *Bessel's equation of order m*. Two linearly independent solutions of this equation are the *Bessel functions of the first and second kind and order m*, denoted by J_m and Y_m, respectively.

3.38 Remarks.   (i) For $m \geq 1$, the J_m satisfy the recurrence relations

$$J_{m-1}(x) + J_{m+1}(x) = \frac{2m}{x} J_m(x),$$

$$J_{m-1}(x) - J_{m+1}(x) = 2J_m'(x),$$

from which we deduce that

$$\left[x^m J_m(x)\right]' = x^m J_{m-1}(x).$$

Similar equalities are satisfied by the Y_m.

(ii) The J_m and Y_m can also be computed by means of the Bessel integrals

$$J_m(x) = \frac{1}{\pi} \int_0^\pi \cos(m\tau - x\sin\tau)\, d\tau = \frac{1}{2\pi} \int_{-\pi}^\pi e^{-i(m\tau - x\sin\tau)}\, d\tau,$$

$$Y_m(x) = \frac{1}{\pi} \int_0^\pi \sin(x\sin\tau - m\tau)\, d\tau$$

$$- \frac{1}{\pi} \int_0^\infty \left[e^{m\tau} + (-1)^m e^{-m\tau}\right] e^{-x\sinh\tau}\, d\tau.$$

(iii) As $x \to 0+$,

$$J_0(x) \to 1, \quad J_m(x) \to 0, \quad m = 1, 2, \ldots,$$

$$Y_m(x) \to -\infty, \quad m = 0, 1, 2, \ldots.$$

(iv) J_m has the Taylor series expansion

$$J_m(x) = \sum_{k=0}^\infty \frac{(-1)^k}{k!(k+m)!} \left(\frac{x}{2}\right)^{2k+m}.$$

(v) The Bessel functions can be defined for any real, and even complex, order m. For example, when the order is a negative integer $-m$, we set

$$J_{-m}(x) = (-1)^m J_m(x).$$

(vi) When the order is an integer, the J_m are generated by the function

$$e^{(x/2)(t-1/t)} = \sum_{m=-\infty}^\infty J_m(x) t^m.$$

The graphs of J_0, J_1 and Y_0, Y_1 are shown in Fig. 3.3.

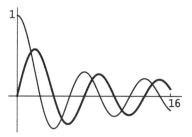 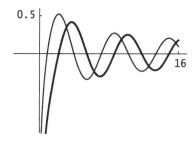

Figure 3.3: Left: J_0 (light line) and J_1 (heavy line). Right: Y_0 (light line) and Y_1 (heavy line).

Going back to (3.17), we can write its general solution as

$$g(\xi) = C_1 J_m(\xi) + C_2 Y_m(\xi), \quad C_1, C_2 = \text{const},$$

from which

$$f(x) = C_1 J_m(\sqrt{\lambda}\, x) + C_2 Y_m(\sqrt{\lambda}\, x).$$

In view of Remark 3.38(iii), the first condition (3.15) implies that $C_2 = 0$. The second one then yields $C_1 J_m(\sqrt{\lambda}\, a) = 0$. Since we want nonzero solutions, we must have

$$J_m(\sqrt{\lambda}\, a) = 0.$$

The function J_m has infinitely many positive zeros, which form a sequence ξ_{mn}, $n = 1, 2, \ldots$, such that $\xi_{mn} \to \infty$ as $n \to \infty$ for each $m = 0, 1, 2, \ldots$; therefore, the singular S–L problem (3.14), (3.15) has the eigenvalue–eigenfunction pairs

$$\lambda_{mn} = \left(\frac{\xi_{mn}}{a}\right)^2, \quad f_{mn}(x) = J_m\left(\frac{\xi_{mn}x}{a}\right), \quad n = 1, 2, \ldots.$$

It turns out that $\{f_{mn}\}_{n=1}^{\infty}$ is a complete system for each $m = 0, 1, 2, \ldots$, and that

$$\int_0^a J_m\left(\frac{\xi_{mn}x}{a}\right) J_m\left(\frac{\xi_{mp}x}{a}\right) x\, dx$$

$$= \begin{cases} 0, & n \neq p, \\ \frac{1}{2} a^2 J_{m+1}^2(\xi_{mp}), & n = p, \end{cases} \quad m = 0, 1, 2, \ldots. \qquad (3.18)$$

These orthogonality (with weight $\sigma(x) = x$) relations allow us to expand any suitable function u in a generalized Fourier series of the form

$$u(x) \sim \sum_{n=1}^{\infty} c_{mn} J_m\left(\frac{\xi_{mn}x}{a}\right) \qquad (3.19)$$

for each $m = 0, 1, 2, \ldots$. To find the coefficients c_{mn}, we follow the procedure described in Remark 3.27(i) with (a, b) replaced by $(0, a)$, c_n by c_{mn}, $f_n(x)$ by $J_m(\xi_{mn}x/a)$, and $\sigma(x)$ by x. Making use of (3.18), in the end we find that

$$c_{mn} = \frac{\int_0^a u(x) J_m(\xi_{mn}x/a) x \, dx}{\frac{1}{2} a^2 J_{m+1}^2(\xi_{mn})}, \quad n = 1, 2, \ldots. \tag{3.20}$$

3.39 Remark. Series (3.19) converges to $u(x)$ at all points x where u is continuous in the interval $(0, a)$. If $m = 0$, the point $x = 0$ is also included. If $m > 0$, then $x = 0$ is included if $u(0) = 0$. The point $x = a$ is included for any $m \geq 0$ if $u(a) = 0$.

3.40 Example. Let

$$u(x) = 2x - 1,$$

and let $a = 1$ and $m = 0$. The values, rounded to the fourth decimal place, of the first five zeros of $J_0(\xi)$ are

$$\xi_{01} = 2.4048, \quad \xi_{02} = 5.5201, \quad \xi_{03} = 8.6537,$$
$$\xi_{04} = 11.7915, \quad \xi_{05} = 14.9309,$$

which, when used in (3.20), generate the coefficients

$$c_{01} = 0.0329, \quad c_{02} = -1.1813, \quad c_{03} = 0.7452,$$
$$c_{04} = -0.7644, \quad c_{05} = 0.6146.$$

Therefore, by (3.19), we have the expansion

$$u(x) \sim 0.0329 \, J_0(2.4048\,x) - 1.1813 \, J_0(5.5201\,x) + 0.7452 \, J_0(8.6537\,x)$$
$$- 0.7644 \, J_0(11.7915\,x) + 0.6146 \, J_0(14.9309\,x) + \cdots. \tag{3.21}$$

We now construct a second expansion for u by taking $m = 1$. The approximate values of the first five zeros of J_1 are

$$\xi_{11} = 3.8317, \quad \xi_{12} = 7.0156, \quad \xi_{13} = 10.1735,$$
$$\xi_{14} = 13.3237, \quad \xi_{15} = 16.4706,$$

so, by (3.20), we arrive at the coefficients

$$c_{11} = 0.3788, \quad c_{12} = -1.3827, \quad c_{13} = 0.4700,$$
$$c_{14} = -0.9199, \quad c_{15} = 0.4248$$

and the corresponding expansion

$$u(x) \sim 0.3788 \, J_1(3.8317\,x) - 1.3827 \, J_1(7.0156\,x) + 0.4700 \, J_1(10.1735\,x)$$
$$- 0.9199 \, J_1(13.3237\,x) + 0.4248 \, J_1(16.4706\,x) + \cdots. \tag{3.22}$$

VERIFICATION WITH MATHEMATICA®. The input (based on (3.21) and (3.22))

```
ApproxJ0 = 0.0329 * BesselJ[0,2.4048 * x]
   - 1.1813 * BesselJ[0,5.5201 * x] + 0.7452 * BesselJ[0,8.6537 * x]
   - 0.7644 * BesselJ[0,11.7915 * x]
   + 0.6146 * BesselJ[0,14.9309 * x];
p0 = Plot[{ApproxJ0, 2 * x - 1}, {x,0,1}, PlotRange -> {-0.1,1.1},
   {-1.2,1.2}}, PlotStyle -> Black, ImageSize -> Scaled[0.25],
   Ticks -> {{1},{-1,1}}, AspectRatio -> 0.9];
ApproxJ1 = 0.3788 * BesselJ[1,3.8317 * x]
   - 1.3827 * BesselJ[1,7.0156 * x] + 0.4700 * BesselJ[1,10.1735 * x]
   - 0.9199 * BesselJ[1,13.3237 * x]
   + 0.4248 * BesselJ[1,16.4706 * x];
p1 = Plot[{ApproxJ1, 2 * x - 1}, {x,0,1}, PlotRange -> {{-0.1,1.1},
   {-1.2,1.2}}, PlotStyle -> Black, ImageSize -> Scaled[0.25],
   Ticks -> {{1},{-1,1}}, AspectRatio -> 0.9];
Show[GraphicsRow[{p0,p1}]]
```

generates the graphical output

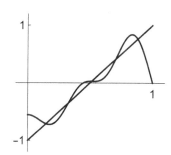 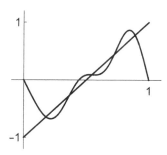

Exercises

Compute the first five terms of the generalized Fourier series expansion for the given function u in the eigenfunctions constructed with the Bessel functions (i) J_0 and (ii) J_1.

1 $u(x) = 1, \quad 0 \le x \le 1.$

2 $u(x) = 2x + 1, \quad 0 \le x \le 1.$

3 $u(x) = 1 - 3x, \quad 0 \le x \le 1.$

4 $u(x) = \begin{cases} 3, & 0 \le x \le 1/2, \\ 1, & 1/2 < x \le 1. \end{cases}$

5 $u(x) = \begin{cases} -1, & 0 \leq x \leq 1/2, \\ 2, & 1/2 < x \leq 1. \end{cases}$ **6** $u(x) = \begin{cases} x+3, & 0 \leq x \leq 1/2, \\ 2, & 1/2 < x \leq 1. \end{cases}$

7 $u(x) = \begin{cases} 1, & 0 \leq x \leq 1/2, \\ 2-2x, & 1/2 < x \leq 1. \end{cases}$ **8** $u(x) = \begin{cases} x-2, & 0 \leq x \leq 1/2, \\ x+1, & 1/2 < x \leq 1. \end{cases}$

Answers to Odd-Numbered Exercises

1 (i) $u(x) \sim 1.6020\, J_0(2.4048\, x) - 1.0463\, J_0(5.5201\, x) + 0.8514\, J_0(8.6537\, x)$
$\qquad - 0.7296\, J_0(11.7915\, x) + 0.6485\, J_0(14.9309\, x) + \cdots .$

   (ii) $u(x) \sim 2.2131\, J_1(3.8317\, x) - 0.5171\, J_1(7.0156\, x) + 1.1046\, J_1(10.1735\, x)$
$\qquad - 0.4550\, J_1(13.3237\, x) + 0.8113\, J_1(16.4706\, x) + \cdots .$

3 (i) $u(x) \sim -0.8503\, J_0(2.4048\, x) + 2.2951\, J_0(5.5201\, x) - 1.5435\, J_0(8.6537\, x)$
$\qquad + 1.5114\, J_0(11.7915\, x) - 1.2461\, J_0(14.9309\, x) + \cdots .$

   (ii) $u(x) \sim -1.6747\, J_1(3.8317\, x) + 2.3326\, J_1(7.0156\, x) - 1.2572\, J_1(10.1735\, x)$
$\qquad + 1.6073\, J_1(13.3237\, x) - 1.0429\, J_1(16.4706\, x) + \cdots .$

5 (i) $u(x) \sim 0.8947\, J_0(2.4048\, x) - 4.0540\, J_0(5.5201\, x) + 2.5517\, J_0(8.6537\, x)$
$\qquad - 0.0663\, J_0(11.7915\, x) + 0.7009\, J_0(14.9309\, x) + \cdots .$

   (ii) $u(x) \sim 2.2028\, J_1(3.8317\, x) - 4.4956\, J_1(7.0156\, x) + 0.8536\, J_1(10.1735\, x)$
$\qquad - 0.1883\, J_1(13.3237\, x) + 1.4431\, J_1(16.4706\, x) + \cdots .$

7 (i) $u(x) \sim 1.2920\, J_0(2.4048\, x) - 0.2378\, J_0(5.5201\, x) - 0.0983\, J_0(8.6537\, x)$
$\qquad + 0.0356\, J_0(11.7915\, x) + 0.0602\, J_0(14.9309\, x) + \cdots .$

   (ii) $u(x) \sim 1.6299\, J_1(3.8317\, x) + 0.4360\, J_1(7.0156\, x) + 0.2091\, J_1(10.1735\, x)$
$\qquad + 0.2123\, J_1(13.3237\, x) + 0.2485\, J_1(16.4706\, x) + \cdots .$

3.4   Legendre Polynomials

Consider the singular S–L problem

$$(1 - x^2)f''(x) - 2xf'(x) + \lambda f(x) = 0, \quad -1 < x < 1, \qquad (3.23)$$

$$f(x),\ f'(x) \text{ bounded as } x \to -1+ \text{ and as } x \to 1-. \qquad (3.24)$$

Since (3.23) can be written as

$$\left[(1 - x^2)f'(x)\right]' + \lambda f(x) = 0,$$

it is clear that $p(x) = 1 - x^2$, $q(x) = 0$, and $\sigma(x) = 1$.

A detailed analysis of problem (3.23), (3.24), which is beyond the scope of this book, shows that its eigenvalues are

$$\lambda_n = n(n+1), \quad n = 0, 1, 2, \ldots,$$

and that the general solution of (3.23) with $\lambda = \lambda_n$ is of the form

$$f_n(x) = C_1 P_n(x) + C_2 Q_n(x), \quad C_1, C_2 = \text{const}, \tag{3.25}$$

where P_n are polynomials of degree n, called the *Legendre polynomials*. They are computed by means of the formula

$$P_n(x) = \frac{1}{2^n} \sum_{k=0}^{s} \frac{(-1)^k}{k!} \frac{(2n-2k)!}{(n-2k)!(n-k)!} x^{n-2k}, \quad n = 0, 1, 2, \ldots,$$

where

$$s = \begin{cases} n/2, & n \text{ even}, \\ (n-1)/2, & n \text{ odd}. \end{cases}$$

The Q_n, called the *Legendre functions of the second kind*, have the representation

$$Q_n(x) = P_n(x) \int \frac{dx}{(1-x^2)P_n^2(x)}, \quad n = 0, 1, 2, \ldots.$$

3.41 Remarks. (i) The P_n are given by the *Rodrigues formula*

$$P_n(x) = \frac{1}{2^n n!} \frac{d^n}{dx^n} (x^2 - 1)^n, \quad n = 0, 1, 2, \ldots;$$

thus, for $n = 0, 1, \ldots 5$,

$$P_0(x) = 1, \quad P_1(x) = x, \quad P_2(x) = \tfrac{1}{2}(-1 + 3x^2),$$

$$P_3(x) = \tfrac{1}{2}(-3x + 5x^3), \quad P_4(x) = \tfrac{1}{8}(3 - 30x^2 + 35x^4),$$

$$P_5(x) = \tfrac{1}{8}(15x - 70x^3 + 63x^5).$$

(ii) The P_n may also be computed by means of the generating function

$$(1 - 2tx + t^2)^{-1/2} = \sum_{n=0}^{\infty} P_n(x)t^n, \quad |x| < 1, \ |t| < 1.$$

(iii) The P_n satisfy the recursive relation

$$(n+1)P_{n+1}(x) - (2n+1)xP_n(x) + nP_{n-1}(x) = 0, \quad n = 1, 2, \ldots.$$

(iv) The set $\{P_n\}_{n=0}^{\infty}$ is orthogonal over $[-1, 1]$; specifically,

$$\int_{-1}^{1} P_n(x)P_m(x)\, dx = \begin{cases} 0, & n \neq m, \\ \dfrac{2}{2n+1}, & n = m. \end{cases} \tag{3.26}$$

(v) $\{P_n\}_{n=0}^{\infty}$ is a complete system.

(vi) $Q_n(x)$ becomes unbounded as $x \to -1+$ and as $x \to 1-$.

The graphs of P_4, P_5 and Q_4, Q_5 are shown in Fig. 3.4.

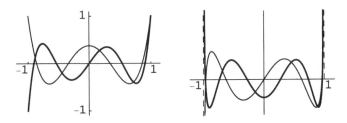

Figure 3.4: Left: P_4 (light line) and P_5 (heavy line). Right: Q_4 (light line) and Q_5 (heavy line).

In view of Remark 3.41(vi), f_n given by (3.25) satisfies the BCs (3.24) only if $C_2 = 0$; therefore, the eigenfunctions of the S–L problem (3.23), (3.24) are

$$f_n(x) = P_n(x), \quad n = 0, 1, 2, \ldots.$$

As before, Remark 3.41(v) allows us to expand any suitable function u in a series of the form

$$u(x) \sim \sum_{n=0}^{\infty} c_n P_n(x), \tag{3.27}$$

where, applying the technique described in Remark 3.27(i) and taking (3.26) into account, we see that the coefficients c_n are computed by means of the formula

$$c_n = \frac{2n+1}{2} \int_{-1}^{1} u(x) P_n(x)\, dx, \quad n = 0, 1, 2, \ldots. \tag{3.28}$$

3.42 Remark. If u is piecewise smooth on $(-1, 1)$, series (3.27) converges to $u(x)$ at all points x where u is continuous, and to $\frac{1}{2}\left[u(x-) + u(x+)\right]$ at the points x where u has a jump discontinuity.

3.43 Example. Consider the function

$$u(x) = \begin{cases} 2x + 1, & -1 \le x \le 0, \\ 3, & 0 < x \le 1. \end{cases}$$

By (3.28) and the expressions of the P_n in Remark 3.41(i), we find the coefficients

$$c_0 = \frac{3}{2}, \quad c_1 = \frac{5}{2}, \quad c_2 = -\frac{5}{8},$$

$$c_3 = -\frac{7}{8}, \quad c_4 = \frac{3}{16}, \quad c_5 = \frac{11}{16},$$

so (3.27) yields the expansion

$$u(x) \sim \tfrac{3}{2} P_0(x) + \tfrac{5}{2} P_1(x) - \tfrac{5}{8} P_2(x) - \tfrac{7}{8} P_3(x)$$

$$+ \tfrac{3}{16} P_4(x) + \tfrac{11}{16} P_5(x) + \cdots . \qquad (3.29)$$

VERIFICATION WITH MATHEMATICA®. The input (based on (3.29))

```
u = Piecewise[{{2*x+1, -1<x<0}, {3, 0<x<1}}];
Approx = (3/2) * LegendreP[0,x] + (5/2) * LegendreP[1,x]
   - (5/8) * LegendreP[2,x] - (7/8) * LegendreP[3,x]
   + (3/16) * LegendreP[4,x] + (11/16) * LegendreP[5,x];
Plot[{Approx, u}, {x, -1,1}, PlotRange -> {{-1.1,1.1},
   {-1.2,3.4}}, PlotStyle -> Black, ImageSize -> Scaled[0.25],
   Ticks -> {{-1,1}, {-1,1,3}}, AspectRatio -> 0.5]
```

generates the graphical output

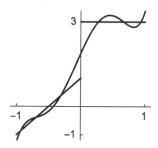

Exercises

Compute the first six terms of the generalized Fourier series expansion for the given function u, in the Legendre polynomials P_n.

1 $u(x) = x^2 - 3x + 4, \quad -1 \le x \le 1.$

2 $u(x) = x^3 + 2x^2 - 3, \quad -1 \le x \le 1.$

3 $u(x) = \begin{cases} -3, & -1 \le x \le 0, \\ 1, & 0 < x \le 1. \end{cases}$
  $\qquad$
4 $u(x) = \begin{cases} 2, & -1 \le x \le 0, \\ -5, & 0 < x \le 1. \end{cases}$

5 $u(x) = \begin{cases} -2, & -1 \le x \le 0, \\ x - 1, & 0 < x \le 1. \end{cases}$
  $\qquad$
6 $u(x) = \begin{cases} 2x + 3, & -1 \le x \le 0, \\ 1, & 0 < x \le 1. \end{cases}$

7 $u(x) = \begin{cases} 2x + 3, & -1 < x \le 0, \\ 1 - 2x, & 0 < x \le 1. \end{cases}$
  $\qquad$
8 $u(x) = \begin{cases} 3x + 1, & -1 \le x \le 0, \\ x + 4, & 0 < x \le 1. \end{cases}$

Answers to Odd-Numbered Exercises

1 $u(x) = (13/3)P_0(x) - 3P_1(x) + (2/3)P_2(x)$.

3 $u(x) \sim -P_0(x) + 3P_1(x) - (7/4)P_3(x) + (11/8)P_5(x) + \cdots$.

5 $u(x) \sim -(5/4)P_0(x) + (5/4)P_1(x) + (5/16)P_2(x) - (7/16)P_3(x)$
$$- (3/32)P_4(x) + (11/32)P_5(x) + \cdots.$$

7 $u(x) \sim P_0(x) - (3/2)P_1(x) - (5/4)P_2(x) + (7/8)P_3(x) + (3/8)P_4(x)$
$$- (11/16)P_5(x) + \cdots.$$

3.5 Spherical Harmonics

A more general form of the singular S–L problem (3.23), (3.24) is obtained if (3.23) is replaced by the *associated Legendre equation*

$$[(1 - x^2)f'(x)]' + \left(\lambda - \frac{m^2}{1 - x^2}\right)f(x) = 0, \quad -1 < x < 1, \tag{3.30}$$

where m is a nonnegative integer. Obviously, here we have

$$p(x) = 1 - x^2, \quad q(x) = -\frac{m^2}{1 - x^2}, \quad \sigma(x) = 1.$$

It can be shown that the eigenvalue–eigenfunction pairs of problem (3.30), (3.24) are

$$\lambda_n = n(n + 1), \quad n = m, m + 1, \ldots,$$

$$f_n(x) = P_n^m(x) = (1 - x^2)^{m/2} \frac{d^m}{dx^m} P_n(x)$$

$$= \frac{(-1)^m}{2^n n!} (1 - x^2)^{m/2} \frac{d^{n+m}}{dx^{n+m}} \left[(x^2 - 1)^n\right]. \tag{3.31}$$

For each $m = 0, 1, \ldots, n$, the P_n^m, called the *associated Legendre functions*, satisfy the orthogonality formula

$$\int_{-1}^{1} P_n^m(x) P_k^m(x) \, dx = \begin{cases} 0, & n \neq k, \\ \dfrac{2(n + m)!}{(2n + 1)(n - m)!}, & n = k. \end{cases} \tag{3.32}$$

The definition of the P_n^m can be extended to negative-integer superscripts by means of the last expression on the right-hand side in (3.31). A straightforward computation shows that

$$P_n^{-m}(x) = \frac{(-1)^{-m}}{2^n n!} (1 - x^2)^{-m/2} \frac{d^{n-m}}{dx^{n-m}} \left[(x^2 - 1)^n\right]$$

$$= (-1)^m \frac{(n - m)!}{(n + m)!} P_n^m(x). \tag{3.33}$$

The spherical harmonics are complex-valued functions (see Section 14.1) that occur in problems formulated in terms of spherical coordinates, which, as is well known, are linked to the Cartesian coordinates x, y, z by the expressions

$$x = r\cos\theta\sin\varphi, \quad y = r\sin\theta\sin\varphi, \quad z = r\cos\varphi.$$

Here, r, $r \geq 0$, is the radius, θ, $0 \leq \theta < 2\pi$, is the azimuthal angle, and φ, $0 \leq \varphi \leq \pi$, is the polar angle. The spherical harmonics are defined by

$$Y_n^m(\theta, \varphi) = P_n^m(\cos\varphi)e^{im\theta}, \quad n = 0, 1, 2, \ldots,$$

where $m = 0, 1, \ldots, n$ and, by Euler's formula,

$$e^{im\theta} = \cos(m\theta) + i\sin(m\theta).$$

In view of (3.32), it is not difficult to verify that

$$\int_0^{2\pi}\int_0^\pi Y_n^m(\theta, \varphi)\bar{Y}_k^l(\theta, \varphi)\sin\varphi\, d\varphi\, d\theta = \begin{cases} \dfrac{4\pi(n+m)!}{(2n+1)(n-m)!}, & m = l,\ n = k, \\ 0 & \text{otherwise,} \end{cases}$$

where the superposed bar denotes complex conjugation. The above equality allows us to normalize the Y_n^m by writing

$$Y_{n,m}(\theta, \varphi) = \left[\frac{(2n+1)(n-m)!}{4\pi(n+m)!}\right]^{1/2} P_n^m(\cos\varphi)e^{im\theta}. \tag{3.34}$$

This is the form by which the spherical harmonics are usually known.
    The $Y_{n,m}$ satisfy the orthonormality formula

$$\int_0^{2\pi}\int_0^\pi Y_{n,m}(\theta, \varphi)\bar{Y}_{k,l}(\theta, \varphi)\sin\varphi\, d\varphi\, d\theta = \begin{cases} 1, & m = l,\ n = k, \\ 0 & \text{otherwise.} \end{cases} \tag{3.35}$$

In view of (3.33), for nonnegative integers m and n we can also define

$$Y_{n,-m}(\theta, \varphi) = (-1)^m \bar{Y}_{n,m}(\theta, \varphi). \tag{3.36}$$

3.44 Example. By the second formula (3.31), (3.34), and (3.36),

$$Y_{0,0}(\theta, \varphi) = \frac{1}{2}\sqrt{\frac{1}{\pi}},$$

$$Y_{1,-1}(\theta, \varphi) = \frac{1}{2}\sqrt{\frac{3}{2\pi}}\, e^{-i\theta}\sin\varphi = \frac{1}{2}\sqrt{\frac{3}{2\pi}}\,\frac{x-iy}{r},$$

$$Y_{1,0}(\theta, \varphi) = \frac{1}{2}\sqrt{\frac{3}{\pi}}\,\cos\varphi = \frac{1}{2}\sqrt{\frac{3}{\pi}}\,\frac{z}{r},$$

$$Y_{1,1}(\theta,\varphi) = -\frac{1}{2}\sqrt{\frac{3}{2\pi}}\, e^{i\theta} \sin\varphi = -\frac{1}{2}\sqrt{\frac{3}{2\pi}}\frac{x+iy}{r},$$

$$Y_{2,-2}(\theta,\varphi) = \frac{1}{4}\sqrt{\frac{15}{2\pi}}\, e^{-2i\theta} \sin^2\varphi = \frac{1}{4}\sqrt{\frac{15}{2\pi}}\frac{(x-iy)^2}{r^2},$$

$$Y_{2,-1}(\theta,\varphi) = \frac{1}{2}\sqrt{\frac{15}{2\pi}}\, e^{-i\theta} \sin\varphi\cos\varphi = \frac{1}{2}\sqrt{\frac{15}{2\pi}}\frac{(x-iy)z}{r^2},$$

$$Y_{2,0}(\theta,\varphi) = \frac{1}{4}\sqrt{\frac{5}{\pi}}\,(3\cos^2\varphi - 1) = \frac{1}{4}\sqrt{\frac{5}{\pi}}\frac{2z^2 - x^2 - y^2}{r^2},$$

$$Y_{2,1}(\theta,\varphi) = -\frac{1}{2}\sqrt{\frac{15}{2\pi}}\, e^{i\theta} \sin\varphi\cos\varphi = -\frac{1}{2}\sqrt{\frac{15}{2\pi}}\frac{(x+iy)z}{r^2},$$

$$Y_{2,2}(\theta,\varphi) = \frac{1}{4}\sqrt{\frac{15}{2\pi}}\, e^{2i\theta} \sin^2\varphi = \frac{1}{4}\sqrt{\frac{15}{2\pi}}\frac{(x+iy)^2}{r^2}.$$

3.45 Remark. Any suitable function u defined on a sphere can be expanded in a spherical harmonics series of the form

$$u(\theta,\varphi) \sim \sum_{n=0}^{\infty} \sum_{m=-n}^{n} c_{n,m} Y_{n,m}(\theta,\varphi), \qquad (3.37)$$

which converges to $u(\theta,\varphi)$ at all points where u is continuous. To find the coefficients $c_{k,l}$, we multiply every term in (3.37) by $\bar{Y}_{k,l}(\theta,\varphi)\sin\varphi$, integrate over $[0,2\pi] \times [0,\pi]$, and take (3.35) into account. This (after k,l are replaced by n,m) leads to

$$c_{n,m} = \int_0^{2\pi}\int_0^{\pi} u(\theta,\varphi)\bar{Y}_{n,m}(\theta,\varphi)\sin\varphi\, d\varphi\, d\theta. \qquad (3.38)$$

3.46 Example. Let
$$v(x,y,z) = x^2 + 2y$$

be defined on the sphere with center at the origin and radius 1. In terms of spherical coordinates, this becomes

$$u(\theta,\varphi) = \cos^2\theta\sin^2\varphi + 2\sin\theta\sin\varphi.$$

Then, using (3.38) and the explicit expressions of the $Y_{n,m}$ given in Example 3.44, we find that

$$c_{0,0} = \frac{2\sqrt{\pi}}{3}, \quad c_{1,0} = 0, \quad c_{1,1} = c_{1,-1} = 2\sqrt{\frac{2\pi}{3}}\,i,$$

$$c_{2,-1} = c_{2,1} = 0, \quad c_{2,0} = -\frac{2}{3}\sqrt{\frac{\pi}{5}}, \quad c_{2,2} = c_{2,-2} = \sqrt{\frac{2\pi}{15}},$$

so, by (3.37),

$$u(\theta, \varphi) = \frac{2\sqrt{\pi}}{3} Y_{0,0}(\theta, \varphi) + 2\sqrt{\frac{2\pi}{3}} i Y_{1,-1}(\theta, \varphi) + 2\sqrt{\frac{2\pi}{3}} i Y_{1,1}(\theta, \varphi)$$

$$+ \sqrt{\frac{2\pi}{15}} Y_{2,-2}(\theta, \varphi) - \frac{2}{3}\sqrt{\frac{\pi}{5}} Y_{2,0}(\theta, \varphi) + \sqrt{\frac{2\pi}{15}} Y_{2,2}(\theta, \varphi). \quad (3.39)$$

This is a terminating series because f is a linear combination of the spherical harmonics listed in Example 3.44.

VERIFICATION WITH MATHEMATICA®. The input (based on (3.39))

```
(2*Pi^(1/2)/3)*SphericalHarmonicY[0,0,φ,θ]]
  +2*((2*Pi/3)^(1/2))*I*SphericalHarmonicY[1,-1,φ,θ]
  +2*((2*Pi/3)^(1/2))as I*SphericalHarmonicY[1,1,φ,θ]
  +((2*Pi/15)^(1/2))*SphericalHarmonicY[2,-2,φ,θ]
  -((2/3)*(Pi/5)^(1/2))*SphericalHarmonicY[2,0,φ,θ]
  +((2*Pi/15)^(1/2))*SphericalHarmonicY[2,2,φ,θ]
  //FullSimplify
```

generates the output $\cos^2 \theta \sin^2 \varphi + 2 \sin \theta \sin \varphi$, which confirms the above statement.

3.47 Example. If the function in Example 3.46 is replaced by

$$v(x, y, z) = \begin{cases} x, & z \geq 0, \\ yz, & z < 0, \end{cases}$$

then its equivalent form in spherical coordinates is

$$u(\theta, \varphi) = \begin{cases} \cos \theta \sin \varphi, & 0 \leq \varphi \leq \pi/2, \\ \sin \theta \sin \varphi \cos \varphi, & \pi/2 \leq \varphi \leq \pi. \end{cases}$$

In this case, (3.38) yields the coefficients

$$c_{0,0} = c_{1,0} = 0, \quad c_{1,1} = -\bar{c}_{1,-1} = -\sqrt{\frac{\pi}{6}}\left(1 + \frac{3}{8}i\right),$$

$$c_{2,-2} = c_{2,0} = c_{2,2} = 0, \quad c_{2,1} = -\bar{c}_{2,-1} = -\sqrt{\frac{\pi}{30}}\left(\frac{15}{8} - i\right),$$

so expansion (3.37) becomes

$$u(\theta, \varphi) \sim \sqrt{\frac{\pi}{6}}\left(1 - \frac{3}{8}i\right) Y_{1,-1}(\theta, \varphi) - \sqrt{\frac{\pi}{6}}\left(1 + \frac{3}{8}i\right) Y_{1,1}(\theta, \varphi)$$

$$+ \sqrt{\frac{\pi}{30}}\left(\frac{15}{8} + i\right) Y_{2,-1}(\theta, \varphi) - \sqrt{\frac{\pi}{30}}\left(\frac{15}{8} - i\right) Y_{2,1}(\theta, \varphi) + \cdots.$$

$$(3.40)$$

VERIFICATION WITH MATHEMATICA®. The input (based on (3.40))

```
u = Piecewise[{{Cos[θ] * Sin[φ], 0<θ<2*Pi && 0<φ<Pi/2},
  {Sin[θ] * Sin[φ] * Cos[φ], 0<θ<2*Pi, && Pi/2<φ<Pi}}];
approx = Plot3D[ComplexExpand[(Pi/6)^(1/2) * (1 - (3/8) * I)
  * SphericalHarmonicY[1, -1, φ, θ] - (Pi/6)^(1/2) * (1 + (3/8) * I)
  * SphericalHarmonicY[1, 1, φ, θ] + (Pi/30)^(1/2) * (15/8 + I)
  * SphericalHarmonicY[2, -1, φ, θ] - (Pi/30)^(1/2) * (15/8 - I)
  * SphericalHarmonicY[2, 1, φ, θ], {φ, 0, Pi}, {θ, 0, 2*Pi},
  ImageSize -> Scaled[0.3], Ticks -> {{0, 2, 4, 6}, {0, 1, 2, 3},
  {-0.8, 0, 0.8}}];
Graphu = Plot3D[u, {φ, 0, Pi}, {θ, 0, 2*Pi}, ImageSize -> Scaled[0.3],
  Ticks -> {{0, 2, 4, 6}, {0, 1, 2, 3}, {-1, 0, 1}}];
GraphicsRow[{Graphu, approx}]
```

generates the graphical output

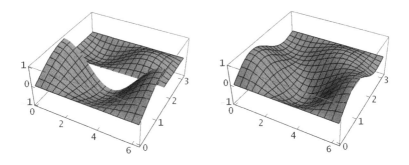

As can be seen, although the approximation is the sum of only the four terms in (3.40), its graph shows reasonable agreement with that of u away from the discontinuity line $\varphi = \pi/2$ of the latter. The approximating function is, of course, continuous across that line.

3.48 Remark. A very general class of what we have called 'suitable' functions in this chapter consists of square-integrable functions, namely, functions u such that

$$\int_D |u(x)|^2 \, dx < \infty,$$

where D is the domain where the functions are defined and x and dx are, respectively, a generic point in D and the element of length/area/volume, as appropriate. If series (3.9), (3.19), (3.27), and (3.37) are written collectively in the form

$$u(x) \sim \sum_{n=1}^{\infty} c_n f_n(x),$$

then they are convergent in the sense that

$$\lim_{N \to \infty} \int_D \left| u(x) - \sum_{n=1}^{N} c_n f_n(x) \right|^2 dx = 0.$$

As already mentioned, when u is sufficiently smooth, the series also converge to $u(x)$ in the classical pointwise sense at every x in D where u is continuous. To keep the notation simple, since the functions of interest to us meet the smoothness requirement, from now on we use the equality sign between them and their generalized Fourier series representations, although, strictly speaking, equality holds only at their points of continuity.

Exercises

The given function v is defined on the sphere with center at the origin and radius 1. Writing it as $u(\theta, \varphi)$ in terms of spherical coordinates, compute the first nine terms of its expansion in the spherical harmonics. (Use the spherical harmonics listed in Example 3.44.)

1 $v(x, y, z) = y^2 - 2z^2$.

2 $v(x, y, z) = 3x^2 + xz$.

3 $v(x, y, z) = \begin{cases} 1, & z \geq 0, \\ 0, & z < 0. \end{cases}$

4 $v(x, y, z) = \begin{cases} -2, & z \geq 0, \\ 1, & z < 0. \end{cases}$

5 $v(x, y, z) = \begin{cases} 2, & z \geq 0, \\ x - y, & z < 0. \end{cases}$

6 $v(x, y, z) = \begin{cases} 2x + z, & z \geq 0, \\ -1, & z < 0. \end{cases}$

7 $v(x, y, z) = \begin{cases} 2y + yz, & z \geq 0, \\ x^2 - 3y, & z < 0. \end{cases}$

8 $v(x, y, z) = \begin{cases} xz + z^2, & z \geq 0, \\ xy, & z < 0. \end{cases}$

Answers to Odd-Numbered Exercises

1 $u(\theta, \varphi) = -(2\sqrt{\pi}/3)Y_{0,0}(\theta, \varphi) - \sqrt{2\pi/15}\, Y_{2,-2}(\theta, \varphi) - (2\sqrt{5\pi}/3)Y_{2,0}(\theta, \varphi)$

$$- \sqrt{2\pi/15}\, Y_{2,2}(\theta, \varphi).$$

3 $u(\theta, \varphi) \sim \sqrt{\pi}\, Y_{0,0}(\theta, \varphi) + (\sqrt{3\pi}/2)Y_{1,0}(\theta, \varphi) + \cdots$.

5 $u(\theta, \varphi) \sim 2\sqrt{\pi}\, Y_{0,0}(\theta, \varphi) + \sqrt{\pi/6}\,(1-i)Y_{1,-1}(\theta, \varphi) + \sqrt{3\pi}\, Y_{1,0}(\theta, \varphi)$

$$- \sqrt{\pi/6}\,(1+i)Y_{1,1}(\theta, \varphi) - (1/8)\sqrt{15\pi/2}\,(1-i)Y_{2,-1}(\theta, \varphi)$$

$$+ (1/8)\sqrt{15\pi/2}\,(1+i)Y_{2,1}(\theta, \varphi) + \cdots .$$

7 $u(\theta, \varphi) \sim (\sqrt{\pi}/3) Y_{0,0}(\theta, \varphi) - (5/8)\sqrt{\pi/6}\, i Y_{1,-1}(\theta, \varphi) - (\sqrt{3\pi}/8) Y_{1,0}(\theta, \varphi)$

$$- (5/8)\sqrt{\pi/6}\, i Y_{1,1}(\theta, \varphi) + \sqrt{\pi/30}\, Y_{2,-2}(\theta, \varphi)$$

$$+ (83/8)\sqrt{\pi/30}\, i Y_{2,-1}(\theta, \varphi) - (1/3)\sqrt{\pi/5}\, Y_{2,0}(\theta, \varphi)$$

$$+ (83/8)\sqrt{\pi/30}\, i Y_{2,1}(\theta, \varphi) + \sqrt{\pi/30}\, Y_{2,2}(\theta, \varphi) + \cdots .$$

CHAPTER 4

Some Fundamental Equations of Mathematical Physics

The great majority of processes and phenomena in the real world are studied by means of mathematical models. An investigation of this kind generally consists of three stages.

(i) The model is set up in terms of mathematical expressions describing the quantitative relationships between the physical quantities involved.

(ii) The equations of the model are solved by means of various mathematical methods.

(iii) The mathematical results are interpreted from a physical point of view in relation to the original process.

In this chapter, we show how three simple but fundamental mathematical models are derived, which involve partial differential equations. These models are important because each of them is representative of an entire class of linear second-order PDEs.

4.1 Definition. A *partial differential equation* is an equation that contains an unknown function of several variables and one or more of its partial derivatives.

To simplify the notation, we will denote the partial derivatives of functions almost exclusively by means of subscripts; thus, for $u = u(x,t)$,

$$u_t \equiv \frac{\partial u}{\partial t}, \quad u_x \equiv \frac{\partial u}{\partial x},$$

$$u_{tt} \equiv \frac{\partial^2 u}{\partial t^2}, \quad u_{xt} \equiv \frac{\partial^2 u}{\partial x \partial t}, \quad u_{xx} \equiv \frac{\partial^2 u}{\partial x^2},$$

and so on.

4.1 The Heat Equation

Heat conduction in a one-dimensional rod. Consider a heat-conducting rod in the shape of a thin cylinder described by the following physical parameters:

L : length;

A : cross-sectional area, assumed constant;

$E(x,t)$: heat energy density (heat energy per unit volume);

$\varphi(x,t)$: heat flux (heat energy per unit area, flowing to the right per unit time);

$q(x,t)$: heat sources or sinks (heat energy per unit volume, generated or lost inside the rod per unit time).

Throughout what follows, we assume that the lateral (cylindrical) surface of the rod is insulated; that is, no heat exchange takes place across it. We also use the generic term 'sources' to describe both sources and sinks.

The physical law we are using to set up the mathematical model in this case is the law of conservation of heat energy, which states that

rate of change of heat energy in body

= heat flow across boundary per unit time

+ heat generated by sources per unit time.

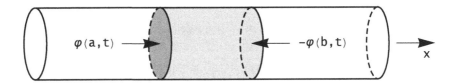

Figure 4.1: An arbitrary segment of rod.

For an arbitrary length of rod between $x = a$ and $x = b$ on the x-axis (see Fig. 4.1), this law translates as

$$\frac{d}{dt} \int_a^b E(x,t)A\,dx = \left[\varphi(a,t) - \varphi(b,t)\right]A + \int_a^b q(x,t)A\,dx$$

$$= -\int_a^b \varphi_x(x,t)\,A\,dx + \int_a^b q(x,t)A\,dx, \quad t > 0.$$

Dividing through by A and moving all the terms to the left-hand side, we find that for $t > 0$ and any a, b, $0 < a < b < L$,

$$\int_a^b \left[E_t(x,t) + \varphi_x(x,t) - q(x,t) \right] dx = 0.$$

In view of the arbitrariness of a and b, this means that

$$E_t(x,t) = -\varphi_x(x,t) + q(x,t), \quad 0 < x < L, \ t > 0. \tag{4.1}$$

We need to introduce further physical parameters for the rod; thus,

$u(x,t)$: temperature;

$c(x)$: specific heat (the heat energy that raises the temperature of one unit of mass by one unit), assumed, for simplicity, to be independent of temperature;

$K_0(x)$: thermal conductivity, also assumed to be independent of temperature;

$\rho(x)$: mass density.

Then the total heat energy in the section of rod between $x = a$ and $x = b$ is

$$\int_a^b E(x,t) A \, dx = \int_a^b c(x)\rho(x)u(x,t) A \, dx, \quad t > 0;$$

as above, this means that

$$E(x,t) = c(x)\rho(x)u(x,t), \quad 0 < x < L, \ t > 0. \tag{4.2}$$

At the same time, according to Fourier's law of heat conduction,

$$\varphi(x,t) = -K_0(x)u_x(x,t), \quad 0 < x < L, \ t > 0. \tag{4.3}$$

Using (4.2) and (4.3), we can now write the conservation of heat energy expressed by (4.1) as

$$c(x)\rho(x)u_t(x,t) = \left(K_0(x)u_x(x,t) \right)_x + q(x,t), \quad 0 < x < L, \ t > 0. \tag{4.4}$$

Assuming that the rod is uniform (c, ρ, and K_0 are constants) and that there are no internal heat sources ($q = 0$), we see that (4.4) reduces to

$$u_t(x,t) = ku_{xx}(x,t), \quad 0 < x < L, \ t > 0, \tag{4.5}$$

where $k = K_0/(c\rho) = \text{const} > 0$ is the *thermal diffusivity* of the rod. Equation (4.5) is called the *heat (diffusion) equation*.

If sources are present in the rod, then this equation is replaced by its nonhomogeneous version

$$u_t(x,t) = ku_{xx}(x,t) + q(x,t), \quad 0 < x < L, \ t > 0,$$

where q now incorporates the constant $1/(c\rho)$.

Initial condition. Since (4.5) is a first-order equation with respect to time, it needs only one initial condition, which is normally taken to be

$$u(x,0) = f(x), \quad 0 < x < L.$$

This means prescribing the initial distribution of temperature in the rod.

Boundary conditions. The equation is of second order with respect to the space variable, so we need two boundary conditions. There are three main types of physically meaningful conditions that are usually prescribed at the 'near' endpoint $x = 0$ and 'far' endpoint $x = L$.

(i) The temperature may be given at one endpoint; for example,

$$u(0,t) = \alpha(t), \quad t > 0.$$

(ii) If the rod is insulated at an endpoint, then the condition must mean that the heat flux there is zero. In view of (4.3), this is equivalent to the derivative u_x being equal to zero; for example,

$$u_x(L,t) = 0, \quad t > 0.$$

More generally, if a (nonzero) heat flux through the endpoint $x = L$ is prescribed, then the above condition is replaced by

$$u_x(L,t) = \beta(t), \quad t > 0.$$

(iii) When one of the endpoints is in contact with another medium, we use Newton's law of cooling, which states that the heat flux at that endpoint is proportional to the difference between the temperature of the rod and the temperature of the external medium; for example,

$$K_0 u_x(0,t) = H[u(0,t) - U(t)], \quad t > 0,$$

where $U(t)$ is the (known) temperature of the external medium and $H > 0$ is the *heat transfer coefficient*. Owing to the convention concerning the direction of the heat flux, at the far endpoint this type of condition becomes

$$-K_0 u_x(L,t) = H[u(L,t) - U(t)], \quad t > 0.$$

4.2 Remarks. (i) Only one boundary condition is prescribed at each endpoint.

(ii) The boundary condition at $x = 0$ may differ from that at $x = L$.

(iii) It is easily verified that the heat equation is linear.

(iv) The one-dimensional heat equation is the simplest example of a so-called *parabolic equation*.

The model. As we have seen, the mathematical model for heat conduction in a rod consists of several elements.

4.3 Definition. A partial differential equation (PDE) and the initial conditions (ICs) and boundary conditions (BCs) associated with it form an *initial boundary value problem* (IBVP). If only initial conditions or boundary conditions are present, then we have an *initial value problem* (IVP) or a *boundary value problem* (BVP), respectively.

4.4 Example. The IBVP modeling heat conduction in a one-dimensional uniform rod with sources, insulated lateral surface, and temperature prescribed at both endpoints is of the form

$$u_t(x,t) = k u_{xx}(x,t) + q(x,t), \quad 0 < x < L, \ t > 0, \qquad \text{(PDE)}$$

$$u(0,t) = \alpha(t), \quad u(L,t) = \beta(t), \quad t > 0, \qquad \text{(BCs)}$$

$$u(x,0) = f(x), \quad 0 < x < L, \qquad \text{(IC)}$$

where q, α, β, and f are given functions.

4.5 Example. If the near endpoint is insulated and the far one is kept in a medium of constant zero temperature, and if the rod contains no sources, then the corresponding IBVP is

$$u_t(x,t) = k u_{xx}(x,t), \quad 0 < x < L, \ t > 0, \qquad \text{(PDE)}$$

$$u_x(0,t) = 0, \quad u_x(L,t) + h u(L,t) = 0, \quad t > 0, \qquad \text{(BCs)}$$

$$u(x,0) = f(x), \quad 0 < x < L, \qquad \text{(IC)}$$

where f is a given function and h is a known positive constant.

4.6 Definition. A *classical solution* of an IBVP is a function u that satisfies the PDE, BCs, and ICs pointwise everywhere in the region where the problem is formulated. For brevity, in what follows we refer to such functions simply as 'solutions'. It is obvious that an IBVP has solutions in this sense only if the data functions have a certain degree of smoothness.

4.7 Example. Suppose that q, α, β, and f in Example 4.4 are continuous functions. Then a solution of that IBVP has the following properties:

(i) It is continuously differentiable with respect to t and twice continuously differentiable with respect to x at all points in the domain (semi-infinite strip) in the (x,t)-plane defined by

$$G = \{(x,t) : 0 < x < L, \ t > 0\},$$

and satisfies the PDE at all points in G.

(ii) Its continuity extends to the 'spatial' boundary lines of G; that is, to the two half-lines

$$\partial G_x = \{(x,t) : x = 0, \ t > 0\} \cup \{(x,t) : x = L, \ t > 0\},$$

and it satisfies the appropriate BC at every point on ∂G_x.

(iii) Its continuity also extends to the 'temporal' boundary line of G, namely
$$\partial G_t = \{(x, t) : 0 < x < L, \ t = 0\},$$
and it satisfies the IC at all points on ∂G_t.

4.8 Remarks. (i) We note that nothing is said in Example 4.7 about the behavior of u at the 'corner' points $(0, 0)$ and $(L, 0)$ of G. If, for example,
$$\lim_{t \to 0} u(0, t) = \lim_{x \to 0} u(x, 0),$$
that is, if
$$\lim_{t \to 0} \alpha(t) = \lim_{x \to 0} f(x),$$

then u can be defined at $(0, 0)$ by the common value of the above limits. If, on the other hand, the two limits are distinct, then u cannot be defined at $(0, 0)$ in the classical sense. In the applications that follow, we do not concern ourselves with these two 'corner' points.

(ii) Suppose that f has a jump discontinuity at a point $(x_0, 0)$ on ∂G_t. In this case, it is impossible to find a solution u that satisfies the IC at this point in the classical sense; however, we might find a type of solution that does so in some 'average' way. We call this a *weak solution*.

(iii) The definition of a solution can be generalized in the obvious manner to IBVPs with more than two independent variables.

4.9 Theorem (Uniqueness). *If the functions α, β, and f are sufficiently smooth to ensure that u, u_t, u_x, and u_{xx} are continuous in G and up to the boundary of G, including the two 'corner' points, then the IBVP in Example 4.4 has at most one solution.*

Proof. Suppose that there are two solutions u_1 and u_2 of the problem in question. Then, because of linearity, it is obvious that
$$u = u_1 - u_2$$

is a solution of the fully homogeneous IBVP; that is, for any arbitrarily fixed number $T > 0$, the function u satisfies
$$u_t(x, t) = k u_{xx}(x, t), \quad 0 < x < L, \ 0 < t < T,$$
$$u(0, t) = 0, \quad u(L, t) = 0, \quad 0 < t < T,$$
$$u(x, 0) = 0, \quad 0 < x < L.$$

Multiplying the PDE by u, integrating over $[0, L]$, and using the BCs, the smoothness properties of the solutions for $0 \leq x \leq L$ and $0 \leq t \leq T$, and integration by parts, we find that

$$0 = \int\limits_0^L (uu_t - kuu_{xx})\, dx$$

$$= \int\limits_0^L \left[\tfrac{1}{2}(u^2)_t + ku_x^2\right] dx - k\left[uu_x\right]_{x=0}^{x=L}$$

$$= \int\limits_0^L \left[\tfrac{1}{2}(u^2)_t + ku_x^2\right] dx;$$

therefore,

$$\frac{1}{2}\frac{d}{dt}\int\limits_0^L u^2\, dx = -k\int\limits_0^L u_x^2\, dx \le 0, \quad 0 \le t \le T,$$

which means that the function

$$W(t) = \int\limits_0^L u^2(x,t)\, dx, \quad 0 \le t \le T,$$

is nonincreasing. Since $W(t) \ge 0$ and $W(0) = 0$ (because of the IC satisfied by u), this is possible only if $W(t) = 0$, $0 \le t \le T$. In view of the nonnegative integrand in the definition of W above and the arbitrariness of $T > 0$, we then conclude that $u = 0$; in other words, the solutions u_1 and u_2 coincide. □

4.10 Remarks.   (i) Theorem 4.9 states that if the IBVP in Example 4.4 has a solution, then that solution is unique. The existence issue is resolved in Chapter 5, where we actually construct the solution.

(ii) The properties required of the solution in Theorem 4.9 are stronger than those mentioned in Example 4.7.

(iii) Uniqueness can also be proved for the solutions of IBVPs with other types of BCs.

(iv) We can easily convince ourselves that the change of variables

$$\xi = \frac{x}{L}, \quad \tau = \frac{k}{L^2} t,$$

$$u(x,t) = u\left(L\xi, \frac{L^2\tau}{k}\right) = v(\xi, \tau)$$

reduces the heat equation to the form

$$v_\tau(\xi, \tau) = v_{\xi\xi}(\xi, \tau), \quad 0 < \xi < 1, \ \tau > 0.$$

Hence, without loss of generality, in most of our applications we consider this simpler version, with x, t, and u in place of ξ, τ, and v, respectively.

Exercises

1 Show that Theorem 4.9 also holds for the IBVP in Example 4.4 with the BCs replaced by

(i) $u_x(0,t) - hu(0,t) = \alpha(t)$, $u(L,t) = \beta(t)$, $t > 0$;

(ii) $u_x(0,t) = \alpha(t)$, $u_x(L,t) + hu(L,t) = \beta(t)$, $t > 0$,

where α and β are given functions and $h = \text{const} > 0$.

2 Verify the statement in Remark 4.10(iv).

4.2   The Laplace Equation

The higher-dimensional heat equation. For simplicity, we study the case of two space dimensions. Consider a heat-conducting body (thin plate) occupying a finite region D in the (x,y)-plane, bounded by a smooth, simple, closed curve ∂D, and let R be an arbitrary element of D with a smooth boundary ∂R (see Fig. 4.2). We adopt the same notation for the physical parameters of this body as in Section 4.1, and assume that the upper and lower faces of the plate are insulated.

Heat flow depends on direction, so here the heat flux is a vector $\vec{\varphi}$, which we can write in the form

$$\vec{\varphi} = \text{normal component } + \text{ tangential component}$$
$$= (\vec{\varphi} \cdot \vec{n})\vec{n} + (\vec{\varphi} \cdot \vec{\tau})\vec{\tau}, \quad |\vec{n}| = |\vec{\tau}| = 1,$$

where $\vec{n}$ and $\vec{\tau}$ are the unit normal and tangent vectors to ∂R, directed as shown in Fig. 4.2. Clearly, the tangential component makes no contribution to the heat exchange between R and the rest of the body.

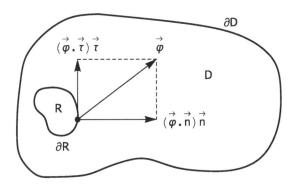

Figure 4.2: A two-dimensional body configuration.

As in the case of a rod, the conservation of heat energy states that

rate of change of heat energy
$\qquad$ = heat flow across boundary per unit time
$\qquad\qquad$ + heat generated by sources per unit time.

This translates mathematically as

$$\frac{d}{dt}\int_R E(x,y,t)\,dA = -\int_{\partial R} \vec{\varphi}(x,y,t)\cdot\vec{n}(x,y)\,ds + \int_R q(x,y,t)\,dA,$$

where dA and ds are, respectively, the elements of area and arc length. Taking (4.2) into account and applying the divergence theorem

$$\int_R (\operatorname{div}\vec{\varphi})(x,y,t)\,dA = \int_{\partial R}\vec{\varphi}(x,y,t)\cdot\vec{n}(x,y)\,ds,$$

we can rewrite the above equality as

$$\int_R \big[c(x,y)\rho(x,y)u_t(x,y,t) + (\operatorname{div}\vec{\varphi})(x,y,t) - q(x,y,t)\big]\,dA = 0.$$

In view of the arbitrariness of R, we conclude that for (x,y) in D and $t>0$,

$$c(x,y)\rho(x,y)u_t(x,y,t) = -(\operatorname{div}\vec{\varphi})(x,y,t) + q(x,y,t). \qquad (4.6)$$

Once again, we now use Fourier's law of heat conduction, which in two dimensions is

$$\vec{\varphi}(x,y,t) = -K_0(x,y)(\operatorname{grad} u)(x,y,t);$$

consequently, (4.6) becomes

$$c(x,y)\rho(x,y)u_t(x,y,t) = \operatorname{div}\big(K_0(x,y)(\operatorname{grad} u)(x,y,t)\big) + q(x,y,t).$$

For a uniform body (c, ρ, and K_0 are constants) with no internal sources ($q = 0$), the last formula reduces to

$$u_t(x,y,t) = k(\operatorname{div}\operatorname{grad} u)(x,y,t), \quad k = \frac{K_0}{c\rho} = \text{const} > 0,$$

or

$$u_t(x,y,t) = k(\Delta u)(x,y,t), \quad (x,y) \text{ in } D, \ t>0, \qquad (4.7)$$

where the differential operator Δ, defined for smooth functions $v(x,y)$ by

$$(\Delta v)(x,y) = v_{xx}(x,y) + v_{yy}(x,y),$$

is called the *Laplacian*. Equality (4.7) is the two-dimensional heat (diffusion) equation.

If the body contains sources, then (4.7) is replaced by its nonhomogeneous counterpart

$$u_t(x, y, t) = k(\Delta u)(x, y, t) + q(x, y, t), \quad (x, y) \text{ in } D, \ t > 0,$$

where, as before, q now incorporates the factor $1/(c\rho)$.

Initial condition. Here, we have

$$u(x, y, 0) = f(x, y), \quad (x, y) \text{ in } D,$$

which represents the initial distribution of temperature in the body.

Boundary conditions. The main types of BCs are similar in nature to those for a rod.

(i) When the temperature is prescribed on the boundary,

$$u(x, y, t) = \alpha(x, y, t), \quad (x, y) \text{ on } \partial D, \ t > 0.$$

This is called a *Dirichlet boundary condition*.

(ii) If the flux through the boundary is prescribed, then

$$(\text{grad } u)(x, y, t) \cdot \vec{n}(x, y) = \beta(x, y, t), \quad (x, y) \text{ on } \partial D, \ t > 0,$$

or, equivalently,

$$u_n(x, y, t) = \beta(x, y, t), \quad (x, y) \text{ on } \partial D, \ t > 0,$$

where $\vec{n}$ is the unit outward normal to ∂D and $u_n = \partial u/\partial n$. This is called a *Neumann boundary condition*. In particular, when the boundary is insulated, we have

$$u_n(x, y, t) = 0, \quad (x, y) \text{ on } \partial D, \ t > 0.$$

(iii) Newton's law of cooling takes the form

$$-K_0 u_n(x, y, t) = H\big[u(x, y, t) - U(x, y, t)\big], \quad (x, y) \text{ on } \partial D, \ t > 0.$$

This is referred to as a *Robin boundary condition*.

Sometimes, we may have one type of condition prescribed on some part of the boundary, and another type on the remaining part.

Equilibrium temperature. An equilibrium (steady-state) temperature is a time-independent solution $u = u(x, y)$ of (4.7); in other words, it is a function satisfying the *Laplace equation*

$$(\Delta u)(x, y) = 0, \quad (x, y) \text{ in } D,$$

and a time-independent boundary condition, for example,

$$u(x, y) = \alpha(x, y), \quad (x, y) \text{ on } \partial D.$$

If steady-state sources $q(x, y)$ are present in the body, then the equilibrium temperature satisfies the *Poisson equation*

$$(\Delta u)(x, y) = -\frac{1}{k} q(x, y), \quad (x, y) \text{ in } D.$$

In what follows, when we discuss the Poisson equation—that is, the nonhomogeneous Laplace equation—we omit the factor $-1/k$ on the right-hand side, regarding it as incorporated in the source term q.

An equilibrium temperature may, or may not, exist.

4.11 Remark. The Poisson equation also arises in electrostatics. According to Maxwell's equations, a time-independent, two-dimensional electric field (u, v) satisfies

$$v_x - u_y = 0, \quad u_x + v_y = \rho,$$

where ρ is the charge density. The first equation shows that $-u\,dx - v\,dy$ is the differential of a function φ, called the *electric potential*; that is,

$$d\varphi = \varphi_x dx + \varphi_y dy = -u dx - v dy,$$

which means that $\varphi_x = -u$ and $\varphi_y = -v$. If we replace this in the second equation, we arrive at

$$\Delta\varphi = \varphi_{xx} + \varphi_{yy} = -\rho.$$

4.12 Remarks.   (i) Problems in three space variables are formulated similarly, with $u = u(x, y, z, t)$.

(ii) In polar coordinates $x = r\cos\theta$, $y = r\sin\theta$, for $u = u(r, \theta)$ we have

$$\Delta u = r^{-1}(ru_r)_r + r^{-2}u_{\theta\theta} = u_{rr} + r^{-1}u_r + r^{-2}u_{\theta\theta}.$$

(iii) In cylindrical coordinates $x = r\cos\theta$, $y = r\sin\theta$, z, the Laplacian of $u = u(r, \theta, z)$ takes the form

$$\Delta u = r^{-1}(ru_r)_r + r^{-2}u_{\theta\theta} + u_{zz}.$$

(iv) In circularly (axially) symmetric problems, the function u is independent of θ, so

$$\Delta u = r^{-1}(ru_r)_r + u_{zz} = u_{rr} + r^{-1}u_r + u_{zz}.$$

(v) In spherical coordinates $x = r\cos\theta\sin\varphi$, $y = r\sin\theta\sin\varphi$, $z = r\cos\varphi$, for $u = u(r, \theta, \varphi)$ we have

$$\Delta u = r^{-2}(r^2u_r)_r + (r^2\sin^2\varphi)^{-1}u_{\theta\theta} + (r^2\sin\varphi)^{-1}((\sin\varphi)u_\varphi)_\varphi.$$

(vi) It is obvious that the Laplace and Poisson equations are linear.

(vii) The Laplace equation is the simplest example of a so-called *elliptic equation*.

4.13 Example. The equilibrium temperature distribution in a thin, uniform finite plate with sources, insulated upper and lower faces, and prescribed temperature on the boundary is modeled by the BVP

$$(\Delta u)(x, y) = q(x, y), \quad (x, y) \text{ in } D, \tag{PDE}$$

$$u(x, y) = \alpha(x, y), \quad (x, y) \text{ on } \partial D, \tag{BC}$$

where ∂D is a simple closed curve. The Laplacian is written either in Cartesian or in polar coordinates, depending on the geometry of the plate. When polar coordinates are used, the nature of the ensuing PDE may require us to consider additional 'boundary' conditions, suggested by the physics of the process.

4.14 Remark. A (classical) solution of the BVP in Example 4.13 has the following properties:

(i) It is twice continuously differentiable in D and satisfies the PDE at every point in D.

(ii) It is continuous up to the boundary ∂D of D and satisfies the BC at every point of ∂D.

As mentioned in Remark 4.8(ii), a boundary data function with discontinuities may generate a less smooth solution.

4.15 Theorem. *If u is a solution of the BVP in Example 4.13, then u attains its maximum and minimum values on the boundary ∂D of D.*

Proof. We anticipate a result established in Section 5.3, according to which (see Remark 5.20) the temperature at the center of a circular disc is equal to the average of the temperature on its boundary circle.

Suppose that the maximum of the solution u occurs at a point P inside D. Regarding P as the center of a small disk lying within D, we deduce that the value of u at P is the average of all the values of u on the boundary circle of that disc, and so it cannot be greater than all of those values. We have thus arrived at a contradiction, which implies that u must attain its maximum on the boundary ∂D. Considering $v = -u$, we also conclude that u attains its minimum on ∂D.

This assertion is known in the literature as the *maximum principle* for the Laplace equation. □

4.16 Corollary. *If a solution u of the BVP in Example 4.13 is identically zero on ∂D, then u is also identically zero in D.*

Proof. By Theorem 4.15, the maximum and minimum of u are zero, since they occur at points on ∂D. Hence, u is zero in D. □

4.17 Theorem (Uniqueness). *The BVP in Example 4.13 has at most one solution.*

Proof. Suppose that there are two solutions u_1 and u_2. Because of the linearity of the PDE and BC, the difference $u = u_1 - u_2$ is a solution of the fully homogeneous BVP

$$(\Delta u)(x, y) = 0, \quad (x, y) \text{ in } D,$$
$$u(x, y) = 0, \quad (x, y) \text{ on } \partial D;$$

therefore, by Corollary 4.16, $u = 0$, which means that u_1 and u_2 are the same function. □

4.18 Definition. A solution of a BVP (IVP, IBVP) is said to *depend continuously on the data* (or to be *stable*) if a small variation in the data (BCs, ICs, nonhomogeneous term in the PDE) induces only a small variation in the solution.

4.19 Theorem. *The solution of the BVP in Example 4.13 depends continuously on the boundary data.*

Proof. Consider the BVPs

$$(\Delta u)(x, y) = q(x, y), \quad (x, y) \text{ in } D,$$
$$u(x, y) = \alpha(x, y), \quad (x, y) \text{ on } \partial D,$$

and

$$(\Delta v)(x, y) = q(x, y), \quad (x, y) \text{ in } D,$$
$$v(x, y) = \alpha(x, y) + \varepsilon(x, y), \quad (x, y) \text{ on } \partial D,$$

where ε is a small perturbation of the boundary function α. Then, taking the linearity of the PDE and BC into account, we see that $w = u - v$ satisfies

$$(\Delta w)(x, y) = 0, \quad (x, y) \text{ in } D,$$
$$w(x, y) = \varepsilon(x, y), \quad (x, y) \text{ on } \partial D.$$

By Theorem 4.15, for all (x, y) in D,

$$\min_{(x,y) \text{ on } \partial D} \varepsilon(x, y) \leq w(x, y) = u(x, y) - v(x, y) \leq \max_{(x,y) \text{ on } \partial D} \varepsilon(x, y);$$

that is, the perturbation of the solution u is small. □

4.20 Remark. A BVP (IVP, IBVP) is said to be *well posed* if it has a unique solution that depends continuously on the data. In the following chapters, we construct a solution for the BVP in Example 4.13, which, according to Theorem 4.17, is the only solution of that problem. Furthermore, by Theorem 4.19, this solution depends continuously on the boundary data, so the Dirichlet problem for the Laplace equation is well posed.

Exercises

1 Show that any (classical) solution of the Poisson equation $\Delta u = q$ in D which is continuously differentiable, rather than merely continuous, up to the boundary ∂D satisfies

$$\int_D (u_x^2 + u_y^2)\, dA + \int_D uq\, dA = \int_{\partial D} uu_n\, ds.$$

Using this formula, show that

   (i) each of the Dirichlet and Robin BVPs for the Poisson equation in D has at most one solution of this type;

  (ii) any two such solutions of the Neumann BVP for the Poisson equation in D differ by a constant.

4.3 The Wave Equation

Vibrating string. Consider the motion of a tightly stretched elastic string, in which the horizontal displacement of the points of the string is negligible (see Fig. 4.3).

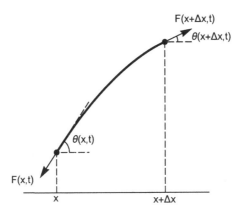

Figure 4.3: An arbitrary small segment of an elastic string.

We define the following physical parameters for the string:

     L : length;

$u(x,t)$: vertical displacement;

$\rho_0(x)$: mass density (mass per unit length);

$F(x,t)$: tension;

$q(x,t)$: vertical component of the body force per unit mass.

Assuming that the vertical displacement from the equilibrium position (the segment $[0, L]$ along the x-axis) is small and writing Newton's second law

$$\text{force} = \text{mass} \times \text{acceleration}$$

in vertical projection for the string segment between two close points corresponding to x and $x + \Delta x$, we obtain the approximate equality

$$\rho_0(x)(\Delta x)u_{tt}(x,t) \cong F(x + \Delta x, t) \sin \left(\theta(x + \Delta x, t) \right)$$
$$- F(x,t) \sin \left(\theta(x,t) \right) + \rho_0(x)(\Delta x)q(x,t).$$

If we divide by Δx and then let $\Delta x \to 0$, we arrive at the equality

$$\rho_0(x)u_{tt}(x,t) = \left[F(x,t) \sin \left(\theta(x,t) \right) \right]_x + \rho_0(x)q(x,t). \tag{4.8}$$

For a small angle θ, the slope of the string satisfies

$$u_x = \tan \theta = \frac{\sin \theta}{\cos \theta} = \frac{\sin \theta}{1 - \frac{1}{2}\theta^2 + \cdots} \cong \sin \theta,$$

so we can rewrite (4.8) in the form

$$\rho_0(x)u_{tt}(x,t) = \left(F(x,t)u_x(x,t) \right)_x + \rho_0(x)q(x,t). \tag{4.9}$$

Assuming that the string is perfectly elastic (that is, when θ is small we may take $F(x,t) = F_0 = \text{const}$) and homogeneous ($\rho_0 = \text{const}$), and that the body force is negligible compared to tension ($q = 0$), we see that (4.9) becomes

$$u_{tt}(x,t) = c^2 u_{xx}(x,t), \quad 0 < x < L, \ t > 0, \tag{4.10}$$

where $c^2 = F_0/\rho_0$. This is the one-dimensional *wave equation*, in which c has the dimensions of velocity. If the body force is not negligible, then (4.10) is replaced by

$$u_{tt}(x,t) = c^2 u_{xx}(x,t) + q(x,t), \quad 0 < x < L, \ t > 0.$$

Initial conditions. Since (4.10) is of second order with respect to time, we need two initial conditions. These usually are

$$u(x,0) = f(x), \quad u_t(x,0) = g(x), \quad 0 < x < L;$$

that is, we prescribe the initial position and velocity of the points in the string.

Boundary conditions. As in the case of the heat equation, there are three main types of BCs for (4.10) which are physically meaningful.

(i) The displacement may be prescribed at an endpoint; for example,

$$u(L, t) = \beta(t).$$

(ii) An endpoint may be free (no vertical tension):

$$F_0 \sin \theta \cong F_0 \tan \theta = F_0 u_x = 0,$$

so we may have a condition of the form

$$u_x(0, t) = 0, \quad t > 0.$$

(iii) An endpoint may have an elastic attachment, described by

$$F_0 u_x(0, t) = k u(0, t) \quad \text{or} \quad F_0 u_x(L, t) = -k u(L, t), \quad t > 0,$$

where $k = \text{const} > 0$.

4.21 Remarks. (i) There is an obvious analogy between the BCs associated with the wave equation and those associated with the heat equation.

(ii) The condition prescribed at one endpoint may be different from that prescribed at the other.

(iii) It is clear that the wave equation is linear.

(iv) The one-dimensional wave equation is the simplest example of a linear second-order *hyperbolic equation*.

The model. As we have seen, the vibrations of an elastic string are governed by the wave equation and appropriate BCs and ICs.

4.22 Example. The IBVP for a vibrating string with fixed endpoints when the effect of the body force is taken into account is of the form

$$u_{tt}(x, t) = c^2 u_{xx}(x, t) + q(x, t), \quad 0 < x < L, \ t > 0, \tag{PDE}$$

$$u(0, t) = 0, \quad u(L, t) = 0, \quad t > 0, \tag{BCs}$$

$$u(x, 0) = f(x), \quad u_t(x, 0) = g(x), \quad 0 < x < L, \tag{ICs}$$

where q, f, and g are given functions.

4.23 Example. If the string in Example 4.22 has an elastic attachment at its near endpoint, a free far endpoint, and a negligible body force, then the corresponding IBVP is

$$u_{tt}(x, t) = c^2 u_{xx}(x, t), \quad 0 < x < L, \ t > 0, \tag{PDE}$$

$$u_x(0, t) - k u(0, t) = 0, \quad u_x(L, t) = 0, \quad t > 0, \tag{BCs}$$

$$u(x, 0) = f(x), \quad u_t(x, 0) = g(x), \quad 0 < x < L, \tag{ICs}$$

where k is a known positive constant and f and g are prescribed functions.

4.24 Remark. Let the domain G and its boundary lines ∂G_x and ∂G_t be as defined in Example 4.7. A (classical) solution u of the IBVP in Example 4.22 has the following properties:

(i) It is twice continuously differentiable with respect to x and t in G and satisfies the PDE at every point in G.

(ii) It is continuous up to ∂G_x and satisfies the BCs at every point on ∂G_x.

(iii) It is continuous together with its first-order time derivative up to ∂G_t and satisfies the ICs at every point on ∂G_t.

As in the case of the heat equation and the Laplace equation, the presence of jump discontinuities in the data functions leads to a solution with reduced smoothness.

4.25 Theorem (Uniqueness). *If the functions q, f, and g are such that u and all its first-order and second-order derivatives are continuous up to the boundary of G, including the 'corner' points $(0,0)$ and $(L,0)$, then the IBVP in Example 4.22 has at most one solution.*

Proof. Suppose that the given IBVP has two solutions u_1 and u_2. Then their difference $u = u_1 - u_2$ satisfies the fully homogeneous IBVP

$$u_{tt}(x,t) = c^2 u_{xx}(x,t), \quad 0 < x < L, \ t > 0,$$

$$u(0,t) = 0, \quad u(L,t) = 0, \quad t > 0,$$

$$u(x,0) = 0, \quad u_t(x,0) = 0, \quad 0 < x < L.$$

Multiplying the PDE by u_t, integrating the result with respect to x over $[0,L]$ and with respect to t over $[0,T]$, where $T > 0$ is an arbitrarily fixed number, using integration by parts, and taking the BCs and the smoothness properties of u into account, we see that

$$0 = \int_0^T \int_0^L (u_{tt}u_t - c^2 u_{xx}u_t)\, dx\, dt$$

$$= \int_0^T \int_0^L (u_{tt}u_t + c^2 u_x u_{xt})\, dx\, dt - c^2 \int_0^T [u_x u_t]_{x=0}^{x=L}\, dt$$

$$= \tfrac{1}{2} \int_0^T \int_0^L (u_t^2 + c^2 u_x^2)_t\, dx\, dt = \tfrac{1}{2} \int_0^L [u_t^2 + c^2 u_x^2]_{t=0}^{t=T}\, dx;$$

hence, for any $T > 0$,

$$\int_0^L [u_t^2(x,T) + c^2 u_x^2(x,T)]\, dx = \int_0^L [u_t^2(x,0) + c^2 u_x^2(x,0)]\, dx,$$

which implies that

$$V(t) = \int_0^L \left[u_t^2(x,t) + c^2 u_x^2(x,t) \right] dx = \kappa = \text{const} \geq 0, \quad t \geq 0.$$

From the ICs and the smoothness properties of u, it follows that $V(0) = 0$, so $\kappa = 0$; in other words, $V = 0$. Since the integrand in V is nonnegative, we see that this is possible if and only if

$$u_t(x,t) = 0, \quad u_x(x,t) = 0, \quad 0 \leq x \leq L, \ t \geq 0.$$

Consequently, u is constant in G and on its boundary lines. Since u is zero on these lines, we conclude that $u = 0$; thus, u_1 and u_2 coincide. □

4.26 Remarks. (i) As for the heat and Laplace equations, in what follows we construct a solution for the IBVP mentioned in Example 4.22. By Theorem 4.25, this will be the only solution of that problem.

(ii) Uniqueness can also be proved for the solutions of IBVPs with other types of BCs.

(iii) The substitution

$$\xi = \frac{1}{L} x, \quad \tau = \frac{c}{L} t, \quad u(x,t) = u(L\xi, L\tau/c) = v(\xi, \tau)$$

reduces the wave equation to

$$v_{\tau\tau}(\xi, \tau) = v_{\xi\xi}(\xi, \tau), \quad 0 < \xi < 1, \ \tau > 0.$$

It is mostly this simpler form that we use in applications.

Exercises

1 Show that Theorem 4.25 also holds for the IBVP in Example 4.22 with $h = \text{const} > 0$ and the BCs replaced by

(i) $u_x(0,t) - hu(0,t) = 0, \quad u(L,t) = 0, \quad t > 0;$
(ii) $u_x(0,t) = 0, \quad u_x(L,t) + hu(L,t) = 0, \quad t > 0.$

2 Verify the statement in Remark 4.26(iii).

4.4 Other Equations

Below is a list of other linear partial differential equations that occur in important mathematical models. Their classification as parabolic, hyperbolic, or elliptic is explained in Chapter 11.

Brownian motion. The function $u(x, t)$ that specifies the probability that a particle undergoing one-dimensional motion in a fluid is located at point x at time t satisfies the second-order parabolic PDE

$$u_t(x, t) = a u_{xx}(x, t) - b u_x(x, t),$$

where the coefficients a, $b > 0$ are related to the average displacement of the particle per unit time and the variance of the observed displacement around the average.

Diffusion–convection problems. The one-dimensional case is described by the second-order parabolic PDE

$$u_t(x, t) = k u_{xx}(x, t) - a u_x(x, t) + b u(x, t),$$

where $u(x, t)$ is the temperature and the coefficients k, $a > 0$ and b are expressed in terms of the physical properties of the medium, the heat flow rate, and the strength of the source.

When $a = 0$ and $b > 0$, the above equation describes a diffusion process with a chain reaction.

Stock market prices. The price $V(S, t)$ of a derivative on the stock market is the solution of the second-order parabolic PDE

$$V_t(S, t) + \tfrac{1}{2} \sigma^2 S^2 V_{SS}(S, t) + r S V_S(S, t) - r V(S, t) = 0,$$

where S is the price of the stock, σ is the volatility of the stock, r is the continuously compounded risk-free interest per annum, and time t is measured in years. This is known as the *Black–Scholes equation*.

Evolution of a quantum state. In quantum mechanics, the wave function $\psi(x, t)$ that characterizes the one-dimensional motion of a particle under the influence of a potential $V(x)$ satisfies the *Schrödinger equation*

$$i\hbar \psi_t(x, t) = -\frac{\hbar^2}{2m} \psi_{xx}(x, t) + V(x)\psi(x, t).$$

This is a second-order parabolic PDE where $\hbar$ is the reduced Planck constant, m is the mass of the particle, and $i^2 = -1$.

Motion of a quantum scalar field. The wave function $\psi(x, y, z, t)$ of a free moving particle in relativistic quantum mechanics is the solution of the (second-order hyperbolic) *Klein–Gordon equation*

$$\psi_{tt}(x, y, z, t) = c^2 \Delta \psi(x, y, z, t) - \frac{m^2 c^4}{\hbar^2} \psi(x, y, z, t),$$

where c is the speed of light.

The one-dimensional version of this equation is of the form

$$u_{tt}(x, t) = c^2 u_{xx}(x, t) - a u(x, t),$$

where $a = \text{const} > 0$.

Loss transmission line. The current and voltage in the propagation of signals on a telegraph line satisfy the second-order hyperbolic PDE

$$au_{tt}(x,t) + bu_t(x,t) + cu(x,t) = u_{xx}(x,t),$$

where the coefficients a, $b > 0$ and $c \geq 0$ are expressed in terms of the resistance, inductance, capacitance, and conductance characterizing the line. This is known as the *telegraph* (or *telegrapher's*) *equation*.

Dissipative waves. In the one-dimensional case, the propagation of such waves is governed by a second-order hyperbolic PDE of the form

$$u_{tt}(x,t) + au_t(x,t) + bu(x,t) = c^2 u_{xx}(x,t) - du_x(x,t),$$

where the coefficients a, b, $d \geq 0$ are not all zero and $c > 0$.

Transverse vibrations of a rod. The deflection $u(x,t)$ of a generic point in the rod satisfies the fourth-order PDE

$$u_{tt}(x,t) + c^2 u_{xxxx}(x,t) = 0,$$

where c is a physical constant related to the rigidity of the rod material.

Scattered waves. The *Helmholtz equation*

$$\Delta u(x,y,z) + k^2 u(x,y,z) = 0,$$

where $k = $ const, plays an important role in the study of scattering of acoustic, electromagnetic, and elastic waves. It is a second-order elliptic PDE.

The equation

$$\Delta u(x,y,z) - k^2 u(x,y,z) = 0$$

is called the *modified Helmholtz equation.*

Steady-state convective heat. In the two-dimensional case, the temperature distribution $u(x,y)$ governing this process satisfies a second-order elliptic PDE of the form

$$\Delta u(x,y) - au_x(x,y) - bu_y(x,y) + cu(x,y) = 0,$$

where the coefficients a, $b \geq 0$, not both zero, and c are related to the thermal properties of the medium, the heat flow rates in the x and y directions, and the strength of the heat source.

Plane problems in continuum mechanics. The Airy stress function in plane elasticity and the stream function in the slow flow of a viscous incompressible fluid are solutions of the elliptic fourth-order *biharmonic equation*

$$\Delta\Delta u(x,y) = u_{xxxx}(x,y) + 2u_{xxyy}(x,y) + u_{yyyy}(x,y) = 0.$$

Plane transonic flow. The transonic flow of a compressible gas is described by the *Euler–Tricomi equation*

$$u_{xx}(x, y) = x u_{yy}(x, y),$$

where $u(x, y)$ is a function of speed. This is a second-order PDE of mixed type: it is hyperbolic for $x > 0$, elliptic for $x < 0$, and parabolic for $x = 0$.

Mathematical biology. The *Fisher equation*

$$u_t(x, t) = D u_{xx}(x, t) + r u(x, t) \big[1 - u(x, t) \big],$$

where $r, D = \text{const} > 0$, models the advance of advantageous genes in a population. This is a nonlinear equation of parabolic type.

Fluid dynamics. The propagation of waves on a free-moving fluid surface is governed by the nonlinear *Boussinesq equation*

$$\eta_{tt}(x, t) - gh\eta_{xx}(x, t) - gh\left(\frac{3}{2h} \eta^2(x, t) + \frac{h^2}{3} \eta_{xx}(x, t) \right)_{xx} = 0,$$

where $\eta = \eta(x, t)$ is the free surface elevation, $h = \text{const}$ is the fluid depth, and g is the acceleration of gravity.

4.27 Remark. In the rest of the book we assume, without explicit mention, that all functions representing nonhomogeneous terms in PDEs, ICs, and BCs are prescribed and 'reasonably behaved', in the sense that they possess the degree of smoothness required by the implementation of the solution procedure.

Exercises

In **1–4**, find real numbers α and β such that the function substitution

$$u(x, t) = e^{\alpha x + \beta t} v(x, t)$$

reduces the given diffusion–convection equation to the heat equation for v.

1 $u_t(x, t) = 2u_{xx}(x, t) - 3u_x(x, t) - u(x, t).$

2 $u_t(x, t) = u_{xx}(x, t) - 6u_x(x, t) - 2u(x, t).$

3 $u_t(x, t) = \frac{1}{2} u_{xx}(x, t) - 4u_x(x, t) - 2u(x, t).$

4 $u_t(x, t) = 3u_{xx}(x, t) - 2u_x(x, t) - 3u(x, t).$

In **5–8**, determine if there are real numbers α and β such that the function substitution $u(x, t) = e^{\alpha x + \beta t} v(x, t)$ in the given dissipative wave equation eliminates (i) both first-order derivatives; (ii) the unknown function and its first-order t-derivative; (iii) the unknown function and its first-order x-derivative.

(In all cases, the coefficients of the reduced equation must satisfy the restrictions mentioned in Section 4.4.) When such a substitution exists, write out the reduced equation.

5 $u_{tt}(x,t) + u_t(x,t) + u(x,t) = u_{xx}(x,t) - 2u_x(x,t).$

6 $u_{tt}(x,t) + 3u_t(x,t) + u(x,t) = 2u_{xx}(x,t) - u_x(x,t).$

7 $u_{tt}(x,t) + 2u_t(x,t) + \frac{1}{2}u(x,t) = 2u_{xx}(x,t) - \sqrt{2}\,u_x(x,t).$

8 $u_{tt}(x,t) + 2u_t(x,t) + 2u(x,t) = u_{xx}(x,t) - u_x(x,t).$

In **9–12**, find real numbers α and β such that the function substitution

$$u(x,y) = e^{\alpha x + \beta y} v(x,y)$$

reduces the given steady-state convective heat equation to the Helmholtz (modified Helmholtz) equation. In each case, write out the reduced equation.

9 $u_{xx}(x,y) + u_{yy}(x,y) - 2u_x(x,y) - 4u_y(x,y) + 7u(x,y) = 0.$

10 $u_{xx}(x,y) + u_{yy}(x,y) - u_x(x,y) - 2u_y(x,y) - u(x,y) = 0.$

11 $u_{xx}(x,y) + u_{yy}(x,y) - 3u_x(x,y) - u_y(x,y) + 2u(x,y) = 0.$

12 $u_{xx}(x,y) + u_{yy}(x,y) - 2u_x(x,y) - 3u_y(x,y) + 4u(x,y) = 0.$

Answers to Odd-Numbered Exercises

1 $\alpha = 3/4,\ \beta = -17/8.$

3 $\alpha = 4,\ \beta = -10.$

5 (i) $\alpha = 1,\ \beta = -1/2,\ v_{tt}(x,t) + (7/4)v(x,t) = v_{xx}(x,t);$

   (ii) $\alpha = 1 - \sqrt{7}/2,\ \beta = -1/2,\ v_{tt}(x,t) = v_{xx}(x,t) - \sqrt{7}\,v_x(x,t).$

7 (iii) $\alpha = \sqrt{2}/4,\ \beta = -1/2,\ v_{tt}(x,t) + v_t(x,t) = 2v_{xx}(x,t).$

9 $\alpha = 1,\ \beta = 2,\ v_{xx}(x,y) + v_{yy}(x,y) + 2v(x,y) = 0.$

11 $\alpha = 3/2,\ \beta = 1/2,\ v_{xx}(x,y) + v_{yy}(x,y) - (1/2)v(x,y) = 0.$

CHAPTER 5

The Method of Separation of Variables

Separation of variables is one of the oldest and most efficient solution techniques for a certain class of PDE problems. Below, we show it in application to initial boundary value problems for the heat and wave equations, and to boundary value problems for the Laplace equation.

5.1 The Heat Equation

We investigate several cases, identified in terms of the BCs prescribed in the mathematical model.

5.1.1 Rod with Zero Temperature at the Endpoints

Consider the IBVP

$$u_t(x,t) = ku_{xx}(x,t), \quad 0 < x < L,\ t > 0, \tag{PDE}$$

$$u(0,t) = 0, \quad u(L,t) = 0, \quad t > 0, \tag{BCs}$$

$$u(x,0) = f(x), \quad 0 < x < L, \tag{IC}$$

where $f \neq 0$. We remark that the PDE and BCs are linear and homogeneous, and seek a solution of the form

$$u(x,t) = X(x)T(t).$$

Substituting into the PDE, we obtain the equality

$$X(x)T'(t) = kX''(x)T(t).$$

It is clear that neither X nor T can be the zero function: if either of them were, then so would u, which is impossible, because the zero solution does not

99

satisfy the nonhomogeneous IC. Hence, we may divide the preceding equality by $kX(x)T(t)$ to arrive at

$$\frac{1}{k}\frac{T'(t)}{T(t)} = \frac{X''(x)}{X(x)}.$$

Since the left-hand side above is a function of t and the right-hand side a function of x alone, this equality is possible if and only if both sides are equal to one and the same constant, say, $-\lambda$. In other words, we must have

$$\frac{1}{k}\frac{T'(t)}{T(t)} = \frac{X''(x)}{X(x)} = -\lambda,$$

where λ is called the *separation constant*. This leads to separate equations for the functions X and T:

$$X''(x) + \lambda X(x) = 0, \quad 0 < x < L, \tag{5.1}$$

$$T'(t) + \lambda k T(t) = 0, \quad t > 0. \tag{5.2}$$

From the first BC we see that

$$u(0,t) = X(0)T(t) = 0 \quad \text{for all } t > 0.$$

As $T \neq 0$, it follows that

$$X(0) = 0. \tag{5.3}$$

Similarly, the second BC yields

$$X(L) = 0. \tag{5.4}$$

We are now ready to find X and T. We want nonzero functions X that satisfy the regular Sturm–Liouville problem (5.1), (5.3), (5.4); in other words, the eigenfunctions X_n corresponding to the eigenvalues λ_n, both computed in Example 3.16 (see also Appendix A.2):

$$\lambda_n = \left(\frac{n\pi}{L}\right)^2, \quad X_n(x) = \sin\frac{n\pi x}{L}, \quad n = 1, 2, \ldots. \tag{5.5}$$

For each λ_n, from (5.2) we find the corresponding time component

$$T_n(t) = e^{-k(n\pi/L)^2 t}, \quad n = 1, 2, \ldots. \tag{5.6}$$

We have taken the arbitrary constant of integration equal to 1 since these functions are used in the next stage with arbitrary numerical coefficients.

Combining (5.5) and (5.6), we conclude that all functions of the form

$$u_n(x,t) = X_n(x)T_n(t) = \sin\frac{n\pi x}{L}e^{-k(n\pi/L)^2 t}, \quad n = 1, 2, \ldots,$$

satisfy both the PDE and the BCs. By the principle of superposition, so does any finite linear combination

$$\sum_{n=1}^{N} b_n u_n(x,t) = \sum_{n=1}^{N} b_n \sin \frac{n\pi x}{L} e^{-k(n\pi/L)^2 t}, \tag{5.7}$$

where b_n are arbitrary numbers. The IC is satisfied by such an expression if

$$f(x) = \sum_{n=1}^{N} b_n u_n(x,0) = \sum_{n=1}^{N} b_n \sin \frac{n\pi x}{L}.$$

But this is impossible unless f is a finite linear combination of the eigenfunctions, which, in general, is not the case. This implies that (5.7) is not a good representation for the solution $u(x,t)$ of the IBVP. However, we recall (see Section 2.2) that if f is piecewise smooth, then it can be written as an *infinite* linear combination of the eigenfunctions (its Fourier sine series):

$$f(x) = \sum_{n=1}^{\infty} b_n \sin \frac{n\pi x}{L}, \quad 0 < x < L, \tag{5.8}$$

where, by (2.12),

$$b_n = \frac{2}{L} \int_0^L f(x) \sin \frac{n\pi x}{L} \, dx, \quad n = 1, 2, \dots. \tag{5.9}$$

Therefore, the solution is given by the infinite series

$$u(x,t) = \sum_{n=1}^{\infty} b_n \sin \frac{n\pi x}{L} e^{-k(n\pi/L)^2 t} \tag{5.10}$$

with the coefficients b_n computed from (5.9).

5.1 Example. Consider the IBVP

$$u_t(x,t) = u_{xx}(x,t), \quad 0 < x < 1, \ t > 0,$$
$$u(0,t) = 0, \quad u(1,t) = 0, \quad t > 0,$$
$$u(x,0) = \sin(3\pi x) - 2\sin(5\pi x), \quad 0 < x < 1.$$

Here $k = 1$, $L = 1$, and the function on the right-hand side in the IC is a finite linear combination of the eigenfunctions. Using (5.8) and Theorem 3.26(ii), we find that

$$b_3 = 1, \quad b_5 = -2, \quad b_n = 0 \ (n \neq 3, 5);$$

so, by (5.10), the solution of the problem is

$$u(x,t) = \sin(3\pi x)e^{-9\pi^2 t} - 2\sin(5\pi x)e^{-25\pi^2 t}.$$

Equally, this result can be obtained from (5.9) and (5.10).

VERIFICATION WITH MATHEMATICA®. The input

```
u = Sin[3*Pi*x] *E^(-9*Pi^2*t) - 2*Sin[5*Pi*x]
  *E^(-25*Pi^2*t);
{D[u,t] - D[u,x,x], u/.x->0, u/.x->1, (u/.t->0)
  - (Sin[3*Pi*x] - 2*Sin[5*Pi*x])} //FullSimplify
```

generates the output $\{0,0,0,0\}$, which confirms that the PDE, BCs, and IC are satisfied.

5.2 Example. To find the solution of the IBVP

$$u_t(x,t) = u_{xx}(x,t), \quad 0 < x < 1, \ t > 0,$$
$$u(0,t) = 0, \quad u(1,t) = 0, \quad t > 0,$$
$$u(x,0) = x, \quad 0 < x < 1,$$

we first use (5.9) with $f(x) = x$ and $L = 1$, and integration by parts, to compute the coefficients b_n:

$$b_n = 2 \int_0^1 x \sin(n\pi x)\, dx = (-1)^{n+1}\frac{2}{n\pi}, \quad n = 1, 2, \ldots;$$

then, by (5.10),

$$u(x,t) = \sum_{n=1}^{\infty}(-1)^{n+1}\frac{2}{n\pi}\sin(n\pi x)\,e^{-n^2\pi^2 t}$$

is the solution of the IBVP.

VERIFICATION WITH MATHEMATICA®. The input

```
un = (-1)^(n+1) * (2/(n*Pi)) * Sin[n*Pi*x] *E^(-n^2*Pi^2*t);
Simplify[{D[un,t] - D[un,x,x], un/.x->0, un/.x->1}, n ∈ Integers]
Plot[{Sum[un/.t->0, {n,1,7}], x}, {x,0,1},
  PlotRange->{{-0.1,1.1}, {-0.2,1.1}}, PlotStyle->Black,
  ImageSize->Scaled[0.2], Ticks->{{1},{1}}, AspectRatio->0.9]
```

generates the output $\{0,0,0\}$ and the graph

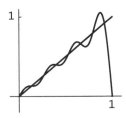

which confirm that the PDE and BCs are satisfied and that the truncation (after 7 terms) of the series solution at $t = 0$ is a good approximation of the IC function.

Exercises

Use separation of variables to solve the IBVP

$$u_t(x,t) = u_{xx}(x,t), \quad 0 < x < 1, \ t > 0,$$
$$u(0,t) = 0, \quad u(1,t) = 0, \quad t > 0,$$
$$u(x,0) = f(x), \quad 0 < x < 1$$

with function f as indicated.

1 $f(x) = \sin(2\pi x) - 3\sin(6\pi x)$.　　**2** $f(x) = 3\sin(\pi x) - \sin(4\pi x)$.

3 $f(x) = -2$.　　**4** $f(x) = 1 - 2x$.

5 $f(x) = 2x + 1$.

6 $f(x) = \begin{cases} 0, & 0 < x \leq 1/2, \\ 2, & 1/2 < x < 1. \end{cases}$

7 $f(x) = \begin{cases} x, & 0 < x \leq 1/2, \\ 0, & 1/2 < x < 1. \end{cases}$

8 $f(x) = \begin{cases} 2, & 0 < x \leq 1/2, \\ x - 1, & 1/2 < x < 1. \end{cases}$

Answers to Odd-Numbered Exercises

1 $u(x,t) = \sin(2\pi x)\, e^{-4\pi^2 t} - 3\sin(6\pi x)\, e^{-36\pi^2 t}$.

3 $u(x,t) = \sum_{n=1}^{\infty} [(-1)^n - 1](4/(n\pi))\sin(n\pi x)\, e^{-n^2\pi^2 t}$.

5 $u(x,t) = \sum_{n=1}^{\infty} [1 - (-1)^n 3](2/(n\pi))\sin(n\pi x)\, e^{-n^2\pi^2 t}$.

7 $u(x,t) = \sum_{n=1}^{\infty} [(2/(n^2\pi^2))\sin(n\pi/2) - (1/(n\pi))\cos(n\pi/2)]$
$$\times \sin(n\pi x)\, e^{-n^2\pi^2 t}.$$

5.1.2　Rod with Insulated Endpoints

The heat conduction problem for a uniform rod with insulated endpoints is described by the IBVP

$$u_t(x,t) = k u_{xx}(x,t), \quad 0 < x < L, \ t > 0, \qquad \text{(PDE)}$$
$$u_x(0,t) = 0, \quad u_x(L,t) = 0, \quad t > 0, \qquad \text{(BCs)}$$
$$u(x,0) = f(x), \quad 0 < x < L. \qquad \text{(IC)}$$

Since the PDE and BCs are linear and homogeneous, we again seek a solution of the form $u(x,t) = X(x)T(t)$ with X, $T \neq 0$. Just as in Subsection 5.1.1, from the PDE and BCs we now find that X and T satisfy, respectively,

$$X''(x) + \lambda X(x) = 0, \quad 0 < x < L,$$
$$X'(0) = 0, \quad X'(L) = 0$$

and

$$T'(t) + k\lambda T(t) = 0, \quad t > 0,$$

where λ is the separation constant.

The nonzero solutions X are the eigenfunctions of the above Sturm–Liouville problem, which were computed in Example 3.17:

$$\lambda_0 = 0, \quad X_0(x) = \tfrac{1}{2},$$
$$\lambda_n = \left(\frac{n\pi}{L}\right)^2, \quad X_n(x) = \cos\frac{n\pi x}{L}, \quad n = 1, 2, \ldots. \tag{5.11}$$

Integrating the equation satisfied by T with $\lambda = \lambda_n$, $n = 0, 1, 2, \ldots$, we obtain the associated time components

$$T_n(t) = e^{-k(n\pi/L)^2 t}, \quad n = 0, 1, 2, \ldots. \tag{5.12}$$

In view of (5.11), (5.12), and the argument used in the preceding case, we now expect the solution of the IBVP to have the series representation

$$u(x,t) = \tfrac{1}{2} a_0 + \sum_{n=1}^{\infty} a_n \cos\frac{n\pi x}{L} e^{-k(n\pi/L)^2 t}, \tag{5.13}$$

where each term satisfies the PDE and the BCs. The IC is satisfied if

$$u(x,0) = f(x) = \tfrac{1}{2} a_0 + \sum_{n=1}^{\infty} a_n \cos\frac{n\pi x}{L}, \quad 0 < x < L.$$

This shows that the a_n are the Fourier cosine series coefficients of f, given by (2.15):

$$a_n = \frac{2}{L} \int_0^L f(x) \cos\frac{n\pi x}{L}\, dx, \quad n = 0, 1, 2, \ldots. \tag{5.14}$$

Therefore, (5.13) with the a_n computed by means of (5.14) is the solution of the IBVP.

5.3 Example. Consider the IBVP

$$u_t(x,t) = u_{xx}(x,t), \quad 0 < x < 1,\ t > 0,$$
$$u_x(0,t) = 0, \quad u_x(1,t) = 0, \quad t > 0,$$
$$u(x,0) = 4 - 3\cos(2\pi x) + 2\cos(3\pi x), \quad 0 < x < 1.$$

With $k = 1$, $L = 1$, and the IC function written as a finite linear combination of the eigenfunctions, we apply formulas (5.11)–(5.13) to find that

$$a_0 = 8, \quad a_2 = -3, \quad a_3 = 2, \quad a_n = 0 \ (n \neq 0, 2, 3);$$

therefore, the solution of the given IBVP is

$$u(x, t) = 4 - 3\cos(2\pi x)e^{-4\pi^2 t} + 2\cos(3\pi x)e^{-9\pi^2 t}.$$

VERIFICATION WITH MATHEMATICA®. The input

```
u = 4 - 3 * Cos [2 * Pi * x] * E^ ( - 4 * Pi^2 * t) +2 * Cos [3 * Pi * x]
   * E^ ( - 9 * Pi^2 * t);
{D[u,t] - D[u,x,x], D[u,x] /. x -> 0, D[u,x] /. x -> 1, (u /. t -> 0)
   - (4 - 3 * Cos [2 * Pi * x] +2 * Cos [3 * Pi * x]}} // FullSimplify
```

generates the output $\{0, 0, 0, 0\}$, which confirms that the PDE, BCs, and IC are satisfied.

5.4 Example. The solution of the IBVP

$$u_t(x, t) = u_{xx}(x, t), \quad 0 < x < 1, \ t > 0,$$
$$u_x(0, t) = 0, \quad u_x(1, t) = 0, \quad t > 0,$$
$$u(x, 0) = x, \quad 0 < x < 1$$

is obtained from (5.13) and (5.14) with

$$L = 1, \quad f(x) = x.$$

Specifically,

$$a_0 = 2 \int_0^1 x \, dx = 1,$$

$$a_n = 2 \int_0^1 x \cos(n\pi x) \, dx = [(-1)^n - 1] \frac{2}{n^2 \pi^2}, \quad n = 1, 2, \ldots;$$

so, as in Example 5.2,

$$u(x, t) = \frac{1}{2} + \sum_{n=1}^{\infty} [(-1)^n - 1] \frac{2}{n^2 \pi^2} \cos(n\pi x) e^{-n^2 \pi^2 t}.$$

VERIFICATION WITH MATHEMATICA®. The input

```
un = ((-1)^n-1) * (2/(n^2*Pi^2)) * Cos[n*Pi*x]
    *E^(-n^2*Pi^2*t);
Simplify[{D[un,t] -D[un,x,x], D[un,x] /.x->0, D[un,x] /.x->1},
    n ∈ Integers]
Plot[{(1/2)+Sum[un/.t->0, {n,1,2}], x}, {x,0,1},
    PlotRange -> {{-0.1,1.1}, {-0.2,1.1}}, PlotStyle -> Black,
    ImageSize -> Scaled[0.2], Ticks -> {{1},{1}}, AspectRatio -> 0.9]
```

generates the output $\{0, 0, 0\}$ and the graph

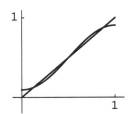

Exercises

Use separation of variables to solve the IBVP

$$u_t(x,t) = u_{xx}(x,t), \quad 0 < x < 1, \ t > 0,$$
$$u_x(0,t) = 0, \quad u_x(1,t) = 0, \quad t > 0,$$
$$u(x,0) = f(x), \quad 0 < x < 1$$

with function f as indicated.

1 $f(x) = 3 - 2\cos(4\pi x).$

2 $f(x) = \cos(2\pi x) - 3\cos(3\pi x).$

3 $f(x) = 2 - 3x.$

4 $f(x) = 3 - 2x.$

5 $f(x) = \begin{cases} -2, & 0 < x \le 1/2, \\ 0, & 1/2 < x < 1. \end{cases}$

6 $f(x) = \begin{cases} 3, & 0 < x \le 1/2, \\ -1, & 1/2 < x < 1. \end{cases}$

7 $f(x) = \begin{cases} 0, & 0 < x \le 1/2, \\ 2x, & 1/2 < x < 1. \end{cases}$

8 $f(x) = \begin{cases} 1, & 0 < x \le 1/2, \\ x+1, & 1/2 < x < 1. \end{cases}$

Answers to Odd-Numbered Exercises

1 $u(x,t) = 3 - 2\cos(4\pi x)\, e^{-16\pi^2 t}$.

3 $u(x,t) = 1/2 + \sum\limits_{n=1}^{\infty} [1 - (-1)^n](6/(n^2\pi^2))\cos(n\pi x)\, e^{-n^2\pi^2 t}$.

5 $u(x,t) = -1 - \sum\limits_{n=1}^{\infty} (4/(n\pi))\sin(n\pi/2)\cos(n\pi x)\, e^{-n^2\pi^2 t}$.

7 $u(x,t) = 3/4 + \sum\limits_{n=1}^{\infty} \{-(2/(n\pi))\sin(n\pi/2) + (4/(n^2\pi^2))[(-1)^n - \cos(n\pi/2)]\}$
$$\times \cos(n\pi x)\, e^{-n^2\pi^2 t}.$$

5.1.3 Rod with Mixed Boundary Conditions

The method of separation of variables can also be used in the case where one endpoint is held at zero temperature, while the other one is insulated.

5.5 Example. Consider the IBVP

$$u_t(x,t) = u_{xx}(x,t), \quad 0 < x < 1,\ t > 0,$$
$$u(0,t) = 0, \quad u_x(1,t) = 0, \quad t > 0,$$
$$u(x,0) = 3\sin\frac{\pi x}{2} - \sin\frac{3\pi x}{2}, \quad 0 < x < 1.$$

Since the PDE and BCs are linear and homogeneous, we seek a solution of the form

$$u(x,t) = X(x)T(t), \quad X,\,T \neq 0.$$

Proceeding as in the other cases, from the PDE and BCs we find that X and T satisfy, respectively,

$$X''(x) + \lambda X(x) = 0, \quad 0 < x < 1,$$
$$X(0) = 0, \quad X'(1) = 0 \tag{5.15}$$

and

$$T'(t) + \lambda T(t) = 0, \quad t > 0,$$

where λ is the separation constant.

Form Example 3.19 with $L = 1$ it follows that the eigenvalues and eigenfunctions of the regular S–L problem for X are

$$\lambda_n = \frac{(2n-1)^2\pi^2}{4}, \quad X_n(x) = \sin\frac{(2n-1)\pi x}{2}, \quad n = 1, 2, \dots,$$

which means that the corresponding time components are

$$T_n(t) = e^{-(2n-1)^2\pi^2 t/4}, \quad n = 1, 2, \dots.$$

Consequently, the functions

$$u_n(x,t) = X_n(x)T_n(t) = \sin \frac{(2n-1)\pi x}{2} e^{-(2n-1)^2\pi^2 t/4}, \quad n = 1,2,\ldots$$

satisfy the PDE and the BCs. Considering the usual arbitrary linear combination of all the (countably many) functions u_n—that is,

$$u(x,t) = \sum_{n=1}^{\infty} c_n u_n(x,t) = \sum_{n=1}^{\infty} c_n \sin \frac{(2n-1)\pi x}{2} e^{-(2n-1)^2\pi^2 t/4}, \quad (5.16)$$

we see that the IC is satisfied if

$$u(x,0) = 3\sin\frac{\pi x}{2} - \sin\frac{3\pi x}{2} = \sum_{n=1}^{\infty} c_n \sin\frac{(2n-1)\pi x}{2}, \quad (5.17)$$

which leads to

$$c_1 = 3, \quad c_2 = -1, \quad c_n = 0 \ (n \neq 1,2).$$

Hence, the solution of the IBVP is

$$u(x,t) = 3\sin\frac{\pi x}{2} e^{-\pi^2 t/4} - \sin\frac{3\pi x}{2} e^{-9\pi^2 t/4}.$$

VERIFICATION WITH MATHEMATICA®. The input

```
u = 3 * Sin[Pi * x/2] * E^(-Pi^2 * t/4) - Sin[3 * Pi * x/2]
    * E^(-9 * Pi^2 * t/4);
{D[u,t] - D[u,x,x] // Simplify, u/.x->0, D[u,x] /.x->1,
    (u/.t->0) - 3 * Sin[Pi * x/2] + Sin[3 * Pi * x/2]}
```

generates the output $\{0,0,0,0\}$, as expected.

5.6 Example. The IBVP

$$u_t(x,t) = u_{xx}(x,t), \quad 0 < x < 1, \ t > 0,$$
$$u(0,t) = 0, \quad u_x(1,t) = 0, \quad t > 0,$$
$$u(x,0) = 1, \quad 0 < x < 1$$

is treated in the same way, except that here, (5.16) and the IC yield

$$u(x,0) = 1 = \sum_{n=1}^{\infty} c_n \sin\frac{(2n-1)\pi x}{2}.$$

In other words, the c_n are the coefficients of the generalized Fourier series for the function

$$f(x) = 1, \quad 0 \le x \le 1$$

in the eigenfunctions X_n of the S–L problem mentioned in Example 5.5. These coefficients are computed by means of (3.10) with

$$u(x) = 1, \quad \sigma(x) = 1, \quad f_n(x) = \sin\frac{(2n-1)\pi x}{2},$$

and direct calculation shows that

$$\int_0^1 \sin^2\frac{(2n-1)\pi x}{2}\, dx = \frac{1}{2}, \quad n = 1, 2, \ldots,$$

so

$$c_n = 2\int_0^1 \sin\frac{(2n-1)\pi x}{2}\, dx = \frac{4}{(2n-1)\pi}, \quad n = 1, 2, \ldots.$$

Thus, the solution of the IBVP has the series representation

$$u(x,t) = \sum_{n=1}^{\infty} \frac{4}{(2n-1)\pi} \sin\frac{(2n-1)\pi x}{2}\, e^{-(2n-1)^2\pi^2 t/4}.$$

VERIFICATION WITH MATHEMATICA®. The input

```
un = (4/((2*n-1)*Pi))*Sin[(2*n-1)*Pi*x/2]*E^(-(2*n-1)^2
   *Pi^2*t/4);
Simplify[{D[un,t] - D[un,x,x], un/.x->0, D[un,x]/.x->1},
   n∈Integers]
Plot[{Sum[un/.t->0, {n,1,7}], 1}, {x,0,1},
   PlotRange->{{-0.1,1.1}, {-0.2,1.3}}, PlotStyle->Black,
   ImageSize->Scaled[0.2], Ticks->{{1},{1}}, AspectRatio->0.9]
```

generates the output $\{0,0,0\}$ and the graph

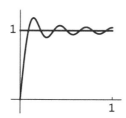

5.7 Example. The solution of the IBVP

$$u_t(x,t) = u_{xx}(x,t), \quad 0 < x < 1, \ t > 0,$$
$$u_x(0,t) = 0, \quad u(1,t) = 0, \quad t > 0,$$
$$u(x,0) = x, \quad 0 < x < 1$$

is constructed just as in Example 5.6, with the difference that in this case the eigenvalue–eigenfunction pairs are (see Example 3.20 with $L = 1$)

$$\lambda_n = \frac{(2n-1)^2\pi^2}{4}, \quad X_n(x) = \cos\frac{(2n-1)\pi x}{2}, \quad n = 1, 2, \ldots.$$

Since

$$\int_0^1 \cos^2\frac{(2n-1)\pi x}{2}\, dx = \frac{1}{2}, \quad n = 1, 2, \ldots,$$

the coefficients of the generalized Fourier series for the IC function, given by (3.10) with

$$u(x) = x, \quad \sigma(x) = 1, \quad k = 1, \quad L = 1,$$

are

$$c_n = 2\int_0^1 x\cos\frac{(2n-1)\pi x}{2}\, dx$$

$$= (-1)^{n+1}\frac{4}{(2n-1)\pi} - \frac{8}{(2n-1)^2\pi^2}, \quad n = 1, 2, \ldots.$$

Consequently, the solution of the IBVP is

$$u(x,t) = \sum_{n=1}^{\infty}\left[(-1)^{n+1}\frac{4}{(2n-1)\pi} - \frac{8}{(2n-1)^2\pi^2}\right]$$

$$\times \cos\frac{(2n-1)\pi x}{2}\, e^{-(2n-1)^2\pi^2 t/4}.$$

VERIFICATION WITH MATHEMATICA®. The input

```
un = (((-1)^(n+1) * 4/((2*n-1) * Pi)) - (8/((2*n-1)^2 * Pi^2)))
    * Cos[(2*n-1) * Pi * x/2] * E^(-(2*n-1)^2 * Pi^2 * t/4);
Simplify[{D[un,t] - D[un,x,x], D[un,x] /.x->0, un/.x->1},
    n ∈ Integers]
Plot[{Sum[un/.t->0, {n,1,7}], 1}, {x,0,1},
    PlotRange -> {{-0.1,1.1}, {-0.2,1.1}}, PlotStyle -> Black,
    ImageSize -> Scaled[0.2], Ticks -> {{1},{1}}, AspectRatio -> 0.9]
```

generates the output $\{0, 0, 0\}$ and the graph

Exercises

Use separation of variables to solve the IBVP consisting of the PDE and IC

$$u_t(x,t) = u_{xx}(x,t), \quad 0 < x < 1, \ t > 0,$$
$$u(x,0) = f(x), \quad 0 < x < 1,$$

with BCs and function f as indicated.

1 $u(0,t) = 0, \quad u_x(1,t) = 0, \quad f(x) = 3\sin(\pi x/2) - \sin(5\pi x/2).$

2 $u(0,t) = 0, \quad u_x(1,t) = 0, \quad f(x) = -3.$

3 $u(0,t) = 0, \quad u_x(1,t) = 0, \quad f(x) = 2 + x.$

4 $u(0,t) = 0, \quad u_x(1,t) = 0, \quad f(x) = \begin{cases} 0, & 0 < x \le 1/2, \\ -1, & 1/2 < x < 1. \end{cases}$

5 $u_x(0,t) = 0, \quad u(1,t) = 0, \quad f(x) = 2\cos(5\pi x/2).$

6 $u_x(0,t) = 0, \quad u(1,t) = 0, \quad f(x) = 4.$

7 $u_x(0,t) = 0, \quad u(1,t) = 0, \quad f(x) = 2x - 3.$

8 $u_x(0,t) = 0, \quad u(1,t) = 0, \quad f(x) = \begin{cases} 2 - x, & 0 < x \le 1/2, \\ 1, & 1/2 < x < 1. \end{cases}$

Answers to Odd-Numbered Exercises

1 $u(x,t) = 3\sin(\pi x/2)\, e^{-\pi^2 t/4} - \sin(5\pi x/2)\, e^{-25\pi^2 t/4}.$

3 $u(x,t) = \sum_{n=1}^{\infty} [8/((2n-1)\pi) + (-1)^{n+1} 8/((2n-1)^2 \pi^2)]$
$$\times \sin((2n-1)\pi x/2)\, e^{-(2n-1)^2 \pi^2 t/4}.$$

5 $u(x,t) = 2\cos(5\pi x/2)\, e^{-25\pi^2 t/4}.$

7 $u(x,t) = \sum_{n=1}^{\infty} [(-1)^n 4/((2n-1)\pi) - 16/((2n-1)^2 \pi^2)]$
$$\times \cos((2n-1)\pi x/2)\, e^{-(2n-1)^2 \pi^2 t/4}.$$

5.1.4 Rod with an Endpoint in a Zero-Temperature Medium

Consider the problem of heat flow in a uniform rod without internal sources, when the near endpoint is kept at zero temperature and the far endpoint is kept in open air of zero temperature. The corresponding IBVP (see

Section 4.1) is

$$u_t(x,t) = k u_{xx}(x,t), \quad 0 < x < L, \ t > 0, \tag{PDE}$$

$$u(0,t) = 0, \quad u_x(L,t) + h u(L,t) = 0, \quad t > 0, \tag{BCs}$$

$$u(x,0) = f(x), \quad 0 < x < L, \tag{IC}$$

where $h = \text{const} > 0$.

Since the PDE and BCs are linear and homogeneous, we use separation of variables and seek a solution of the form

$$u(x,t) = X(x)T(t), \quad X, T \neq 0.$$

In the usual way, from the PDE and BCs we find that X satisfies the regular S–L problem

$$\begin{aligned} X''(x) + \lambda X(x) &= 0, \quad 0 < x < L, \\ X(0) &= 0, \quad X'(L) + hX(L) = 0, \end{aligned} \tag{5.18}$$

and that T is the solution of the equation

$$T'(t) + k\lambda T(t) = 0, \quad t > 0.$$

The eigenvalue–eigenfunction pairs of (5.18) were computed in Example 3.21:

$$\lambda_n = \left(\frac{\zeta_n}{L}\right)^2, \quad X_n(x) = \sin\frac{\zeta_n x}{L}, \quad n = 1, 2, \ldots,$$

where ζ_n are the positive roots of the equation

$$\tan\zeta = -\frac{\zeta}{hL}.$$

Hence,

$$T_n(t) = e^{-k(\zeta_n/L)^2 t}, \quad n = 1, 2, \ldots,$$

so we expect the solution of the IBVP to have a series representation of the form

$$u(x,t) = \sum_{n=1}^{\infty} c_n X_n(x) T_n(t) = \sum_{n=1}^{\infty} c_n \sin\frac{\zeta_n x}{L} e^{-k(\zeta_n/L)^2 t}. \tag{5.19}$$

The coefficients c_n are found from the IC, according to which

$$u(x,0) = f(x) = \sum_{n=1}^{\infty} c_n \sin\frac{\zeta_n x}{L}.$$

Since the eigenfunctions of (5.18) are orthogonal on $[0, L]$, the c_n are computed by means of (3.10):

$$c_n = \frac{\int\limits_0^L f(x)\sin(\zeta_n x/L)\,dx}{\int\limits_0^L \sin^2(\zeta_n x/L)\,dx}, \quad n = 1, 2, \ldots. \tag{5.20}$$

Therefore, the solution of the IBVP is (5.19) with the c_n given by (5.20).

5.8 Example. Consider the IBVP

$$u_t(x,t) = u_{xx}(x,t), \quad 0 < x < 1, \ t > 0,$$

$$u(0,t) = 0, \quad u_x(1,t) + u(1,t) = 0, \quad t > 0,$$

$$u(x,0) = \begin{cases} 0, & 0 < x \le 1/2, \\ 1, & 1/2 < x < 1. \end{cases}$$

From Examples 3.21 and (3.24) with $L = h = 1$, we know that the eigenvalue–eigenfunction pairs associated with this problem are

$$\lambda_n = \zeta_n^2, \quad X_n(x) = \sin(\zeta_n x), \quad n = 1, 2, \ldots,$$

where, rounded to the fourth decimal place,

$$\zeta_1 = 2.0288, \quad \zeta_2 = 4.9132, \quad \zeta_3 = 7.9787, \quad \zeta_4 = 11.0855, \quad \zeta_5 = 14.2074.$$

Using (5.20) and $f(x) = u(x,0)$, we now compute the approximate coefficients

$$c_1 = 0.8001, \quad c_2 = -0.3813, \quad c_3 = -0.1326,$$

$$c_4 = 0.1160, \quad c_5 = 0.1053.$$

Consequently, by (5.19), the solution has the expansion

$$u(x,t) = 0.8001 \sin(2.0288\,x)e^{-4.1159\,t} - 0.3813\sin(4.9132\,x)e^{-24.1393\,t}$$

$$- 0.1326\sin(7.9787\,x)e^{-63.6591\,t}$$

$$+ 0.1160\sin(11.0855\,x)e^{-122.8890\,t}$$

$$+ 0.1053\sin(14.2074\,x)e^{-201.8510\,t} + \cdots. \tag{5.21}$$

VERIFICATION WITH MATHEMATICA®. Each term in (5.21) is an approximation of the corresponding term in (5.19), so it is not expected to satisfy the PDE and BCs exactly. However, computation shows that the error for the truncation consisting of the terms shown in (5.21) is less than 10^{-3}, and that the graph of the function represented by this truncation at $t = 0$ is close to that of $u(x,0)$. Specifically, the input

```
Approx = 0.8001 * Sin[2.0288 * x] * E^(-4.1159 * t) - 0.3813
  * Sin[4.9132 * x] * E^(-24.1393 * t) - 0.1326 * Sin[7.9787 * x]
  * E^(-63.6591 * t) + 0.1160 * Sin[11.0855 * x] * E^(-122.8890 * t)
  + 0.1053 * Sin[14.2074 * x] * E^(-201.8510 * t);
Chop[Simplify[{D[Approx,t] - D[Approx,x,x], Approx /. x ->0,
  (D[Approx,x] /. x -> 1) + Approx /. x->1}, n ∈ Integers],0.001]
Plot[{Approx /. t -> 0, Piecewise[{{0, 0<x<1/2},{1, 1/2<x<1}}]},
```

$\{x,0,1\}$, PlotRange $-> \{\{-0.1,1.1\}, \{-0.2,1.1\}\}$,
PlotStyle $->$ Black, ImageSize $->$ Scaled$[0.2]$,
Ticks $-> \{\{0.5,1\}, \{1\}\}$, AspectRatio $-> 0.9]$

generates the output $\{0,0,0\}$ and the graph

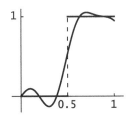

Exercises

Use separation of variables to solve the IBVP

$$u_t(x,t) = u_{xx}(x,t), \quad 0 < x < 1, \ t > 0,$$
$$u_x(0,t) = 0, \quad u_x(1,t) + hu(1,t) = 0, \quad t > 0,$$
$$u(x,0) = f(x), \quad 0 < x < 1$$

with coefficient h and function f as indicated. Compute only the first five terms of the series solution, rounding all numbers to the fourth decimal place.

1 $h = 1$, $f(x) = x + 1$.   **2** $h = 2$, $f(x) = 1 - 2x$.

3 $h = 1/2$, $f(x) = \begin{cases} 1, & 0 < x \le 1/2, \\ -2, & 1/2 < x < 1. \end{cases}$

4 $h = 1$, $f(x) = \begin{cases} x, & 0 < x \le 1/2, \\ 1 - x, & 1/2 < x < 1. \end{cases}$

Answers to Odd-Numbered Exercises

1 $u(x,t) = 1.9184 \sin(2.0288\,x)e^{-4.1159\,t} + 0.1573 \sin(4.9132\,x)e^{-24.1393\,t}$

$\quad + 0.3389 \sin(7.9787\,x)e^{-63.6591\,t} + 0.1307 \sin(11.0855\,x)e^{-122.8890\,t}$

$\quad + 0.1696 \sin(14.2074\,x)e^{-201.8510\,t} + \cdots.$

3 $u(x,t) = -1.2889 \sin(1.8366\,x)e^{-3.3731\,t} + 1.3966 \sin(4.8158\,x)e^{-23.1923\,t}$

$\quad + 0.7337 \sin(7.9171\,x)e^{-62.6797\,t} - 0.1945 \sin(11.0408\,x)e^{-121.9000\,t}$

$\quad - 0.1625 \sin(14.1724\,x)e^{-200.8580\,t} + \cdots.$

5.1.5 Thin Uniform Circular Ring

Physically, the ring of circumference $2L$ shown in Fig. 5.1 can be regarded as a rod of length $2L$ that, for continuity reasons, has the same temperature and heat flux at the two endpoints $x = -L$ and $x = L$.

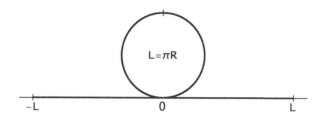

Figure 5.1: Representation of a circular ring as a rod.

Thus, the corresponding IBVP is

$$u_t(x,t) = ku_{xx}(x,t), \quad -L < x < L, \ t > 0, \tag{PDE}$$

$$u(-L,t) = u(L,t), \quad u_x(-L,t) = u_x(L,t), \quad t > 0, \tag{BCs}$$

$$u(x,0) = f(x), \quad -L < x < L. \tag{IC}$$

The PDE and BCs are homogeneous, so, as before, we seek a solution of the form $u(x,t) = X(x)T(t)$ with X, $T \neq 0$. Then from the PDE and BCs we find by the standard argument that

$$\begin{aligned} X''(x) + \lambda X(x) = 0, \quad -L < x < L, \\ X(-L) = X(L), \quad X'(-L) = X'(L), \end{aligned} \tag{5.22}$$

and

$$T'(t) + \lambda kT(t) = 0, \quad t > 0. \tag{5.23}$$

The nonzero function X must be a solution of the periodic Sturm–Liouville problem (5.22), which was discussed in Example 3.33. The eigenvalues of this S–L problem are

$$\lambda_0 = 0, \quad \lambda_n = \left(\frac{n\pi}{L}\right)^2, \quad n = 1, 2, \dots,$$

with corresponding eigenfunctions

$$X_0(x) = \tfrac{1}{2}, \quad X_{1n}(x) = \cos\frac{n\pi x}{L}, \quad X_{2n}(x) = \sin\frac{n\pi x}{L}, \quad n = 1, 2, \dots.$$

Therefore, the time components given by (5.23) with $\lambda = \lambda_n$ are

$$T_n(t) = e^{-k(n\pi/L)^2 t}, \quad n = 0, 1, 2, \dots,$$

so we consider a series solution of the form

$$u(x,t) = \tfrac{1}{2}a_0 + \sum_{n=1}^{\infty}\left[a_n X_{1n}(x) + b_n X_{2n}(x)\right]T_n(t)$$

$$= \tfrac{1}{2}a_0 + \sum_{n=1}^{\infty}\left(a_n \cos\frac{n\pi x}{L} + b_n \sin\frac{n\pi x}{L}\right)e^{-k(n\pi/L)^2 t}. \tag{5.24}$$

The IC requires that

$$u(x,0) = f(x)$$

$$= \tfrac{1}{2}a_0 + \sum_{n=1}^{\infty}\left(a_n \cos\frac{n\pi x}{L} + b_n \sin\frac{n\pi x}{L}\right), \quad -L < x < L.$$

As mentioned in Example 3.33, the coefficients of this full Fourier series of f are computed by means of formulas (2.7)–(2.9):

$$a_n = \frac{1}{L}\int_{-L}^{L} f(x)\cos\frac{n\pi x}{L}\,dx, \quad n = 0, 1, 2, \ldots,$$

$$\tag{5.25}$$

$$b_n = \frac{1}{L}\int_{-L}^{L} f(x)\sin\frac{n\pi x}{L}\,dx, \quad n = 1, 2, \ldots.$$

Consequently, the solution of the IBVP is given by (5.24) and (5.25).

5.9 Example. Consider the IBVP

$$u_t(x,t) = u_{xx}(x,t), \quad -1 < x < 1,\ t > 0,$$
$$u(-1,t) = u(1,t), \quad u_x(-1,t) = u_x(1,t), \quad t > 0,$$
$$u(x,0) = x + 1, \quad -1 < x < 1.$$

Here $L = 1$ and $f(x) = x + 1$, so, by (5.25),

$$a_0 = \int_{-1}^{1}(x+1)\,dx = 2,$$

$$a_n = \int_{-1}^{1}(x+1)\cos(n\pi x)\,dx = 0,$$

$$b_n = \int_{-1}^{1}(x+1)\sin(n\pi x)\,dx = (-1)^{n+1}\frac{2}{n\pi}.$$

Hence, by (5.24), the solution of the IBVP is

$$u(x,t) = 1 + \sum_{n=1}^{\infty} (-1)^{n+1} \frac{2}{n\pi} \sin(n\pi x)\, e^{-n^2\pi^2 t}.$$

VERIFICATION WITH MATHEMATICA®. The input

```
un = (-1)^(n+1) * (2/(n*Pi)) * Sin[n*Pi*x] * E^(-n^2*Pi^2*t);
Simplify[{D[un,t] - D[un,x,x], (un /. x -> -1) - (un /. x -> 1),
   (D[un,x] /. x -> -1) - (D[un,x] /. x -> 1)}, n ∈ Integers]
Plot[{1 + Sum[un /. t -> 0, {n,1,5}], x + 1}, {x, -1,1},
   PlotRange -> {{-1.1,1.1}, {-0.2,2.1}}, PlotStyle -> Black,
   ImageSize -> Scaled[0.3], Ticks -> {{-1,1}, {2}},
   AspectRatio -> 0.5]
```

generates the output $\{0,0,0\}$ and the graph

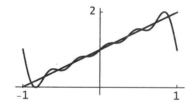

Exercises

Use separation of variables to solve the IBVP

$$u_t(x,t) = u_{xx}(x,t), \quad -1 < x < 1,\ t > 0,$$
$$u(-1,t) = u(1,t), \quad u_x(-1,t) = u_x(1,t), \quad t > 0,$$
$$u(x,0) = f(x), \quad 0 < x < 1,$$

with function f as indicated.

1 $f(x) = 2\sin(2\pi x) - \cos(5\pi x)$. **2** $f(x) = 3x - 2$.

3 $f(x) = \begin{cases} 3, & -1 < x \le 0, \\ 0, & 0 \le x < 1. \end{cases}$ **4** $f(x) = \begin{cases} 2, & -1 < x \le 0, \\ x - 1, & 0 < x < 1. \end{cases}$

Answers to Odd-Numbered Exercises

1 $u(x,t) = 2\sin(2\pi x)\, e^{-4\pi^2 t} - \cos(5\pi x)\, e^{-25\pi^2 t}$.

3 $u(x,t) = 3/2 + \sum_{n=1}^{\infty} [(-1)^n - 1](3/(n\pi)) \sin(n\pi x)\, e^{-n^2\pi^2 t}$.

5.2 The Wave Equation

Generally speaking, the method of separation of variables is applied to IBVPs for the wave equation in much the same way as for the heat equation. Here, however, the time component satisfies a second-order ODE, which introduces a second IC.

5.2.1 Vibrating String with Fixed Endpoints

We know from Section 4.3 that the corresponding IBVP is

$$u_{tt}(x,t) = c^2 u_{xx}(x,t), \quad 0 < x < L, \ t > 0, \tag{PDE}$$

$$u(0,t) = 0, \quad u(L,t) = 0, \quad t > 0, \tag{BCs}$$

$$u(x,0) = f(x), \quad u_t(x,0) = g(x), \quad 0 < x < L. \tag{ICs}$$

Since, as in the earlier problems, the PDE and BCs are linear and homogeneous, we seek a solution of the form

$$u(x,t) = X(x)T(t),$$

where, clearly, neither X nor T can be the zero function. Replacing this in the PDE, we obtain

$$X(x)T''(t) = c^2 X''(x)T(t),$$

which, divided by $c^2 X(x)T(t)$, yields

$$\frac{1}{c^2} \frac{T''(t)}{T(t)} = \frac{X''(x)}{X(x)} = -\lambda, \tag{5.26}$$

where $\lambda = \text{const}$ is the separation constant. At the same time, it can be seen that the BCs lead to $X(0) = 0$ and $X(L) = 0$; hence, X satisfies

$$X''(x) + \lambda X(x) = 0, \quad 0 < x < L,$$

$$X(0) = 0, \quad X(L) = 0.$$

This regular S–L problem was solved in Example 3.16 (see also Appendix A.2), where it was shown that its eigenvalue–eigenfunction pairs are

$$\lambda_n = \left(\frac{n\pi}{L}\right)^2, \quad X_n(x) = \sin\frac{n\pi x}{L}, \quad n = 1, 2, \ldots. \tag{5.27}$$

From (5.26) we deduce that for each eigenvalue λ_n, we obtain a function T_n that satisfies the equation

$$T_n''(t) + c^2 \lambda_n T_n(t) = 0, \quad n = 1, 2, \ldots,$$

whose general solution (since $\lambda_n > 0$) is

$$T_n(t) = b_{1n} \cos\frac{n\pi ct}{L} + b_{2n} \sin\frac{n\pi ct}{L}, \quad b_{1n}, b_{2n} = \text{const.} \tag{5.28}$$

In view of (5.27) and (5.28), we assume that the solution of the IBVP has the series representation

$$u(x,t) = \sum_{n=1}^{\infty} X_n(x) T_n(t)$$

$$= \sum_{n=1}^{\infty} \sin \frac{n\pi x}{L} \left(b_{1n} \cos \frac{n\pi ct}{L} + b_{2n} \sin \frac{n\pi ct}{L} \right). \tag{5.29}$$

Each term of this series satisfies the PDE and the BCs. To satisfy the ICs, we set $t = 0$ in (5.29) and in the expression of $u_t(x,t)$ obtained by differentiating (5.29) term by term:

$$u(x,0) = f(x) = \sum_{n=1}^{\infty} b_{1n} \sin \frac{n\pi x}{L},$$

$$u_t(x,0) = g(x) = \sum_{n=1}^{\infty} b_{2n} \frac{n\pi c}{L} \sin \frac{n\pi x}{L}.$$

We see that the numbers b_{1n} and $b_{2n}(n\pi c)/L$ are the Fourier sine series coefficients of f and g, respectively; so, by (2.12),

$$b_{1n} = \frac{2}{L} \int_0^L f(x) \sin \frac{n\pi x}{L} \, dx, \quad n = 1, 2, \ldots,$$

$$\tag{5.30}$$

$$b_{2n} = \frac{2}{n\pi c} \int_0^L g(x) \sin \frac{n\pi x}{L} \, dx, \quad n = 1, 2, \ldots.$$

Therefore, the solution of the IBVP is given by (5.29) with the b_{1n} and b_{2n} computed by means of (5.30).

5.10 Remark. The terms in the expansion of u are called *normal modes of vibration*. Solutions of this type are called *standing waves*.

5.11 Example. We notice that in the IBVP

$$u_{tt}(x,t) = u_{xx}(x,t), \quad 0 < x < 1, \ t > 0,$$

$$u(0,t) = 0, \quad u(1,t) = 0, \quad t > 0,$$

$$u(x,0) = 3\sin(\pi x) - 2\sin(3\pi x), \quad u_t(x,0) = 4\pi \sin(2\pi x), \quad 0 < x < 1,$$

the functions f and g are linear combinations of the eigenfunctions. Setting $c = 1$ and $L = 1$ in their eigenfunction expansions above, from Theorem 3.26(ii) we deduce that

$$b_{11} = 3, \quad b_{13} = -2, \quad b_{1n} = 0 \ (n \neq 1,3),$$

$$b_{22} = 2, \quad b_{2n} = 0 \ (n \neq 2).$$

Hence, by (5.29), the solution of the IBVP is

$$u(x,t) = 3\sin(\pi x)\cos(\pi t) - 2\sin(3\pi x)\cos(3\pi t) + 2\sin(2\pi x)\sin(2\pi t).$$

VERIFICATION WITH MATHEMATICA®. The input

```
u = 3 * Sin[Pi * x] * Cos[Pi * t] - 2 * Sin[3 * Pi * x] * Cos[3 * Pi * t]
  +2 * Sin[2 * Pi * x] * Sin[2 * Pi * t];
{D[u,t,t] - D[u,x,x], u /. x -> 0, u /. x -> 1,
  (u /. t -> 0) - (3 * Sin[Pi * x] - 2 * Sin[3 * Pi * x]),
  (D[u,t] /. t -> 0) - 4 * Pi * Sin[2 * Pi * x]} // FullSimplify
```

generates the output $\{0,0,0,0,0\}$, which confirms that the PDE, BCs, and ICs are satisfied.

5.12 Example. If the initial conditions in the IBVP in Example 5.11 are replaced by

$$u(x,0) = \begin{cases} x, & 0 < x \le 1/2, \\ 1 - x, & 1/2 < x < 1, \end{cases} \quad u_t(x,0) = 0, \quad 0 < x < 1,$$

then, using (5.30) (with $c = 1$ and $L = 1$) and integration by parts, we find that

$$b_{1n} = 2\left\{ \int_0^{1/2} f(x)\sin(n\pi x)\,dx + \int_{1/2}^1 f(x)\sin(n\pi x)\,dx \right\}$$

$$= 2\left\{ \int_0^{1/2} x\sin(n\pi x)\,dx + \int_{1/2}^1 (1-x)\sin(n\pi x)\,dx \right\} = \frac{4}{n^2\pi^2}\sin\frac{n\pi}{2},$$

$$b_{2n} = \frac{2}{n\pi} \int_0^1 g(x)\sin(n\pi x)\,dx = 0, \quad n = 1, 2, \ldots;$$

so, by (5.29), the solution of the IBVP is

$$u(x,t) = \sum_{n=1}^\infty \frac{4}{n^2\pi^2}\sin\frac{n\pi}{2}\sin(n\pi x)\cos(n\pi t).$$

VERIFICATION WITH MATHEMATICA®. The input

```
un = (4/(n^2APi^2)) * Sin[n * Pi/2] * Sin[n * Pi * x] * Cos[n * Pi * t];
Simplify[{D[un,t,t] - D[un,x,x], un /. x -> 0, un /. x -> 1,
  D[un,t] /. t -> 0}, n ∈ Integers]
```

```
f = Piecewise[{{x, 0<x<1/2}, {1-x, 1/2<x<1}}];
Plot[{Sum[un /. t ->0, {n,1,3}], f}, {x,0,1},
  PlotRange -> {{-0.1,1.1}, {-0.1,0.6}}, PlotStyle -> Black,
  ImageSize -> Scaled[0.3], Ticks -> {{0.5,1}, {0.5}},
  AspectRatio -> 0.9]
```

generates the output $\{0,0,0,0\}$ and the graph

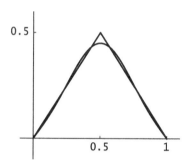

Exercises

Use separation of variables to solve the IBVP

$$u_{tt}(x,t) = u_{xx}(x,t), \quad 0 < x < 1, \ t > 0,$$
$$u(0,t) = 0, \quad u(1,t) = 0, \quad t > 0,$$
$$u(x,0) = f(x), \quad u_t(x,0) = g(x), \quad 0 < x < 1$$

with functions f and g as indicated.

1 $f(x) = -3\sin(2\pi x) + 4\sin(7\pi x), \quad g(x) = \sin(3\pi x).$

2 $f(x) = -1, \quad g(x) = 3\sin(\pi x).$

3 $f(x) = 2\sin(3\pi x), \quad g(x) = 2.$

4 $f(x) = 1, \quad g(x) = x.$

5 $f(x) = \begin{cases} 1, & 0 < x \le 1/2, \\ 2, & 1/2 < x < 1, \end{cases} \quad g(x) = 3\sin(2\pi x).$

6 $f(x) = 1, \quad g(x) = \begin{cases} 2, & 0 < x \le 1/2, \\ -1, & 1/2 < x < 1. \end{cases}$

7 $f(x) = \begin{cases} x, & 0 < x \le 1/2, \\ 1, & 1/2 < x < 1, \end{cases} \quad g(x) = -1.$

8 $f(x) = x + 1$, $g(x) = \begin{cases} -1, & 0 < x \le 1/2, \\ 2x, & 1/2 < x < 1. \end{cases}$

Answers to Odd-Numbered Exercises

1 $u(x,t) = -3\sin(2\pi x)\cos(2\pi t) + 4\sin(7\pi x)\cos(7\pi t)$
$$+ (1/(3\pi))\sin(3\pi x)\sin(3\pi t).$$

3 $u(x,t) = 2\sin(3\pi x)\cos(3\pi t) + \sum\limits_{n=1}^{\infty} [1 - (-1)^n](4/(n^2\pi^2))$
$$\times \sin(n\pi x)\sin(n\pi t).$$

5 $u(x,t) = \sum\limits_{n=1}^{\infty} (2/(n\pi))[1 - (-1)^n 2 + \cos(n\pi/2)]\sin(n\pi x)\cos(n\pi t)$
$$+ (3/(2\pi))\sin(2\pi x)\sin(2\pi t).$$

7 $u(x,t) = \sum\limits_{n=1}^{\infty} \sin(n\pi x)\{(1/(n^2\pi^2))[(-1)^{n+1}2n\pi + n\pi\cos(n\pi/2)$
$$+ 2\sin(n\pi/2)]\cos(n\pi t) + [(-1)^n - 1](2/(n^2\pi^2))\sin(n\pi t)\}.$$

5.2.2 Vibrating String with Free Endpoints

The IBVP corresponding to this case is (see Section 4.3)

$$u_{tt}(x,t) = c^2 u_{xx}(x,t), \quad 0 < x < L,\ t > 0, \tag{PDE}$$

$$u_x(0,t) = 0, \quad u_x(L,t) = 0, \quad t > 0, \tag{BCs}$$

$$u(x,0) = f(x), \quad u_t(x,0) = g(x), \quad 0 < x < L. \tag{ICs}$$

Separating the variables, we are led to the eigenvalue–eigenfunction pairs

$$\lambda_0 = 0, \quad X_0(x) = \tfrac{1}{2},$$

$$\lambda_n = \left(\frac{n\pi}{L}\right)^2, \quad X_n(x) = \cos\frac{n\pi x}{L}, \quad n = 1, 2, \ldots.$$

For $\lambda_0 = 0$, the general solution of the equation satisfied by T is

$$T_0(t) = a_{10} + a_{20}t, \quad a_{10}, a_{20} = \text{const},$$

while for λ_n, $n \ge 1$, the functions T_n are given by (5.28). Hence, we expect the solution of the IBVP to be of the form

$$u(x,t) = \tfrac{1}{2} a_{10} + \tfrac{1}{2} a_{20}t$$

$$+ \sum\limits_{n=1}^{\infty} \cos\frac{n\pi x}{L} \left(a_{1n} \cos\frac{n\pi ct}{L} + a_{2n} \sin\frac{n\pi ct}{L} \right), \tag{5.31}$$

where the a_{1n} and a_{2n}, $n = 0, 1, 2, \ldots$, are constants. Every term in (5.31) satisfies the PDE and the BCs.

To satisfy the ICs, we require that

$$u(x, 0) = f(x) = \tfrac{1}{2} a_{10} + \sum_{n=1}^{\infty} a_{1n} \cos \frac{n\pi x}{L},$$

$$u_t(x, 0) = g(x) = \tfrac{1}{2} a_{20} + \sum_{n=1}^{\infty} a_{2n} \frac{n\pi c}{L} \cos \frac{n\pi x}{L};$$

consequently, by (2.15),

$$a_{10} = \frac{2}{L} \int_0^L f(x)\, dx,$$

$$a_{1n} = \frac{2}{L} \int_0^L f(x) \cos \frac{n\pi x}{L}\, dx, \quad n = 1, 2, \ldots,$$

$$a_{20} = \frac{2}{L} \int_0^L g(x)\, dx,$$

$$a_{2n} = \frac{2}{n\pi c} \int_0^L g(x) \cos \frac{n\pi x}{L}\, dx, \quad n = 1, 2, \ldots,$$

(5.32)

and we conclude that the solution of the IBVP is given by (5.31) with the a_{1n} and a_{2n} computed by means of (5.32).

5.13 Example. To find the solution of the IBVP

$$u_{tt}(x, t) = u_{xx}(x, t), \quad 0 < x < 1,\ t > 0,$$
$$u_x(0, t) = 0, \quad u_x(1, t) = 0, \quad t > 0,$$
$$u(x, 0) = \cos(2\pi x), \quad u_t(x, 0) = -2\pi \cos(\pi x), \quad 0 < x < 1,$$

we remark that $c = 1$ and $L = 1$, and that the initial data functions f and g are linear combinations of the eigenfunctions; thus,

$$a_{12} = 1, \quad a_{1n} = 0 \ (n \neq 2),$$
$$a_{21} = -2, \quad a_{2n} = 0 \ (n \neq 1),$$

so, by (5.31), the solution of the IBVP is

$$u(x, t) = -2\cos(\pi x) \sin(\pi t) + \cos(2\pi x) \cos(2\pi t).$$

VERIFICATION WITH MATHEMATICA®. The input

```
u = - 2 * Cos [Pi * x] * Sin [Pi * t] + Cos [2 * Pi * x] * Cos [2 * Pi * t];
{D[u,t,t] - D[u,x,x], D[u,x] /.x -> 0, D[u,x] /.x -> 1,
  (u /.t -> 0) - Cos [2 * Pi * x], (D[u,t] /.t -> 0) + 2 * Pi * Cos [Pi * x]}
//FullSimplify
```

generates the output $\{0, 0, 0, 0, 0\}$, which confirms that the PDE, BCs, and ICs are satisfied.

5.14 Example. Consider the IBVP

$$u_{tt}(x,t) = u_{xx}(x,t), \quad 0 < x < 1, \ t > 0,$$

$$u_x(0,t) = 0, \quad u_x(1,t) = 0, \quad t > 0,$$

$$u(x,0) = 0, \ 0 < x < 1, \quad u_t(x,0) = \begin{cases} -1, & 1/4 \le x \le 3/4, \\ 0 & \text{otherwise.} \end{cases}$$

Again, here we have $c = 1$ and $L = 1$; so, by (5.32),

$$a_{1n} = 0, \quad n = 0, 1, 2, \ldots,$$

$$a_{20} = 2\left\{ \int_0^{1/4} g(x)\, dx + \int_{1/4}^{3/4} g(x)\, dx + \int_{3/4}^1 g(x)\, dx \right\} = 2\int_{1/4}^{3/4} (-1)\, dx = -1,$$

$$a_{2n} = \frac{2}{n\pi}\left\{ \int_0^{1/4} g(x)\cos(n\pi x)\, dx + \int_{1/4}^{3/4} g(x)\cos(n\pi x)\, dx \right.$$

$$\left. + \int_{3/4}^1 g(x)\cos(n\pi x)\, dx \right\}$$

$$= \frac{2}{n\pi} \int_{1/4}^{3/4} (-1)\cos(n\pi x)\, dx = -\frac{4}{n^2\pi^2}\sin\frac{n\pi}{4}\cos\frac{n\pi}{2}, \quad n = 1, 2, \ldots.$$

Hence, by (5.31), the solution of the IBVP is

$$u(x,t) = -\frac{1}{2}t - \sum_{n=1}^{\infty} \frac{4}{n^2\pi^2}\sin\frac{n\pi}{4}\cos\frac{n\pi}{2}\cos(n\pi x)\sin(n\pi t).$$

VERIFICATION WITH MATHEMATICA®. The input

```
un = - (4/ (n^2 * Pi^2)) * Sin [n * Pi/4] * Cos [n * Pi/2]
  * Cos [n * Pi * x] * Sin [n * Pi * t];
Simplify [{D[un,t,t] - D[un,x,x], D[un,x] /. x -> 0, D[un,x] /. x -> 1,
  un /. t -> 0}, n ∈ Integers]
```

```
f = Piecewise[{{0, 0<x<1/4}, {-1, 1/4<x<3/4}, {0, 3/4<x<1}}];
Plot[{D[-(1/2)*t+Sum[un, {n,1,11}], t] /. t->0, f}, {x,0,1},
  PlotRange->{{-0.1,1.1}, {-1.1,0.1}}, PlotStyle->Black,
  ImageSize->Scaled[0.25], Ticks->{{1},{-1}},
  AspectRatio->0.9]
```

generates the output $\{0, 0, 0, 0\}$ and the graph

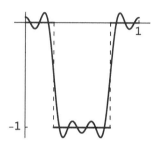

Exercises

Use separation of variables to solve the IBVP

$$u_{tt}(x,t) = u_{xx}(x,t), \quad 0 < x < 1, \ t > 0,$$
$$u_x(0,t) = 0, \quad u_x(1,t) = 0, \quad t > 0,$$
$$u(x,0) = f(x), \quad u_t(x,0) = g(x), \quad 0 < x < 1$$

with functions f and g as indicated.

1 $f(x) = 2 - 3\cos(4\pi x), \quad g(x) = 2\cos(3\pi x).$

2 $f(x) = x - 1, \quad g(x) = 2 - \cos(\pi x).$

3 $f(x) = -3\cos(2\pi x), \quad g(x) = 2x - 1.$

4 $f(x) = x, \quad g(x) = 2x - 1.$

5 $f(x) = \begin{cases} 2, & 0 < x \le 1/2, \\ 3, & 1/2 < x < 1, \end{cases} \quad g(x) = -\cos(3\pi x).$

6 $f(x) = x, \quad g(x) = \begin{cases} 2, & 0 < x \le 1/2, \\ -1, & 1/2 < x < 1. \end{cases}$

7 $f(x) = \begin{cases} x + 1, & 0 < x < 1/2, \\ 2, & 1/2 < x < 1, \end{cases} \quad g(x) = 1 - x.$

8 $f(x) = 2x$, $g(x) = \begin{cases} 2x + 1, & 0 < x \le 1/2, \\ x + 2, & 1/2 < x < 1. \end{cases}$

Answers to Odd-Numbered Exercises

1 $u(x,t) = 2 - 3\cos(4\pi x)\cos(4\pi t) + (2/(3\pi))\cos(3\pi x)\sin(3\pi t)$.

3 $u(x,t) = -3\cos(2\pi x)\cos(2\pi t)$
$$+ \sum_{n=1}^{\infty} [(-1)^n - 1](4/(n^3\pi^3))\cos(n\pi x)\sin(n\pi t).$$

5 $u(x,t) = 5/2 + \sum_{n=1}^{\infty} -(2/(n\pi))\sin(n\pi/2)\cos(n\pi x)\cos(n\pi t)$
$$- (1/(3\pi))\cos(3\pi x)\sin(3\pi t).$$

7 $u(x,t) = 13/8 + t/2 + \sum_{n=1}^{\infty} \cos(n\pi x)\{(1/(n^2\pi^2))[2\cos(n\pi/2)$
$$- n\pi\sin(n\pi/2) - 2]\cos(n\pi t)$$
$$+ [1 - (-1)^n](2/(n^3\pi^3))\sin(n\pi t)\}.$$

5.2.3 Vibrating String with Other Types of Boundary Conditions

By way of illustration, we consider a string with one free and one fixed endpoints, and then a string where one endpoint is fixed and the other has an elastic attachment.

5.15 Example. Consider the IBVP

$$u_{tt}(x,t) = u_{xx}(x,t), \quad 0 < x < 1, \ t > 0,$$
$$u(0,t) = 0, \quad u_x(1,t) = 0, \quad t > 0,$$
$$u(x,0) = \sin\frac{\pi x}{2} + 2\sin\frac{5\pi x}{2}, \quad u_t(x,0) = -2\pi\sin\frac{\pi x}{2}, \quad 0 < x < 1.$$

Seeking the solution in the form $u(x,t) = X(x)T(t)$ with $X, T \ne 0$ and replacing in the PDE and BCs, we arrive at the equation

$$T''(t) + \lambda T(t) = 0, \quad t > 0$$

and the regular S–L problem

$$X''(x) + \lambda X(x) = 0, \quad 0 < x < 1,$$
$$X(0) = 0, \quad X'(1) = 0.$$

The eigenvalue–eigenfunction pairs of the latter are (see Example 5.5)

$$\lambda_n = \frac{(2n-1)^2\pi^2}{4}, \quad X_n(x) = \sin\frac{(2n-1)\pi x}{2}, \quad n = 1, 2, \ldots,$$

so

$$T_n(t) = c_{1n}\cos\frac{(2n-1)\pi t}{2} + c_{2n}\sin\frac{(2n-1)\pi t}{2}, \quad n = 1, 2, \ldots.$$

Then

$$u(x,t) = \sum_{n=1}^{\infty}\sin\frac{(2n-1)\pi x}{2}\left\{c_{1n}\cos\frac{(2n-1)\pi t}{2} + c_{2n}\sin\frac{(2n-1)\pi t}{2}\right\},$$

and the ICs produce the equalities

$$u(x,0) = \sum_{n=1}^{\infty}c_{1n}\sin\frac{(2n-1)\pi x}{2} = \sin\frac{\pi x}{2} + 2\sin\frac{5\pi x}{2},$$

$$u_t(x,0) = \sum_{n=1}^{\infty}\frac{(2n-1)\pi}{2}c_{2n}\sin\frac{(2n-1)\pi x}{2} = -2\pi\sin\frac{\pi x}{2};$$

therefore,

$$c_{1n} = \begin{cases} 1, & n = 1, \\ 2, & n = 3, \\ 0, & n \neq 1, 2, \end{cases} \qquad c_{2n} = \begin{cases} -4, & n = 1, \\ 0, & n \neq 1. \end{cases}$$

This means that the solution of the given IBVP is

$$u(x,t) = \sin\frac{\pi x}{2}\left(\cos\frac{\pi t}{2} - 4\sin\frac{\pi t}{2}\right) + 2\sin\frac{5\pi x}{2}\cos\frac{5\pi t}{2}.$$

VERIFICATION WITH MATHEMATICA®. The input

```
u = Sin[Pi*x/2] * (Cos[Pi*t/2] - 4*Sin[Pi*t/2]) +
    2*Sin[5*Pi*x/2] * Cos[5*Pi*t/2];
{f,g} = {Sin[Pi*x/2] + 2*Sin[5*Pi*x/2], -2*Pi*Sin[Pi*x/2];
{D[u,t,t] - D[u,x,x], u/.x->0, D[u,x]/.x->1, (u/.t->0) - f,
    (D[u,t]/.t->0) - g}//Simplify
```

generates the output $\{0, 0, 0, 0, 0\}$.

5.16 Example. To apply the method of separation of variables to the IBVP

$$u_{tt}(x,t) = u_{xx}(x,t), \quad 0 < x < 1, \; t > 0,$$

$$u_x(0,t) = 0, \quad u(1,t) = 0, \quad t > 0,$$

$$u(x,0) = \begin{cases} 2, & 0 < x \le 1/2, \\ 1, & 1/2 < x < 1, \end{cases} \quad u_t(x,0) = \cos\frac{3\pi x}{2}, \quad 0 < x < 1,$$

we start, as usual, by seeking the solution in the form $u(x,t) = X(x)T(t)$ with $X, T \neq 0$. Replaced in the PDE and BCs, this yields the equation

$$T''(t) + \lambda T(t) = 0, \quad t > 0$$

and the regular S–L problem

$$X''(x) + \lambda X(x) = 0, \quad 0 < x < 1,$$
$$X'(0) = 0, \quad X(1) = 0.$$

The eigenvalue–eigenfunction pairs of the latter are (see Example 5.6 or Appendix A.2)

$$\lambda_n = \frac{(2n-1)^2 \pi^2}{4}, \quad X_n(x) = \cos \frac{(2n-1)\pi x}{2}, \quad n = 1, 2, \ldots,$$

so the equation for T generates the solutions

$$T_n(t) = c_{1n} \cos \frac{(2n-1)\pi t}{2} + c_{2n} \sin \frac{(2n-1)\pi t}{2}, \quad n = 1, 2, \ldots.$$

Following the general procedure, we now write

$$u(x,t) = \sum_{n=1}^{\infty} \cos \frac{(2n-1)\pi x}{2} \left\{ c_{1n} \cos \frac{(2n-1)\pi t}{2} + c_{2n} \sin \frac{(2n-1)\pi t}{2} \right\}$$

and use the ICs to determine the coefficients c_{1n} and c_{2n}.

The first IC leads to

$$u(x,0) = \sum_{n=1}^{\infty} c_{1n} \cos \frac{(2n-1)\pi x}{2}.$$

Since, according to Theorem 3.26(ii), the eigenfunctions X_n are orthogonal on the interval $0 \leq x \leq 1$, we multiply the above equation by $\cos((2m-1)\pi x/2)$, integrate the new equality term by term from 0 to 1, and arrive at

$$c_{1n} = \frac{4}{(2n-1)\pi} \left[(-1)^{n+1} + \sin \frac{(2n-1)\pi}{4} \right].$$

The second IC is

$$\sum_{n=1}^{\infty} \frac{(2n-1)\pi}{2} c_{2n} \cos \frac{(2n-1)\pi x}{2} = \cos \frac{3\pi x}{2},$$

so

$$c_{22} = \frac{2}{3\pi}, \quad c_{2n} = 0 \ (n \neq 2).$$

Hence, the solution of the IBVP is

$$u(x,t) = \sum_{n=1}^{\infty} \frac{4}{(2n-1)\pi}\left[(-1)^{n+1} + \sin\frac{(2n-1)\pi}{4}\right]$$

$$\times \cos\frac{(2n-1)\pi x}{2}\cos\frac{(2n-1)\pi t}{2} + \frac{2}{3\pi}\cos\frac{3\pi x}{2}\sin\frac{3\pi t}{2}.$$

VERIFICATION WITH MATHEMATICA®. The input

```
un = (4/((2*n-1)*Pi)) * ((-1)^(n+1) + Sin[(2*n-1)*Pi/4])
   * Cos[(2*n-1)*Pi*x/2] * Cos[(2*n-1)*Pi*t/2];
AddTerm = (2/(3*Pi)) * Cos[3*Pi*x/2] * Sin[3*Pi*t/2];
Simplify[{D[un + AddTerm,t,t] - D[un + AddTerm,x,x],
   D[un + AddTerm,x] /. x->0, (un + AddTerm) /. x->1,
   (D[(un + AddTerm),t] /. t->0) - Cos[3*Pi*x/2]}, n∈Integers]
f = Piecewise[{{2, 0<x<1/2}, {1, 1/2<x<1}}];
Plot[{AddTerm + Sum[un, {n,1,11}] /. t->0, f}, {x,0,1},
   PlotRange -> {{-0.1,1.1}, {-0.1,2.1}}, PlotStyle -> Black,
   ImageSize -> Scaled[0.25], Ticks -> {{0.5,1},{1,2}},
   AspectRatio -> 0.9]
```

generates the output $\{0,0,0,0\}$ and the graph

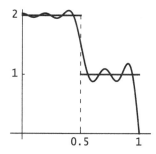

5.17 Example. In the IBVP

$$u_{tt}(x,t) = u_{xx}(x,t), \quad 0 < x < 1, \ t > 0,$$
$$u(0,t) = 0, \quad u_x(1,t) + u(1,t) = 0, \quad t > 0,$$
$$u(x,0) = \begin{cases} 1, & 0 < x \le 1/2, \\ 0, & 1/2 < x < 1, \end{cases} \quad u_t(x,0) = 1, \ 0 < x < 1,$$

the substitution $u(x,t) = X(x)T(t)$, $X, T \ne 0$, leads to the equation

$$T''(t) + \lambda T(t) = 0, \quad t > 0,$$

and the regular S–L problem

$$X''(x) + \lambda X(x) = 0, \quad 0 < x < 1,$$
$$X(0) = 0, \quad X'(1) + X(1) = 0.$$

As shown in Example 3.21, the eigenvalues and eigenfunctions of the latter, for $h = L = 1$, are, respectively, $\lambda_n = \zeta_n^2$ and $X_n(x) = \sin(\zeta_n x)$, $n = 1, 2, \ldots$, where the first five numbers ζ_n, computed in Example 3.29 with rounding at the fourth decimal place, are

$$\zeta_1 = 2.0288, \quad \zeta_2 = 4.9132, \quad \zeta_3 = 7.9787,$$
$$\zeta_4 = 11.0855, \quad \zeta_5 = 14.2074.$$

Then

$$T_n(t) = c_{1n} \cos(\zeta_n t) + c_{2n} \sin(\zeta_n t), \quad n = 1, 2, \ldots,$$

so the series form of the solution is

$$u(x, t) = \sum_{n=1}^{\infty} \sin(\zeta_n x) \big[c_{1n} \cos(\zeta_n t) + c_{2n} \sin(\zeta_n t) \big].$$

Using this series and the ICs, and constructing the generalized Fourier series expansions

$$u(x, 0) = \sum_{n=1}^{\infty} a_n \sin(\zeta_n x), \quad u_t(x, 0) = \sum_{n=1}^{\infty} b_n \sin(\zeta_n x),$$

we find that $c_{1n} = a_n$ and $c_{2n} = b_n / \zeta_n$. Completing the calculation, in the end we arrive at the approximate solution

$$
\begin{aligned}
u(x, t) = {} & \sin(2.0288\, x)[0.3891 \cos(2.0288\, t) + 0.5862 \sin(2.0288\, t)] \\
& + \sin(4.9132\, x)[0.6947 \cos(4.9132\, t) + 0.0638 \sin(4.9132\, t)] \\
& + \sin(7.9787\, x)[0.4102 \cos(7.9787\, t) + 0.0348 \sin(7.9787\, t)] \\
& + \sin(11.0855\, x)[0.0469 \cos(11.0855\, t) + 0.0147 \sin(11.0855\, t)] \\
& + \sin(14.2074\, x)[0.0446 \cos(14.2074\, t) + 0.0106 \sin(14.2074\, t)] + \cdots .
\end{aligned}
$$

VERIFICATION WITH MATHEMATICA®. The input

```
Approx = Sin[2.0288 * x] * (0.3891 * Cos[2.0288 * t] + 0.5862
    * Sin[2.0288 * t]) + Sin[4.9132 * x] * (0.6947 * Cos[4.9132 * t]
    + 0.0638 * Sin[4.9132 * t]) + Sin[7.9787 * x] * (0.4102
    * Cos[7.9787 * t] + 0.0348 * Sin[7.9787 * t]) + Sin[11.0855 * x]
    * (0.046 * Cos[11.0855 * t] + 0.0147 * Sin[11.0855 * t])
    + Sin[14.2074 * x] * (0.0446 * Cos[14.2074 * t] + 0.0106
```

```
      *Sin[14.2074*t]);
f=Piecewise[{{1, 0<x<1/2}, {0, 1/2<x<1}}];
Chop[FullSimplify[{D[Approx,t,t]-D[Approx,x,x], Approx/. x->0,
    (D[Approx,x]/. x->1)+Approx/. x->1}, n∈Integers], 0.001]
p1=Plot[{Approx/. t->0, f}, {x,0,1}, PlotRange->{{-0.1,1.1},
    {-0.1,1.3}}, PlotStyle->Black, ImageSize->Scaled[0.2],
    Ticks->{{0.5,1},{1}}, AspectRatio->0.9];
p2=Plot[{D[Approx,t]/. t->0, 1}, {x,0,1},
    PlotRange->{{-0.1,1.1}, {-0.1,1.3}}, PlotStyle->Black,
    ImageSize->Scaled[0.2], Ticks->{{1},{1}}, AspectRatio->0.9];
Show[GraphicsRow[{p1,p2}]]
```

generates the output $\{0, 0, 0\}$ and the graphs

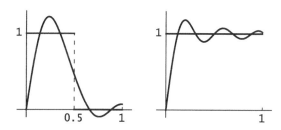

Exercises

Use separation of variables to solve the IBVP consisting of the PDE and ICs

$$u_{tt}(x, t) = u_{xx}(x, t), \quad 0 < x < 1, \ t > 0,$$
$$u(x, 0) = f(x), \quad u_t(x, 0) = g(x), \quad 0 < x < 1$$

with BCs and functions f and g as indicated. Where appropriate, round the numbers to the fourth decimal place.

1 $u(0, t) = 0$, $u_x(1, t) = 0$, $f(x) = 2\sin(\pi x/2) - \sin(5\pi x/2)$,
   $g(x) = -4\pi\sin(\pi x/2)$.

2 $u(0, t) = 0$, $u_x(1, t) = 0$, $f(x) = 1$, $g(x) = x$.

3 $u_x(0, t) = 0$, $u(1, t) = 0$, $f(x) = -2\cos(3\pi x/2)$,
   $g(x) = \pi\big[3\cos(\pi x/2) - 12\sin(3\pi x/2)\big]$.

4 $u_x(0, t) = 0$, $u(1, t) = 0$, $f(x) = x - 1$, $g(x) = 1$.

5 $u(0,t) = 0, \quad u_x(1,t) + u(1,t) = 0, \quad f(x) = 1, \quad g(x) = x.$

6 $u(0,t) = 0, \quad u_x(1,t) + u(1,t) = 0, \quad f(x) = x - 1,$

$$g(x) = \begin{cases} 0, & 0 < x \le 1/2, \\ 2, & 1/2 < x < 1. \end{cases}$$

Answers to Odd-Numbered Exercises

1 $u(x,t) = 2\sin(\pi x/2)\big[\cos(\pi t/2) - 4\sin(\pi t/2)\big] - \sin(5\pi x/2)\cos(5\pi t/2).$

3 $u(x,t) = 6\cos(\pi x/2)\sin(\pi t/2) - 2\cos(3\pi x/2)\big[\cos(3\pi t/2) + 4\sin(3\pi t/2)\big].$

5 $u(x,y) = \sin(2.0288\,x)[1.1892\cos(2.0288\,t) + 0.3594\sin(2.0288\,t)]$

$\qquad\qquad + \sin(4.9132\,x)[0.3134\cos(4.9132\,t) - 0.0318\sin(4.9132\,t)]$

$\qquad\qquad + \sin(7.9787\,x)[0.2775\cos(7.9787\,t) + 0.0077\sin(7.9787\,t)]$

$\qquad\qquad + \sin(11.0855\,x)[0.1629\cos(11.0855\,t) - 0.0029\sin(11.0855\,t)]$

$\qquad\qquad + \sin(14.2074\,x)[0.1499\cos(14.2074\,t) + 0.0014\sin(14.2074\,t)]$

$\qquad\qquad\qquad\qquad\qquad\qquad\qquad\qquad\qquad + \cdots .$

5.3   The Laplace Equation

In this section, we discuss the solution of two types of problems: one formulated in terms of Cartesian coordinates, and one where the use of polar coordinates is more appropriate.

5.3.1   Laplace Equation in a Rectangle

Consider the equilibrium temperature in a uniform rectangle

$$D = \{(x,y) : 0 \le x \le L,\ 0 \le y \le K\},\ L, K = \text{const},$$

with no sources and with temperature prescribed on the boundary.

The corresponding problem is (see Section 4.2)

$$u_{xx}(x,y) + u_{yy}(x,y) = 0, \quad 0 < x < L,\ 0 < y < K,$$
$$u(0,y) = f_1(y), \quad u(L,y) = f_2(y), \quad 0 < y < K,$$
$$u(x,0) = g_1(x), \quad u(x,K) = g_2(x), \quad 0 < x < L.$$

Although this is a boundary value problem, it turns out that the method of separation of variables can still be applied. However, since here the BCs are nonhomogeneous, we need to do some additional preliminary work.

By the principle of superposition (see Theorem 1.22), we can seek the solution of the above BVP as

$$u(x, y) = u_1(x, y) + u_2(x, y),$$

where both u_1 and u_2 satisfy the PDE, u_1 satisfies the BCs with $f_1(y) = 0$, $f_2(y) = 0$, and u_2 satisfies the BCs with $g_1(x) = 0$, $g_2(x) = 0$; thus, the BVP for u_1 is

$$(u_1)_{xx}(x, y) + (u_1)_{yy}(x, y) = 0, \quad 0 < x < L, \ 0 < y < K, \qquad \text{(PDE)}$$

$$u_1(0, y) = 0, \quad u_1(L, y) = 0, \quad 0 < y < K,$$
$$u_1(x, 0) = g_1(x), \quad u_1(x, K) = g_2(x). \qquad \text{(BCs)}$$

We seek a solution of this BVP of the form

$$u_1(x, y) = X(x)Y(y).$$

Clearly, for the same reason as in the case of the heat and wave equations, neither X nor Y can be the zero function. Replacing in the PDE, we obtain

$$X''(x)Y(y) + X(x)Y''(y) = 0,$$

from which, on division by $X(x)Y(y)$, we find that

$$\frac{X''(x)}{X(x)} = -\frac{Y''(y)}{Y(y)} = -\lambda,$$

where $\lambda = \text{const}$ is the separation constant.

The two homogeneous BCs yield

$$X(0)Y(y) = 0, \quad X(L)Y(y) = 0, \quad 0 < y < K.$$

Since, as already mentioned, $Y(y)$ cannot be identically equal to zero, we conclude that

$$X(0) = 0, \quad X(L) = 0.$$

Hence, X is the solution of the regular Sturm–Liouville problem

$$X''(x) + \lambda X(x) = 0, \quad 0 < x < L,$$
$$X(0) = 0, \quad X(L) = 0.$$

From Example 3.16 we know that the eigenvalue–eigenfunction pairs of this problem are

$$\lambda_n = \left(\frac{n\pi}{L}\right)^2, \quad X_n(x) = \sin\frac{n\pi x}{L}, \quad n = 1, 2, \ldots. \qquad (5.33)$$

Then for each $n = 1, 2, \ldots$, we seek functions Y_n that satisfy

$$Y_n''(y) - \left(\frac{n\pi}{L}\right)^2 Y_n(y) = 0, \quad 0 < y < K.$$

By Remark 1.4(i), the general solution of the equation for Y_n can be written as

$$Y_n(y) = C_{1n} \sinh \frac{n\pi(y - K)}{L} + C_{2n} \sinh \frac{n\pi y}{L}, \quad C_{1n}, C_{2n} = \text{const.} \quad (5.34)$$

We have chosen this combination of hyperbolic functions because when the nonhomogeneous BCs are applied, one term vanishes at $y = 0$ and the other at $y = K$, which is very convenient for the computation of the coefficients C_{1n} and C_{2n}.

In view of (5.33) and (5.34), it seems reasonable to expect u_1 to be of the form

$$u_1(x, y) = \sum_{n=1}^{\infty} X_n(x) Y_n(y)$$

$$= \sum_{n=1}^{\infty} \sin \frac{n\pi x}{L} \left[C_{1n} \sinh \frac{n\pi(y - K)}{L} + C_{2n} \sinh \frac{n\pi y}{L} \right], \quad (5.35)$$

where each term satisfies the PDE and the two homogeneous BCs. To satisfy the two remaining BCs, we must have

$$u_1(x, 0) = g_1(x)$$

$$= \sum_{n=1}^{\infty} \left(-C_{1n} \sinh \frac{n\pi K}{L} \right) \sin \frac{n\pi x}{L} = \sum_{n=1}^{\infty} b_{1n} \sin \frac{n\pi x}{L},$$

$$u_1(x, K) = g_2(x)$$

$$= \sum_{n=1}^{\infty} \left(C_{2n} \sinh \frac{n\pi K}{L} \right) \sin \frac{n\pi x}{L} = \sum_{n=1}^{\infty} b_{2n} \sin \frac{n\pi x}{L};$$

in other words, we need the Fourier sine series coefficients b_{1n} and b_{2n} of g_1 and g_2, respectively. By (2.12), for $n = 1, 2, \ldots$,

$$b_{1n} = -C_{1n} \sinh \frac{n\pi K}{L} = \frac{2}{L} \int_0^L g_1(x) \sin \frac{n\pi x}{L} \, dx,$$

$$b_{2n} = C_{2n} \sinh \frac{n\pi K}{L} = \frac{2}{L} \int_0^L g_2(x) \sin \frac{n\pi x}{L} \, dx;$$

therefore,

$$C_{1n} = -\frac{b_{1n}}{\sinh(n\pi K/L)} = -\frac{2}{L} \operatorname{csch} \frac{n\pi K}{L} \int_0^L g_1(x) \sin \frac{n\pi x}{L} \, dx, \quad (5.36)$$

$$C_{2n} = \frac{b_{2n}}{\sinh(n\pi K/L)} = \frac{2}{L} \operatorname{csch} \frac{n\pi K}{L} \int_0^L g_2(x) \sin \frac{n\pi x}{L} \, dx. \quad (5.37)$$

This means that the function $u_1(x, y)$ is given by (5.35) with the coefficients C_{1n} and C_{2n} computed by means of (5.36) and (5.37).

The procedure for finding u_2 is similar, with the obvious modifications.

5.18 Example. For the BVP

$$u_{xx}(x, y) + u_{yy}(x, y) = 0, \quad 0 < x < 1, \ 0 < y < 2,$$
$$u(0, y) = 0, \quad u(1, y) = 0, \quad 0 < y < 2,$$
$$u(x, 0) = x, \quad u(x, 2) = 3\sin(\pi x), \quad 0 < x < 1,$$

we have

$$L = 1, \ K = 2, \ f_1(y) = 0, \ f_2(y) = 0, \ g_1(x) = x, \ g_2(x) = 3\sin(\pi x).$$

First, integrating by parts, we see that

$$\int_0^1 x\sin(n\pi x)\, dx = (-1)^{n+1}\frac{1}{n\pi};$$

so, by (5.36),

$$C_{1n} = (-1)^n \frac{2}{n\pi}\operatorname{csch}(2n\pi), \quad n = 1, 2, \dots.$$

Second, using (5.37), we easily convince ourselves that

$$C_{21} = 3\operatorname{csch}(2\pi), \quad C_{2n} = 0, \quad n = 2, 3, \dots.$$

Consequently, by (5.35), the solution of the given BVP is

$$u(x, y) = \sum_{n=1}^{\infty}(-1)^n\frac{2}{n\pi}\operatorname{csch}(2n\pi)\sin(n\pi x)\sinh\left(n\pi(y - 2)\right)$$
$$+ 3\operatorname{csch}(2\pi)\sin(\pi x)\sinh(\pi y).$$

VERIFICATION WITH MATHEMATICA®. The input

```
un = ( - 1)^n * (2/(n * Pi)) * Csch[2 * n * Pi] * Sin[n * Pi * x]
    * Sinh[n * Pi * (y - 2)];
AddTerm = 3 * Csch[2 * Pi] * Sin[Pi * x] * Sinh[Pi * y];
Simplify[{D[un + AddTerm, x, x] + D[un + AddTerm, y, y],
    (un + AddTerm) /. x -> 0, (un + AddTerm) /. x -> 1,
    ((un + AddTerm) /. y -> 2) - 3 * Sin[Pi * x]}, n ∈ Integers]
Plot [{AddTerm + Sum[un, {n, 1, 9}] /. y -> 0, x}, {x, 0, 1},
    PlotRange * {{ - 0.1, 1.1}, { - 0.1, 1.2}}, PlotStyle -> Black,
```

```
ImageSize -> Scaled[0.25], Ticks -> {{1},{1}},
AspectRatio -> 0.9]
```

generates the output $\{0,0,0,0\}$ and the graph

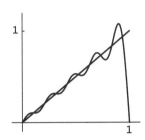

5.19 Remark. As mentioned in Remark 1.4(i) and Example 1.6, the general solution of Y_n should be expressed in the most convenient form for the given nonhomogeneous BCs. We have already indicated the best choice when $u_1(x,0)$ and $u_1(x,K)$ are prescribed. The other combinations are as follows:

(i) If $(u_1)_y(x,0) = g_1(x)$, $(u_1)_y(x,K) = g_2(x)$, then

$$Y_n(y) = C_{1n} \cosh \frac{n\pi(y-K)}{L} + C_{2n} \cosh \frac{n\pi y}{L}.$$

(ii) If $u_1(x,0) = g_1(x)$, $(u_1)_y(x,K) = g_2(x)$, then

$$Y_n(y) = C_{1n} \cosh \frac{n\pi(y-K)}{L} + C_{2n} \sinh \frac{n\pi y}{L}.$$

(iii) If $(u_1)_y(x,0) = g_1(x)$, $u_1(x,K) = g_2(x)$, then

$$Y_n(y) = C_{1n} \cosh \frac{n\pi y}{L} + C_{2n} \sinh \frac{n\pi(y-K)}{L}.$$

The order of the hyperbolic functions above has been chosen to induce subscript symmetry in (5.36) and (5.37).

Exercises

Use separation of variables to solve the BVP consisting of the PDE

$$u_{xx}(x,y) + u_{yy}(x,y) = 0, \quad 0 < x < 1, \ 0 < y < 2$$

and BCs as indicated.

1 $u(0,y) = 0$, $u(1,y) = 0$, $u(x,0) = 3\sin(2\pi x)$, $u(x,2) = \sin(3\pi x)$.

2 $u_x(0,y) = 2\sin(\pi y)$, $u_x(1,y) = -\sin(2\pi y)$, $u(x,0) = 0$, $u(x,2) = 0$.

3 $u_x(0,y) = y - 2$, $u(1,y) = 3\cos(2\pi y)$, $u_y(x,0) = 0$, $u_y(x,2) = 0$.

4 $u_x(0, y) = 0$, $u_x(1, y) = 0$, $u(x, 0) = 2 - \cos(\pi x)$, $u_y(x, 2) = x$.

5 $u_x(0, y) = 0$, $u(1, y) = 0$,

$$u(x, 0) = \begin{cases} 1, & 0 < x \le 1/2, \\ 2, & 1/2 < x < 1, \end{cases} \quad u_y(x, 2) = 3 \cos(\pi x/2).$$

6 $u_x(0, y) = -\sin(3\pi y/4)$, $u_x(1, y) = \begin{cases} 0, & 0 < y \le 1, \\ 1, & 1 < y < 2, \end{cases}$

$u(x, 0) = 0$, $u_y(x, 2) = 0$.

Answers to Odd-Numbered Exercises

1 $u(x, y) = -3\operatorname{csch}(4\pi)\sin(2\pi x)\sinh(2\pi(y - 2))$

$$+ \operatorname{csch}(6\pi)\sin(3\pi x)\sinh(3\pi y).$$

3 $u(x, y) = 3\operatorname{sech}(2\pi)\cosh(2\pi x)\cos(2\pi y) + 1 - x$

$$+ \sum_{n=1}^{\infty} [(-1)^n - 1](8/(n^3\pi^3))\operatorname{sech}(n\pi/2)$$
$$\times \sinh(n\pi(x - 1)/2)\cos(n\pi y/2).$$

5 $u(x, y) = (6/\pi)\operatorname{sech}\pi\cos(\pi x/2)\sinh(\pi y/2)$

$$+ \sum_{n=1}^{\infty} (4/(2n - 1)\pi)[(-1)^{n+1}2 - \sin((2n - 1)\pi/4)]$$
$$\times \operatorname{sech}((2n - 1)\pi)\cos((2n - 1)\pi x/2)$$
$$\times \cosh((2n - 1)\pi(y - 2)/2).$$

5.3.2 Laplace Equation in a Circular Disc

Consider the equilibrium temperature in a uniform circular disc

$$D = \{(r, \theta) : 0 \le r < \alpha, \ -\pi < \theta \le \pi\}$$

with no sources, under the condition of prescribed temperature on the boundary $r = \alpha = $ const. Because of the geometry of the body, it is advisable that we write the modeling BVP in terms of polar coordinates r and θ with the pole at the center of the disc. Thus, recalling the form of the Laplacian in polar coordinates given in Remark 4.12(ii), we have

$$u_{rr}(r, \theta) + r^{-1}u_r(r, \theta) + r^{-2}u_{\theta\theta}(r, \theta) = 0,$$

$$0 < r < \alpha, \ -\pi < \theta < \pi, \qquad \text{(PDE)}$$

$$u(\alpha, \theta) = f(\theta), \quad -\pi < \theta \le \pi. \qquad \text{(BC)}$$

In Subsection 5.3.1, we solved the BVP by fully using the boundary data prescribed at the end-points of the intervals on which the variables x and y were defined. This time, we have a BC given only at $r = 1$, so we need to consider additional 'boundary' conditions, given at the remaining end-points for r and θ, namely $r = 0$, $\theta = -\pi$, and $\theta = \pi$. The form of these conditions is suggested by analytic requirements based on physical considerations: u (the temperature) and u_r (the radial component of the heat flux) in a BVP in a disc with a 'reasonably behaved' function f (see Remark 4.27) are expected to be continuous (therefore bounded) everywhere in the disc. Thus, our additional conditions can be formulated as

$$u(r, \theta), \ u_r(r, \theta) \text{ bounded as } r \to 0+, \quad -\pi < \theta < \pi,$$
$$u(r, -\pi) = u(r, \pi), \quad u_\theta(r, -\pi) = u_\theta(r, \pi), \quad 0 < r < \alpha.$$

The PDE and all the BCs except $u(\alpha, \theta) = f(\theta)$ are homogeneous; consequently, it seems advisable to seek a solution of the form

$$u(r, \theta) = R(r)\Theta(\theta), \quad R, \Theta \neq 0. \tag{5.38}$$

Then the last two (periodic) BCs become

$$\Theta(-\pi)R(r) = \Theta(\pi)R(r), \quad \Theta'(-\pi)R(r) = \Theta'(\pi)R(r), \quad 0 < r < \alpha,$$

from which, since $R \neq 0$, we deduce that

$$\Theta(-\pi) = \Theta(\pi), \quad \Theta'(-\pi) = \Theta'(\pi).$$

In view of (5.38), the PDE can be written as

$$\left[R''(r) + r^{-1}R'(r) \right]\Theta(\theta) + r^{-2}R(r)\Theta''(\theta) = 0,$$

which, on division by $r^{-2}R(r)\Theta(\theta)$ and in accordance with the usual argument, yields

$$\frac{\Theta''(\theta)}{\Theta(\theta)} = -\frac{r^2 R''(r) + r R'(r)}{R(r)} = -\lambda, \tag{5.39}$$

where $\lambda = \text{const}$ is the separation constant. Then Θ is a solution of the periodic Sturm–Liouville problem

$$\Theta''(\theta) + \lambda\Theta(\theta) = 0, \quad -\pi < \theta < \pi,$$
$$\Theta(-\pi) = \Theta(\pi), \quad \Theta'(-\pi) = \Theta'(\pi).$$

This problem was discussed in Example 3.33. Hence, for $L = \pi$, the eigenvalues and their corresponding eigenfunctions are, respectively,

$$\lambda_0 = 0, \quad \Theta_0(\theta) = \tfrac{1}{2},$$
$$\lambda_n = n^2, \quad \Theta_{1n}(\theta) = \cos(n\theta), \quad \Theta_{2n}(\theta) = \sin(n\theta), \quad n = 1, 2, \ldots.$$

From the second equality in (5.39) we see that for each $n = 0, 1, 2, \ldots$, the corresponding radial component R_n satisfies

$$r^2 R_n''(r) + r R_n'(r) - n^2 R_n = 0, \quad 0 < r < \alpha. \tag{5.40}$$

This is a Cauchy–Euler equation, for which, if $n \neq 0$, we seek solutions of the form (see Section 1.4)

$$R_n(r) = r^p, \quad p = \text{const}.$$

Replacing in (5.40), we find that

$$[p(p-1) + p - n^2]r^p = 0, \quad 0 < r < \alpha,$$

or $p^2 - n^2 = 0$, with roots $p_1 = n$ and $p_2 = -n$. Therefore, the general solution of (5.40) in this case is

$$R_n(r) = C_1 r^n + C_2 r^{-n}, \quad C_1, C_2 = \text{const}. \tag{5.41}$$

If $n = 0$, then (5.40) reduces to $(r R_0'(r))' = 0$, so

$$r R_0'(r) = \bar{C}_2 = \text{const};$$

in turn, this leads to $R_0'(r) = r^{-1} \bar{C}_2$, and then to the general solution

$$R_0(r) = \bar{C}_1 + \bar{C}_2 \ln r, \quad \bar{C}_1, \bar{C}_2 = \text{const}. \tag{5.42}$$

From (5.38) and the condition of boundedness at $r = 0$ we deduce that all the $R_n(0)$ must be finite. As (5.41) and (5.42) indicate, this happens only if $C_2 = 0$ and $\bar{C}_2 = 0$; in other words,

$$R_0(r) = \bar{C}_1, \quad R_n(r) = C_1 r^n, \quad n = 1, 2, \ldots.$$

Taking $\bar{C}_1 = 1$ and $C_1 = 1$, we write

$$R_0(r) = 1, \quad R_n(r) = r^n, \quad n = 1, 2, \ldots.$$

In accordance with the separation of variables procedure, we expect the solution of the IBVP to have a series representation of the form

$$u(r, \theta) = a_0 R_0(r) \Theta_0(\theta) + \sum_{n=1}^{\infty} R_n(r) \big[a_n \Theta_{1n}(\theta) + b_n \Theta_{2n}(\theta) \big]$$

$$= \tfrac{1}{2} a_0 + \sum_{n=1}^{\infty} r^n \big[a_n \cos(n\theta) + b_n \sin(n\theta) \big], \tag{5.43}$$

where each term satisfies the PDE and the three homogeneous BCs. The unknown numerical coefficients a_0, a_n, and b_n are determined from the remaining (nonhomogeneous) condition:

$$u(\alpha, \theta) = f(\theta) = \tfrac{1}{2} a_0 + \sum_{n=1}^{\infty} \alpha^n \big[a_n \cos(n\theta) + b_n \sin(n\theta) \big],$$
$$-\pi < \theta \leq \pi. \qquad (5.44)$$

This is the full Fourier series for f with $L = \pi$ and coefficients $a_0/2$, $\alpha^n a_n$, and $\alpha^n b_n$; so, by (2.7)–(2.9),

$$a_0 = \frac{1}{\pi} \int_{-\pi}^{\pi} f(\theta) \, d\theta,$$

$$a_n = \frac{1}{\pi \alpha^n} \int_{-\pi}^{\pi} f(\theta) \cos(n\theta) \, d\theta, \quad n = 1, 2, \ldots, \qquad (5.45)$$

$$b_n = \frac{1}{\pi \alpha^n} \int_{-\pi}^{\pi} f(\theta) \sin(n\theta) \, d\theta, \quad n = 1, 2, \ldots.$$

Consequently, the solution of the BVP is given by (5.43) with the a_0, a_n, and b_n computed by means of (5.45).

5.20 Remark. From (5.43) and (5.45) it follows that

$$u(0, \theta) = \tfrac{1}{2} a_0 = \frac{1}{2\pi} \int_{-\pi}^{\pi} f(\theta) \, d\theta;$$

in other words, the temperature at the center of the disk is equal to the average of the temperature on the boundary circle. This result, already used in the proof of Theorem 4.15, is called the *mean value property* for the Laplace equation.

5.21 Example. In the BVP

$$(\Delta u)(r, \theta) = 0, \quad 0 < r < 2, \; -\pi < \theta < \pi,$$
$$u(2, \theta) = 1 + 8 \sin \theta - 32 \cos(4\theta), \quad -\pi < \theta < \pi$$

we have

$$\alpha = 2, \quad f(\theta) = 1 + 8 \sin \theta - 32 \cos(4\theta).$$

Since f is a linear combination of the eigenfunctions associated with this problem, we use (5.44) and Theorem 3.26(ii) to deduce that

$$\tfrac{1}{2} a_0 = 1, \quad 2^4 a_4 = -32, \quad 2 b_1 = 8,$$

while the remaining coefficients are zero; therefore,

$$a_0 = 2, \quad a_4 = -2, \quad a_n = 0 \; (n \neq 0, 4),$$
$$b_1 = 4, \quad b_n = 0 \; (n \neq 1),$$

which means that, by (5.43), the solution of the given BVP is

$$u(r, \theta) = 1 + 4r \sin \theta - 2r^4 \cos(4\theta).$$

VERIFICATION WITH MATHEMATICA®. The input

```
u = 1 + 4 * r * Sin[θ] - 2 * r^4 * Cos[4 * θ];
{D[u,r,r] + r^(-1) * D[u,r] + r^(-2) * D[u,θ,θ],
 (u /. r -> 2) - 1 - 8 * Sin[θ] + 32 * Cos[4 * θ]}
// FullSimplify
```

generates the output $\{0, 0\}$, which confirms that the function u satisfies both the PDE and the BC.

Exercises

Use separation of variables to solve the BVP

$$u_{rr}(r, \theta) + r^{-1}u_r(r, \theta) + r^{-2}u_{\theta\theta}(r, \theta) = 0,$$

$$0 < r < \alpha, \quad -\pi < \theta < \pi,$$

$$u(r, \theta), \ u_r(r, \theta) \text{ bounded as } r \to 0+, \quad u(\alpha, \theta) = f(\theta),$$

$$-\pi < \theta < \pi,$$

$$u(r, -\pi) = u(r, \pi), \quad u_\theta(r, -\pi) = u_\theta(r, \pi), \quad 0 < r < \alpha$$

with constant α and function f as indicated.

1 $\alpha = 1/2$, $f(\theta) = 3 - 4\cos(3\theta)$.

2 $\alpha = 3$, $f(\theta) = 2\cos(4\theta) + \sin(2\theta)$.

3 $\alpha = 2$, $f(\theta) = \begin{cases} -1, & -\pi < \theta \le 0, \\ 3, & 0 < \theta < \pi. \end{cases}$

4 $\alpha = 1$, $f(\theta) = \begin{cases} 1, & -\pi < \theta \le 0, \\ \theta, & 0 < \theta < \pi. \end{cases}$

Answers to Odd-Numbered Exercises

1 $u(r, \theta) = 3 - 32r^3 \cos(3\theta).$

3 $u(r, \theta) = 1 + \sum_{n=1}^{\infty} [1 - (-1)^n](1/(2^{n-2}n\pi))r^n \sin(n\theta).$

5.4   Other Equations

The method of separation of variables can also be applied to problems involving some of the equations mentioned in Section 4.4. Before we illustrate it in specific cases, we need to make a useful observation.

5.22 Remark. It is easy to show that the method of separation of variables applied to the generic IBVP

$$\alpha u_{tt}(x,t) + \beta u_t(x,t) = A u_{xx}(x,t) + B u_x(x,t) + C u(x,t),$$
$$0 < x < L, \ t > 0,$$
$$u(0,t) = 0, \quad u(L,t) = 0, \quad t > 0,$$
$$u(x,0) = f(x), \quad u_t(x,0) = g(x), \quad 0 < x < L,$$

where α, β are not both zero and $A \neq 0$, leads to the regular S–L problem

$$AX''(x) + BX'(x) + CX(x) + \lambda X(x) = 0, \quad 0 < x < L,$$
$$X(0) = 0, \quad X(L) = 0.$$

Dividing the equation by A, we bring it to the form in Remark 3.25, with

$$a = \frac{B}{A}, \quad b = \frac{C}{A}, \quad c = \frac{1}{A},$$

so the eigenvalues and eigenfunctions associated with the given IBVP are (after simple algebraic manipulation)

$$\lambda_n = \frac{n^2 \pi^2 A}{L^2} + \frac{B^2}{4A} - C, \quad X_n(x) = e^{-(B/(2A))x} \sin \frac{n \pi x}{L}, \quad n = 1, 2, \ldots.$$

Additionally,

$$\sigma(x) = \frac{1}{A} e^{(B/A)x}.$$

These results remain valid also for a BVP in the variables x, y.

5.23 Example. Consider the diffusion–convection IBVP

$$u_t(x,t) = 2u_{xx}(x,t) - u_x(x,t), \quad 0 < x < 1, \ t > 0, \qquad \text{(PDE)}$$
$$u(0,t) = 0, \quad u(1,t) = 0, \quad t > 0, \qquad \text{(BCs)}$$
$$u(x,0) = -2e^{x/4} \sin(3\pi x), \quad 0 < x < 1. \qquad \text{(IC)}$$

Applying the arguments developed earlier in this chapter, we seek the solution in the form $u(x,t) = X(x)T(t)$ with $X, T \neq 0$ and, from the PDE and BCs, deduce that the components X and T satisfy

$$2X''(x) - X'(x) + \lambda X(x) = 0, \quad 0 < x < 1,$$
$$X(0) = 0, \quad X(1) = 0,$$

(5.46)

and

$$T'(t) + \lambda T(t) = 0, \quad t > 0. \tag{5.47}$$

The regular Sturm–Liouville problem (5.46) is solved by the method used in Example 3.22, and, according to Remark 5.22 with

$$A = 2, \quad B = -1, \quad C = 0, \quad L = 1,$$

its eigenvalues and eigenfunctions are

$$\lambda_n = 2n^2\pi^2 + \tfrac{1}{8}, \quad X_n(x) = e^{x/4}\sin(n\pi x), \quad n = 1, 2, \ldots.$$

Then (5.47) becomes

$$T'_n(t) + \left(2n^2\pi^2 + \tfrac{1}{8}\right)T_n(t) = 0,$$

with general solution

$$T_n(t) = c_n e^{-(2n^2\pi^2 + 1/8)t}, \quad c_n = \text{const}.$$

Combining the X_n and T_n in the usual way, we find that the function

$$u(x,t) = \sum_{n=1}^{\infty} c_n e^{x/4 - (2n^2\pi^2 + 1/8)t} \sin(n\pi x) \tag{5.48}$$

satisfies both the PDE and the BCs.

The coefficients c_n are found from the IC; thus, for $t = 0$ we must have

$$u(x,0) = \sum_{n=1}^{\infty} c_n e^{x/4} \sin(n\pi x) = -2e^{x/4}\sin(3\pi x),$$

from which

$$c_3 = -2, \quad c_n = 0 \ (n \neq 3),$$

so the solution, given by (5.48), of the IBVP is

$$u(x,t) = -2e^{x/4 - (18\pi^2 + 1/8)t}\sin(3\pi x).$$

VERIFICATION WITH MATHEMATICA®. The input

```
u = - 2 * E^ (x/4 - (144 * Pi^2 + 1) * t/8) * Sin[3 * Pi * x];
FullSimplify[{D[u,t] - 2 * D[u,x,x] + D[u,x],
  {u /. x -> 0, u /. x -> 1}, (u /. t -> 0)
  + 2 * E^ (x/4) * Sin[3 * Pi * x]}]
```

generates the output $\{0, \{0, 0\}, 0\}$.

5.24 Example. Consider the dissipative wave propagation problem

$$u_{tt}(x,t) + 2u_t(x,t) = u_{xx}(x,t) - 3u_x(x,t), \quad 0 < x < 1, \ t > 0,$$

$$u(0,t) = 0, \quad u(1,t) = 0, \quad t > 0,$$

$$u(x,0) = -e^{3x/2}\sin(2\pi x), \quad u_t(x,0) = 2e^{3x/2}\sin(\pi x), \quad 0 < x < 1.$$

Writing $u(x,t) = X(x)T(t)$, X, $T \neq 0$, and proceeding as in Example 5.23, we find that

$$X''(x) - 3X'(x) + \lambda X(x) = 0, \quad 0 < x < 1,$$
$$X(0) = 0, \quad X(1) = 0, \tag{5.49}$$

and

$$T''(t) + 2T' + \lambda T(t) = 0, \quad t > 0. \tag{5.50}$$

By Remark 5.22 with $A = 1$, $B = -3$, $C = 0$, and $L = 1$, the regular S–L problem (5.49) yields the eigenvalues and eigenfunctions

$$\lambda_n = n^2\pi^2 + \tfrac{9}{4}, \quad X_n(x) = e^{3x/2}\sin(n\pi x), \quad n = 1, 2, \dots.$$

Then, with λ replaced by λ_n, we find that the roots of the characteristic equation associated with (5.50) are $-1 \pm i\alpha_n$, where

$$\alpha_n = \sqrt{n^2\pi^2 + \tfrac{5}{4}}, \quad n = 1, 2, \dots,$$

so

$$T_n(t) = e^{-t}\big[c_{1n}\cos(\alpha_n t) + c_{2n}\sin(\alpha_n t)\big], \quad c_{1n}, c_{2n} = \text{const}.$$

Putting together the X_n and T_n, we now see that the function

$$u(x,t) = \sum_{n=1}^{\infty} e^{3x/2-t}\sin(n\pi x)\big[c_{1n}\cos(\alpha_n t) + c_{2n}\sin(\alpha_n t)\big] \tag{5.51}$$

satisfies both the PDE and the BCs. We now set $t = 0$ in (5.51) and in its t-derivative and compute the coefficients c_{1n} and c_{2n} from the ICs:

$$u(x,0) = \sum_{n=1}^{\infty} c_{1n}e^{3x/2}\sin(n\pi x) = -e^{3x/2}\sin(2\pi x),$$

$$u_t(x,0) = \sum_{n=1}^{\infty}\big(-c_{1n} + \alpha_n c_{2n}\big)e^{3x/2}\sin(n\pi x) = 2e^{3x/2}\sin(\pi x);$$

hence,

$$c_{12} = -1, \quad c_{1n} = 0 \ (n \neq 2),$$
$$-c_{11} + \alpha_1 c_{21} = 2, \quad -c_{1n} + \alpha_n c_{2n} = 0 \ (n \neq 1).$$

From these equalities we find that

$$c_{21} = \frac{2}{\alpha_1}, \quad c_{22} = -\frac{1}{\alpha_2}, \quad c_{2n} = 0 \ (n \neq 1, 2),$$

so the solution of the IBVP, obtained from (5.51), is

$$u(x,t) = e^{3x/2-t}\left[\frac{2}{\alpha_1}\sin(\alpha_1 t)\sin(\pi x) - \cos(\alpha_2 t)\sin(2\pi x)\right.$$
$$\left. - \frac{1}{\alpha_2}\sin(\alpha_2 t)\sin(2\pi x)\right],$$

where

$$\alpha_1 = \sqrt{\pi^2 + \frac{5}{4}}, \quad \alpha_2 = \sqrt{4\pi^2 + \frac{5}{4}}.$$

VERIFICATION WITH MATHEMATICA®. The input

```
{α1,α2}={(Pi^2+5/4)^(1/2),(4*Pi^2+5/4)^(1/2)};
u=FullSimplify[E^(3*x/2-t)*((2/α1)*Sin[α1*t]*Sin[Pi*x]
 -Cos[α2*t]*Sin[2*Pi*x]-(1/α2)*Sin[α2*t]*Sin[2*Pi x])];
FullSimplify[{D[u,t,t]+2*D[u,t]-D[u,x,x]+3*D[u,x],
 u/.x->0,u/.x->1,(u/.t->0)+E^(3*x/2)*Sin[2*Pi*x],
 (D[u,t]/.t->0)-2*E^(3*x/2)*Sin[Pi*x]}]
```

generates the output $\{0,0,0,0,0\}$.

5.25 Example. Consider the two-dimensional steady-state diffusion–convection problem

$$u_{xx}(x,y) + u_{yy}(x,y) - 2u_x(x,y) = 0, \quad 0 < x < 1, \ 0 < y < 2, \qquad \text{(PDE)}$$

$$u(0,y) = 0, \quad u(1,y) = -3\sin(\pi y), \quad 0 < y < 2,$$
$$u(x,0) = 0, \quad u(x,2) = 0, \quad 0 < x < 1. \qquad \text{(BCs)}$$

Seeking the solution in the form $u(x,y) = X(x)Y(y)$ with $X, Y \neq 0$, from the PDE and the last two (homogeneous) BCs it follows that

$$X''(x) - 2X'(x) - \lambda X(x) = 0, \quad 0 < x < 1, \qquad (5.52)$$

and

$$Y''(y) + \lambda Y(y) = 0, \quad 0 < y < 2,$$
$$Y(0) = 0, \quad Y(2) = 0.$$

The regular S–L problem for Y has been solved in Example 3.16 and, with $L = 2$, has the eigenvalues and eigenfunctions

$$\lambda_n = \frac{n^2\pi^2}{4}, \quad Y_n(y) = \sin\frac{n\pi y}{2}, \quad n = 1, 2, \ldots,$$

so (5.52) becomes

$$X_n''(x) - 2X_n'(x) - \frac{n^2\pi^2}{4}X_n(x) = 0,$$

with general solution

$$X_n(x) = c_{1n}e^{s_{1n}x} + c_{2n}e^{s_{2n}x},$$

where

$$s_{1n} = 1 + \frac{1}{2}\sqrt{n^2\pi^2 + 4}, \quad s_{2n} = 1 - \frac{1}{2}\sqrt{n^2\pi^2 + 4}, \quad c_{1n}, c_{2n} = \text{const.}$$

Consequently, the function

$$u(x, y) = \sum_{n=1}^{\infty} \sin\frac{n\pi y}{2}\left(c_{1n}e^{s_{1n}x} + c_{2n}e^{s_{2n}x}\right) \tag{5.53}$$

satisfies the PDE and the last two BCs. The coefficients c_{1n} and c_{2n} are determined from the first two BCs; thus, setting $x = 0$ and then $x = 1$ in (5.53), we find that

$$\sum_{n=1}^{\infty}(c_{1n} + c_{2n})\sin\frac{n\pi y}{2} = 0,$$

$$\sum_{n=1}^{\infty}\left(c_{1n}e^{s_{1n}} + c_{2n}e^{s_{2n}}\right)\sin\frac{n\pi y}{2} = -3\sin(\pi y),$$

so

$$c_{1n} + c_{2n} = 0, \quad n = 1, 2, \ldots,$$

$$c_{12}e^{s_{12}} + c_{22}e^{s_{22}} = -3,$$

$$c_{1n}e^{s_{1n}} + c_{2n}e^{s_{2n}} = 0 \quad (n \neq 2).$$

Then

$$c_{12} = -c_{22} = \frac{3}{e^{s_{22}} - e^{s_{12}}},$$

$$c_{1n} = c_{2n} = 0 \quad (n \neq 2),$$

which, substituted in (5.53), yield the solution of the given BVP:

$$u(x, y) = \frac{3}{e^{s_{22}} - e^{s_{12}}}\left(e^{s_{12}x} - e^{s_{22}x}\right)\sin(\pi y),$$

where

$$s_{12} = 1 + \sqrt{\pi^2 + 1}, \quad s_{22} = 1 - \sqrt{\pi^2 + 1}.$$

VERIFICATION WITH MATHEMATICA®. The input

```
{s12,s22} = {1 + (Pi^2 + 1)^(1/2), 1 - (Pi^2 + 1)^(1/2)};
u = FullSimplify[(3/(E^s22 - E^s12)) * (E^(s12 * x) - E^(s22 * x))
    * Sin[Pi * y]];
FullSimplify[{D[u,x,x] + D[u,y,y] - 2 * D[u,x], u /. x -> 0,
    (u /. x -> 1) + 3 * Sin[Pi * y], (u /. y -> 0), u /. y -> 2}]
```

generates the output $\{0, 0, 0, 0, 0\}$.

Exercises

In **1–4**, use separation of variables to solve the IBVP

$$u_t(x,t) = ku_{xx}(x,t) - au_x(x,t) + bu(x,t), \quad 0 < x < 1, \ t > 0,$$
$$u(0,t) = 0, \quad u(1,t) = 0, \quad t > 0,$$
$$u(x,0) = f(x), \quad 0 < x < 1$$

with numbers k, a, b and function f as indicated.

1 $k = 1$, $a = 1$, $b = 0$, $f(x) = 3e^{x/2}\sin(2\pi x)$.

2 $k = 1$, $a = 2$, $b = 0$, $f(x) = 1$.

3 $k = 1$, $a = 1$, $b = 1$, $f(x) = \begin{cases} 0, & 0 < x \le 1/2, \\ 1, & 1/2 < x < 1. \end{cases}$

4 $k = 2$, $a = 2$, $b = -1$, $f(x) = \begin{cases} 1, & 0 < x \le 1/2, \\ -1, & 1/2 < x < 1. \end{cases}$

In **5–8**, use separation of variables to solve the IBVP

$$u_{tt}(x,t) + au_t(x,t) + bu(x,t) = c^2 u_{xx}(x,t) - du_x(x,t),$$
$$0 < x < 1, \ t > 0,$$
$$u(0,t) = 0, \quad u(1,t) = 0, \quad t > 0,$$
$$u(x,0) = f(x), \quad u_t(x,0) = g(x), \quad 0 < x < 1$$

with numbers a, b, c, d and functions f, g as indicated.

5 $a = 1$, $b = 1$, $c = 2$, $d = 2$, $f(x) = -2e^{x/4}\sin(3\pi x)$,
$g(x) = e^{x/4}\sin(2\pi x)$.

6 $a = 1$, $b = 1$, $c = 1$, $d = 1$, $f(x) = -3$, $g(x) = 1$.

7 $a = 1$, $b = 2$, $c = 1$, $d = 2$, $f(x) = \begin{cases} -1, & 0 < x \le 1/2, \\ 0, & 1/2 < x < 1, \end{cases}$
$g(x) = e^x\sin(2\pi x)$.

8 $a = 2$, $b = 1$, $c = 1/2$, $d = 1$, $f(x) = -3e^{2x}\sin(4\pi x)$,
$g(x) = \begin{cases} 0, & 0 < x \le 1/2, \\ 1, & 1/2 < x < 1. \end{cases}$

In **9–12**, use separation of variables to solve the BVP consisting of the PDE

$$u_{xx}(x,y) + u_{yy}(x,y) - au_x(x,y) - bu_y(x,y) + cu(x,y) = 0,$$
$$0 < x < L, \ 0 < y < K$$

with numbers L, K, a, b, c, and BCs as indicated.

9   $L = 2$, $K = 1$, $a = 0$, $b = 2$, $c = 0$, $u(0, y) = 0$, $u(2, y) = 0$,

    $u(x, 0) = -\sin(\pi x/2)$, $u(x, 1) = 2\sin(\pi x)$.

10   $L = 1$, $K = 2$, $a = 1$, $b = 1$, $c = 0$, $u(0, y) = 0$, $u(1, y) = 0$,

    $u(x, 0) = 1$, $u(x, 2) = -e^{x/2}\sin(\pi x)$.

11   $L = 2$, $K = 1$, $a = 0$, $b = 2$, $c = 0$, $u(0, y) = 3e^y\sin(2\pi y)$,

    $u(2, y) = -2e^y\sin(\pi y)$, $u(x, 0) = 0$, $u(x, 1) = 0$.

12   $L = 1$, $K = 2$, $a = 1$, $b = 1$, $c = 1$, $u(0, y) = 2e^{y/2}\sin(3\pi y/2)$,

$$u(1, y) = \begin{cases} 1, & 0 < y \le 1, \\ 0, & 1 < y < 2, \end{cases} \quad u(x, 0) = 0, \quad u(x, 2) = 0.$$

Answers to Odd-Numbered Exercises

1   $u(x, t) = 3e^{-(1+16\pi^2)t/4 + x/2}\sin(2\pi x)$.

3   $u(x, t) = \sum\limits_{n=1}^{\infty}(1 + 4n^2\pi^2)^{-1}$

$$\times \{(-1)^{n+1}8n\pi + 4e^{1/4}[2n\pi\cos(n\pi/2) + \sin(n\pi/2)]\}$$

$$\times e^{-1/2 + (3/4 - n^2\pi^2)t + x/2}\sin(n\pi x).$$

5   $u(x, t) = (1 + 16\pi^2)^{-1/2}e^{(x-2t)/4}\sin(2\pi x)\sin((1 + 16\pi^2)^{1/2}t)$

$$- e^{(x-2t)/4}\sin(3\pi x)[2\cos((1 + 36\pi^2)^{1/2}t)$$

$$+ (1 + 36\pi^2)^{-1/2}\sin((1 + 36\pi^2)^{1/2}t)].$$

7   $u(x, t) = 2(11 + 16\pi^2)^{-1/2}e^{(2x-t)/2}\sin(2\pi x)\sin((1/2)(11 + 16\pi^2)^{1/2}t)$

$$+ \sum\limits_{n=1}^{\infty} 2(1 + n^2\pi^2)^{-1}(11 + 4n^2\pi^2)^{-1/2}$$

$$\times [-e^{1/2}n\pi + n\pi\cos(n\pi/2) + \sin(n\pi/2)]e^{(2x-t-1)/2}$$

$$\times \sin(n\pi x)[(11 + 4n^2\pi^2)^{1/2}\cos((1/2)(11 + 4n^2\pi^2)^{1/2}t)$$

$$+ \sin((1/2)(11 + 4n^2\pi^2)^{1/2}t)].$$

9   $u(x, y) = e^y[\operatorname{csch}((4 + \pi^2)^{1/2}/2)\sin(\pi x/2)\sinh(((4 + \pi^2)^{1/2}/2)(y - 1))$

$$+ 2e^{-1}\operatorname{csch}((1 + \pi^2)^{1/2})\sin(\pi x)\sinh((1 + \pi^2)^{1/2}y)].$$

11   $u(x, y) = -e^y[3\operatorname{csch}(2(1 + 4\pi^2)^{1/2})\sinh((1 + 4\pi^2)^{1/2}(x - 2))\sin(2\pi y)$

$$+ 2\operatorname{csch}(2(1 + \pi^2)^{1/2})\sinh((1 + \pi^2)^{1/2}x)\sin(\pi y)].$$

5.5 Equations with More than Two Variables

In Section 4.2, we derived a model for heat conduction in a uniform plate, where the unknown temperature function depends on the time variable and two space variables. Here, we consider the two-dimensional analog of the wave equation and solve it by the method of separation of variables in terms of both Cartesian and polar coordinates.

5.5.1 Vibrating Rectangular Membrane

The vibrations of a rectangular membrane with negligible body force and clamped edge are modeled by the IBVP

$$u_{tt}(x,t) = c^2 (\Delta u)(x,y) = c^2 \big[u_{xx}(x,y) + u_{yy}(x,y) \big],$$
$$0 < x < L, \ 0 < y < K, \ t > 0, \tag{PDE}$$

$$u(0,y,t) = 0, \quad u(L,y,t) = 0, \quad 0 < y < K, \ t > 0,$$
$$u(x,0,t) = 0, \quad u(x,K,t) = 0, \quad 0 < x < L, \ t > 0, \tag{BCs}$$

$$u(x,y,0) = f(x,y), \quad u_t(x,y,0) = g(x,y),$$
$$0 < x < L, \ 0 < y < K. \tag{ICs}$$

Since the PDE and BCs are linear and homogeneous, we seek the solution as a product of a function of the space variables and a function of the time variable:

$$u(x,y,t) = S(x,y)T(t), \quad S, T \neq 0. \tag{5.54}$$

Reasoning as in previous similar situations, from the PDE and BCs we find that T satisfies the ODE

$$T''(t) + \lambda c^2 T(t) = 0, \quad t > 0,$$

and that S is a solution of the BVP

$$S_{xx}(x,y) + S_{yy}(x,y) + \lambda S(x,y) = 0,$$
$$0 < x < L, \ 0 < y < K,$$
$$S(0,y) = 0, \quad S(L,y) = 0, \quad 0 < y < K,$$
$$S(x,0) = 0, \quad S(x,K) = 0, \quad 0 < x < L, \tag{5.55}$$

where $\lambda = \text{const}$ is the separation constant.

We notice that the PDE and BCs for S are themselves linear and homogeneous, so we use separation of variables once more and seek a solution for (5.55) of the form

$$S(x,y) = X(x)Y(y), \quad X, Y \neq 0. \tag{5.56}$$

The PDE in (5.55) then becomes

$$X''(x)Y(y) + X(x)Y''(y) = -\lambda X(x)Y(y),$$

which, on division by $X(x)Y(y)$, yields

$$\frac{X''(x)}{X(x)} = -\lambda - \frac{Y''(y)}{Y(y)} = -\mu,$$

where $\mu = \text{const}$ is a second separation constant.

From the above chain of equalities and the BCs in (5.55), we find in the usual way that X and Y are, respectively, solutions of the regular Sturm–Liouville problems

$$\begin{aligned} X''(x) + \mu X(x) &= 0, \quad 0 < x < L, \\ X(0) = 0, \quad X(L) &= 0, \end{aligned} \tag{5.57}$$

and

$$\begin{aligned} Y''(y) + (\lambda - \mu)Y(y) &= 0, \quad 0 < y < K, \\ Y(0) = 0, \quad Y(K) &= 0. \end{aligned} \tag{5.58}$$

The nonzero solutions that we require for (5.57) are the eigenfunctions of this S–L problem, which were computed in Example 3.16 together with the corresponding eigenvalues:

$$\mu_n = \left(\frac{n\pi}{L}\right)^2, \quad X_n(x) = \sin\frac{n\pi x}{L}, \quad n = 1, 2, \dots.$$

For each fixed value of n, the function Y satisfies the S–L problem (5.58) with $\mu = \mu_n$. Since this problem is similar to (5.57), its eigenvalues and eigenfunctions (nonzero solutions) are

$$\lambda_{nm} - \mu_n = \left(\frac{m\pi}{K}\right)^2, \quad Y_{nm}(y) = \sin\frac{m\pi y}{K}, \quad m = 1, 2, \dots.$$

We remark that (5.55) may be regarded as a two-dimensional eigenvalue problem with eigenvalues

$$\lambda_{nm} = \mu_n + \left(\frac{m\pi}{K}\right)^2 = \left(\frac{n\pi}{L}\right)^2 + \left(\frac{m\pi}{K}\right)^2, \quad n, m = 1, 2, \dots,$$

and corresponding eigenfunctions

$$S_{nm}(x, y) = X_n(x)Y_{nm}(y) = \sin\frac{n\pi x}{L}\sin\frac{m\pi y}{K}, \quad n, m = 1, 2, \dots.$$

Going back to the equation satisfied by T, for $\lambda = \lambda_{nm}$ we obtain the functions

$$T_{nm}(t) = a_{nm}\cos\left(\sqrt{\lambda_{nm}}\, ct\right) + b_{nm}\sin\left(\sqrt{\lambda_{nm}}\, ct\right), \quad a_{nm}, b_{nm} = \text{const}.$$

From this, (5.54), and (5.56) we now conclude that the solution of the original IBVP should have a representation of the form

$$u(x, y, t) = \sum_{n=1}^{\infty} \sum_{m=1}^{\infty} S_{nm}(x, y) T_{nm}(t)$$

$$= \sum_{n=1}^{\infty} \sum_{m=1}^{\infty} \sin \frac{n\pi x}{L} \sin \frac{m\pi y}{K}$$

$$\times \left[a_{nm} \cos \left(\sqrt{\lambda_{nm}}\, ct \right) + b_{nm} \sin \left(\sqrt{\lambda_{nm}}\, ct \right) \right]. \tag{5.59}$$

Each term in (5.59) satisfies the PDE and the BCs. To satisfy the first IC, we must have

$$u(x, y, 0) = f(x, y)$$

$$= \sum_{m=1}^{\infty} \left(\sum_{n=1}^{\infty} a_{nm} \sin \frac{n\pi x}{L} \right) \sin \frac{m\pi y}{K}.$$

Multiplying the second equality by $\sin(p\pi y/K)$, $p = 1, 2, \ldots$, integrating the result with respect to y over $[0, K]$, and taking into account formula (2.5) with L replaced by K, we find that

$$\int_0^K f(x, y) \sin \frac{p\pi y}{K}\, dy = \frac{K}{2} \sum_{n=1}^{\infty} a_{np} \sin \frac{n\pi x}{L}.$$

Multiplying this new equality by $\sin(q\pi x/L)$, $q = 1, 2, \ldots$, and integrating with respect to x over $[0, L]$, we arrive at

$$\int_0^L \int_0^K f(x, y) \sin \frac{p\pi y}{K} \sin \frac{q\pi x}{L}\, dy\, dx = \frac{LK}{4} a_{qp},$$

from which, replacing q by n and p by m, we obtain

$$a_{nm} = \frac{4}{LK} \int_0^L \int_0^K f(x, y) \sin \frac{n\pi x}{L} \sin \frac{m\pi y}{K}\, dy\, dx,$$

$$n, m = 1, 2, \ldots. \tag{5.60}$$

The second IC is satisfied if

$$u_t(x, y, 0) = g(x, y)$$

$$= \sum_{m=1}^{\infty} \left(\sum_{n=1}^{\infty} \sqrt{\lambda_{nm}}\, c\, b_{nm} \sin \frac{n\pi x}{L} \right) \sin \frac{m\pi y}{K};$$

as above, this yields

$$b_{nm} = \frac{4}{LK\sqrt{\lambda_{nm}}\, c} \int_0^L \int_0^K g(x,y) \sin\frac{n\pi x}{L} \sin\frac{m\pi y}{K} \, dy\, dx,$$

$$n, m = 1, 2, \ldots. \qquad (5.61)$$

In conclusion, the solution of the IBVP is given by series (5.59) with the coefficients a_{nm} and b_{nm} computed by means of (5.60) and (5.61).

Other homogeneous BCs are treated similarly.

5.26 Example. In the IBVP

$$u_{tt}(x,y,t) = u_{xx}(x,y,t) + u_{yy}(x,y,t), \quad 0 < x < 1, \ 0 < y < 2, \ t > 0,$$
$$u(0,y,t) = 0, \quad u(1,y,t) = 0, \quad 0 < y < 2, \ t > 0,$$
$$u(x,0,t) = 0, \quad u(x,2,t) = 0, \quad 0 < x < 1, \ t > 0,$$
$$u(x,y,0) = \sin(2\pi x)\left[\sin\frac{\pi y}{2} - 2\sin(\pi y)\right], \quad 0 < x < 1, \ 0 < y < 2,$$
$$u_t(x,y,0) = 2\sin(2\pi x)\sin\frac{\pi y}{2}, \quad 0 < x < 1, \ 0 < y < 2,$$

we have $c = 1$, $L = 1$, and $K = 2$; thus, comparing the initial data functions with the double Fourier series expansions of f and g, we see that

$$a_{21} = 1, \quad a_{22} = -2, \quad a_{2m} = 0 \ (m \neq 1, 2),$$
$$\sqrt{\lambda_{21}}\, b_{21} = 2, \quad \sqrt{\lambda_{2m}}\, b_{2m} = 0 \ (m \neq 1).$$

Since $\lambda_{21} = (2\pi)^2 + (\pi/2)^2 = 17\pi^2/4$ and $\lambda_{22} = 5\pi^2$, formula (5.59) yields the solution

$$u(x,y,t) = \sin(2\pi x)\sin\frac{\pi y}{2}\cos\frac{\sqrt{17}\,\pi t}{2}$$
$$- 2\sin(2\pi x)\sin(\pi y)\cos\left(\sqrt{5}\,\pi t\right)$$
$$+ \frac{4}{\sqrt{17}\,\pi}\sin(2\pi x)\sin\frac{\pi y}{2}\sin\frac{\sqrt{17}\,\pi t}{2}.$$

VERIFICATION WITH MATHEMATICA®. The input

```
u = Sin[2 * Pi * x] * Sin[Pi * y/2] * Cos[(17)^(1/2) * Pi * t/2]
  - 2 * Sin[2 * Pi * x] * Sin[Pi * y] * Cos[5^(1/2) * Pi * t]
  + (4/(17^(1/2) * Pi)) * Sin[2 * Pi * x] * Sin[Pi * y/2]
    * Sin[17^(1/2) * Pi * t/2];
FullSimplify[{D[u,t,t] - D[u,x,x] - D[u,y,y], {u /. x -> 0, u /. x -> 1},
  {u /. y -> 0, u /. y -> 2}, {(u /. t -> 0) - Sin[2 * Pi * x]
```

```
    * (Sin [Pi * y/2] - 2 * Sin [Pi * y]), (D[u,t] /. t -> 0)
    - 2 * Sin [2 * Pi * x] * Sin [Pi * y/2]}}]
```
generates the output $\{0, \{0, 0\}, \{0, 0\}\}$.

Exercises

Use separation of variables to solve the IBVP consisting of the PDE
$$u_{tt}(x, y, t) = u_{xx}(x, y, t) + u_{yy}(x, y, t),$$
$$0 < x < 1, \ 0 < y < 1, \ t > 0$$
and BCs and ICs as indicated.

1 $u(0, y, t) = 0, \ u(1, y, t) = 0, \ u(x, 0, t) = 0, \ u(x, 1, t) = 0,$

$u(x, y, 0) = \sin(\pi x) \sin(2\pi y), \ u_t(x, y, 0) = -2 \sin(2\pi x) \sin(\pi y).$

2 $u(0, y, t) = 0, \ u(1, y, t) = 0, \ u(x, 0, t) = 0, \ u(x, 1, t) = 0,$

$u(x, y, 0) = 3 \sin(2\pi x) \sin(\pi y),$

$u_t(x, y, 0) = \sin(3\pi x) [\sin(2\pi y) - 2 \sin(3\pi y)].$

3 $u(0, y, t) = 0, \ u(1, y, t) = 0, \ u_y(x, 0, t) = 0, \ u_y(x, 1, t) = 0,$

$u(x, y, 0) = -\sin(\pi x) \cos(2\pi y), \ u_t(x, y, 0) = 3 \sin(2\pi x).$

4 $u_x(0, y, t) = 0, \ u_x(1, y, t) = 0, \ u(x, 0, t) = 0, \ u(x, 1, t) = 0,$

$u(x, y, 0) = \cos(2\pi x), \ u_t(x, y, 0) = -2 \cos(\pi x) \sin(3\pi y).$

5 $u(0, y, t) = 0, \ u(1, y, t) = 0, \ u(x, 0, t) = 0, \ u(x, 1, t) = 0,$

$u(x, y, 0) = 1, \ u_t(x, y, 0) = xy.$

6 $u(0, y, t) = 0, \ u(1, y, t) = 0, \ u_y(x, 0, t) = 0, \ u_y(x, 1, t) = 0,$

$u(x, y, 0) = (x - 1)y, \ u_t(x, y, 0) = 2xy.$

Answers to Odd-Numbered Exercises

1 $u(x, y, t) = \sin(\pi x) \sin(2\pi y) \cos(\sqrt{5}\,\pi t)$
$$- (2/(\sqrt{5}\,\pi)) \sin(2\pi x) \sin(\pi y) \sin(\sqrt{5}\,\pi t).$$

3 $u(x, y, t) = -\sin(\pi x) \cos(2\pi y) \cos(\sqrt{5}\,\pi t) + (3/(2\pi)) \sin(2\pi x) \sin(2\pi t).$

5 $u(x, y, t) = (4/\pi^2) \sum\limits_{m=1}^{\infty} \sum\limits_{n=1}^{\infty} (1/(mn))\{[1 - (-1)^m][1 - (-1)^n]$
$$\times \cos((m^2 + n^2)^{1/2}\pi t)$$
$$+ (-1)^{m+n}(\pi(m^2 + n^2))^{-1/2} \sin((m^2 + n^2)^{1/2}\pi t)\}$$
$$\times \sin(n\pi x) \sin(m\pi y).$$

5.5.2 Vibrating Circular Membrane

A two-dimensional body of this type is defined in terms of polar coordinates by $0 \le r \le \alpha$, $-\pi < \theta \le \pi$. If the body force is negligible and the boundary $r = \alpha$ is clamped, then the vertical vibrations of the membrane are described by the IBVP

$$u_{tt}(r, \theta, t) = c^2(\Delta u)(r, \theta, t), \quad 0 < r < \alpha, \ -\pi < \theta < \pi, \ t > 0, \qquad \text{(PDE)}$$

$$u(\alpha, \theta, t) = 0, \quad u(r, \theta, t), \ u_r(r, \theta, t) \text{ bounded as } r \to 0+,$$
$$-\pi < \theta < \pi, \ t > 0,$$
$$u(r, -\pi, t) = u(r, \pi, t), \quad u_\theta(r, -\pi, t) = u_\theta(r, \pi, t), \qquad \text{(BCs)}$$
$$0 < r < \alpha, \ t > 0,$$

$$u(r, \theta, 0) = f(r, \theta), \quad u_t(r, \theta, 0) = g(r, \theta),$$
$$0 < r < \alpha, \ -\pi < \theta < \pi. \qquad \text{(ICs)}$$

As in Subsection 5.3.2, the BCs at $r = 0$ and $\theta = -\pi$, π express the continuity (therefore boundedness) of the displacement u and radial and circular components u_r, u_θ of the tension in the membrane for 'reasonably behaved' functions f and g.

Since the PDE and BCs are linear and homogeneous, we try the method of separation of variables and seek a solution of the form

$$u(r, \theta, t) = S(r, \theta)T(t), \quad S, \ T \ne 0. \qquad (5.62)$$

Then, following the standard procedure, from the PDE and BCs we find that S is a solution of the problem

$$(\Delta S)(r, \theta) + \lambda S(r, \theta) = 0, \quad 0 < r < \alpha, \ -\pi < \theta < \pi,$$
$$S(\alpha, \theta) = 0, \quad S(r, \theta), \ S_r(r, \theta) \text{ bounded as } r \to 0+,$$
$$-\pi < \theta < \pi, \qquad (5.63)$$
$$S(r, -\pi) = S(r, \pi), \quad S_\theta(r, -\pi) = S_\theta(r, \pi), \quad 0 < r < \alpha,$$

where $\lambda = \text{const}$ is the separation constant, while T satisfies

$$T''(t) + \lambda c^2 T(t) = 0, \quad t > 0.$$

Since the PDE and BCs for S are linear and homogeneous, we seek S of the form
$$S(r, \theta) = R(r)\Theta(\theta), \quad R, \ \Theta \ne 0. \qquad (5.64)$$

Next, recalling the expression of the Laplacian in polar coordinates (see Remark 4.12(ii)), we write the PDE in (5.63) as

$$\left[R''(r) + r^{-1}R'(r)\right]\Theta(\theta) + r^{-2}R(r)\Theta''(\theta) + \lambda R(r)\Theta(\theta) = 0,$$

which, on division by $r^{-2}R(r)\Theta(\theta)$, yields

$$\frac{\Theta''(\theta)}{\Theta(\theta)} = -\frac{r^2 R''(r) + rR'(r)}{R(r)} - \lambda r^2 = -\mu,$$

where $\mu = \text{const}$ is a second separation constant.

Applying the usual argument to the BCs, we now conclude that R and Θ are solutions of Sturm–Liouville problems. First, Θ satisfies the periodic S–L problem

$$\Theta''(\theta) + \mu\Theta(\theta) = 0, \quad -\pi < \theta < \pi,$$
$$\Theta(-\pi) = \Theta(\pi), \quad \Theta'(-\pi) = \Theta'(\pi),$$

whose eigenvalues and eigenfunctions, computed in Example 3.33, are

$$\mu_0 = 0, \quad \Theta_0(\theta) = \tfrac{1}{2},$$
$$\mu_n = n^2, \quad \Theta_{1n}(\theta) = \cos(n\theta), \quad \Theta_{2n}(\theta) = \sin(n\theta), \quad n = 1, 2, \dots.$$

For each $n = 0, 1, 2, \dots$, we find that R_n is the solution of the S–L problem

$$r^2 R_n''(r) + rR_n'(r) + (\lambda r^2 - n^2)R_n(r) = 0, \quad 0 < r < \alpha,$$
$$R_n(\alpha) = 0, \quad R_n(r), \ R_n'(r) \text{ bounded as } r \to 0+.$$

This is the singular Sturm–Liouville problem (3.14), (3.15) (with m replaced by n) discussed in Section 3.3, which has the countably many eigenvalues and complete set of corresponding eigenfunctions

$$\lambda_{nm} = \left(\frac{\xi_{nm}}{\alpha}\right)^2, \quad R_{nm}(r) = J_n\left(\frac{\xi_{nm}r}{\alpha}\right), \quad n = 0, 1, 2, \dots, \quad m = 1, 2, \dots,$$

where J_n is the Bessel function of the first kind and order n and ξ_{nm} are the zeros of J_n.

Returning to the ODE satisfied by T, for $\lambda = \lambda_{nm} = (\xi_{nm}/\alpha)^2$ we now obtain the solutions

$$T_{nm}(t) = A_{nm}\cos\frac{\xi_{nm}ct}{\alpha} + B_{nm}\sin\frac{\xi_{nm}ct}{\alpha}, \quad A_{nm}, B_{nm} = \text{const}.$$

Suppose, for simplicity, that $g = 0$; then it is clear that the sine term in T_{nm} must vanish. Hence, if we piece together the various components of u according to (5.62) and (5.64), we conclude that the solution of the IBVP should be of the form

$$u(r, \theta, t) = \sum_{n=0}^{\infty}\sum_{m=1}^{\infty} \left[a_{nm}\Theta_{1n}(\theta) + b_{nm}\Theta_{2n}(\theta)\right] R_{nm}(r)T_{nm}(t)$$

$$= \sum_{n=0}^{\infty}\sum_{m=1}^{\infty} \left[a_{nm}\cos(n\theta) + b_{nm}\sin(n\theta)\right] J_n\left(\frac{\xi_{nm}r}{\alpha}\right)\cos\frac{\xi_{nm}ct}{\alpha}.$$

Each term in this sum satisfies the PDE, the BCs, and the second IC. To satisfy the first IC, we need to have

$$u(r,\theta,0) = f(r,\theta) = \sum_{m=1}^{\infty}\left\{\sum_{n=0}^{\infty}\left[a_{nm}\cos(n\theta) + b_{nm}\sin(n\theta)\right]\right\}J_n\left(\frac{\xi_{nm}r}{\alpha}\right).$$

The a_{nm} and b_{nm} are now computed as in other problems of this type, by means of the orthogonality properties of $\cos(n\theta)$, $\sin(n\theta)$, and $J_n(\xi_{nm}r/\alpha)$ expressed by (2.4)–(2.6) with $L = \pi$, and by (3.18).

5.27 Remark. In the circularly symmetric case (that is, when the data functions are independent of θ) we use the solution split

$$u(r,t) = R(r)T(t), \quad R, T \neq 0.$$

Here, the problem reduces to solving Bessel's equation of order zero. In the end, we obtain

$$u(r,t) = \sum_{m=1}^{\infty}\left[a_m\cos\left(\sqrt{\lambda_m}\,ct\right) + b_m\sin\left(\sqrt{\lambda_m}\,ct\right)\right]J_0\left(\sqrt{\lambda_m}\,r\right),$$

where λ_m is the same as $\lambda_{0m} = (\xi_{0m}/\alpha)^2$ in the general problem and the a_m and b_m are determined from the property of orthogonality with weight r on $[0,\alpha]$ of the functions $J_0\left(\sqrt{\lambda_m}\,r\right)$.

5.28 Example. With

$$\alpha = 1, \quad c = 1, \quad f(r,\theta) = 1, \quad g(r,\theta) = 0,$$

the formula in Remark 5.27 yields

$$u(r,0) = \sum_{m=1}^{\infty} a_m J_0\left(\sqrt{\lambda_m}\,r\right) = 1,$$

$$u_t(r,0) = \sum_{m=1}^{\infty} b_m\sqrt{\lambda_m}\,J_0\left(\sqrt{\lambda_m}\,r\right) = 0,$$

so the a_m are the coefficients of the expansion of the constant function 1 in the functions $J_0\left(\sqrt{\lambda_m}\,r\right)$ on $(0,1)$, and $b_m = 0$. Since, with the numbers rounded to the fourth decimal place,

$$1 = 1.6020\,J_0(2.4048\,r) - 1.0648\,J_0(5.5201\,r)$$
$$+ 0.8514\,J_0(8.6537\,r) - 0.7296\,J_0(11.7915\,r)$$
$$+ 0.6485\,J_0(14.9309\,r) - 0.5895\,J_0(18.0711\,r) + \cdots,$$

it follows that

$$a_1 = 1.6020, \quad a_2 = -1.0648, \quad a_3 = 0.8514,$$
$$a_4 = -0.7296, \quad a_5 = 0.6485, \quad a_6 = -0.5895, \quad \ldots,$$

so the solution of the problem is

$$u(r,t) = 1.6020 \cos(2.4048\,t) J_0(2.4048\,r)$$
$$- 1.0648 \cos(5.5201\,t) J_0(5.5201\,r)$$
$$+ 0.8514 \cos(8.6537\,t) J_0(8.6537\,r)$$
$$- 0.7296 \cos(11.7915\,t) J_0(11.7915\,r)$$
$$+ 0.6485 \cos(14.9309\,t) J_0(14.9309\,r)$$
$$- 0.5895 \cos(18.0711\,t) J_0(18.0711\,r) + \cdots.$$

VERIFICATION WITH MATHEMATICA®. The input

```
Approx = 1.6020 * Cos[2.4048 * t] * BesselJ[0,2.4048 * r]
  - 1.0648 * Cos[5.5201 * t] * BesselJ[0,5.5201 * r]
  + 0.8514 * Cos[8.6537 * t] * BesselJ[0,8.6537 * r]]
  - 0.7296 * Cos[11.7915 * t] * BesselJ[0,11.7915 * r]
  + 0.6485 * Cos[14.9309 * t] * BesselJ[0, 14.9309 * r]
  - 0.5895 * Cos[18.0711 * t] * BesselJ[0, 18.0711 * r]
  // FullSimplify
eqn = r * D[Approx,t,t] - r * D[Approx,r,r] - D[Approx,r]
  // FullSimplify
MaxEqn = FindMaximum[{eqn,0<r<1,t>0},{r,t}][[1]]
ApproxIC = Approx /. t->0
MaxIC = FindMaximum[{ApproxIC,0<r<1},r][[1]]
{Chop[MaxEqn],Chop[MaxIC-1,0.2]}
```

generates the output $\{0,0\}$, which confirms that u is an approximate solution of the BVP.

Exercises

Use separation of variables to compute the first five terms of the series solution of the IBVP

$$u_{tt}(r,\theta,t) = u_{rr}(r,\theta,t) + r^{-1}u_r(r,\theta,t) + r^{-2}u_{\theta\theta}(r,\theta,t),$$
$$0 < r < 1, \quad -\pi < \theta < \pi, \quad t > 0,$$
$$u(1,\theta,t) = 0, \quad u(r,\theta,t), \ u_r(r,\theta,t) \text{ bounded as } r \to 0+,$$
$$-\pi < \theta < \pi, \quad t > 0,$$
$$u(r,-\pi,t) = u(r,\pi,t), \quad u_\theta(r,-\pi,t) = u_\theta(r,\pi,t),$$
$$0 < r < 1, \quad t > 0,$$
$$u(r,\theta,0) = f(r,\theta), \quad u_t(r,\theta,0) = g(r,\theta), \quad 0 < r < 1, \quad -\pi < \theta < \pi$$

with functions f and g as indicated.

1 $f(r, \theta) = 2\sin(2\theta), \quad g(r, \theta) = 0.$

2 $f(r, \theta) = 0, \quad g(r, \theta) = -\cos\theta.$

3 $f(r, \theta) = r\sin\theta, \quad g(r, \theta) = 0.$

4 $f(r, \theta) = 0, \quad g(r, \theta) = (r - 1)\cos(2\theta).$

Answers to Odd-Numbered Exercises

1 $u(r, \theta, t) = \big[5.2698\, J_2(5.1356\, r)\cos(5.1356\, t)$

$$- 0.3168\, J_2(8.4172\, r)\cos(8.4172\, t)$$
$$+ 2.6215\, J_2(11.6198\, r)\cos(11.6198\, t)$$
$$- 0.4762\, J_2(14.7960\, r)\cos(14.7960\, t)$$
$$+ 1.9049\, J_2(17.9598\, r)\cos(17.9598\, t) + \cdots \big]\sin(2\theta).$$

3 $u(r, \theta, t) = \big[-0.3382\, J_1(3.8317\, r)\cos(3.8317\, t)$

$$+ 0.1354\, J_1(7.0156\, r)\cos(7.0156\, t)$$
$$- 0.0774\, J_1(10.1735\, r)\cos(10.1735\, t)$$
$$+ 0.0516\, J_1(13.3237\, r)\cos(13.3237\, t)$$
$$- 0.0375\, J_1(16.4706\, r)\cos(16.4706\, t) + \cdots \big]\sin\theta.$$

5.5.3 Equilibrium Temperature in a Solid Sphere

The steady-state distribution of heat inside a homogeneous sphere of radius α when the (time-independent) temperature is prescribed on the surface and no sources are present is the solution of the BVP

$$\Delta u(r, \theta, \varphi) = 0, \quad 0 < r < \alpha, \ 0 < \theta < 2\pi, \ 0 < \varphi < \pi, \qquad \text{(PDE)}$$

$$u(\alpha, \theta, \varphi) = f(\theta, \varphi), \quad u(r, \theta, \varphi), \ u_r(r, \theta, \varphi) \text{ bounded as } r \to 0+,$$
$$0 < \theta < 2\pi, \ 0 < \varphi < \pi,$$

$$u(r, 0, \varphi) = u(r, 2\pi, \varphi), \quad u_\theta(r, 0, \varphi) = u_\theta(r, 2\pi, \varphi),$$
$$0 < r < \alpha, \ 0 < \varphi < \pi, \qquad \text{(BCs)}$$

$$u(r, \theta, \varphi), \ u_\varphi(r, \theta, \varphi) \text{ bounded as } \varphi \to 0+ \text{ and as } \varphi \to \pi-,$$
$$0 < r < \alpha, \ 0 < \theta < 2\pi,$$

where f is a known function. The first boundary condition represents the given surface temperature; the remaining ones have been adjoined on the basis of arguments similar to those used in the case of a uniform circular disc.

In terms of the spherical coordinates r, θ, φ, the above PDE is written as (see Remark 4.12(v))

$$\frac{1}{r^2}(r^2 u_r)_r + \frac{1}{r^2\sin^2\varphi} u_{\theta\theta} + \frac{1}{r^2\sin\varphi}((\sin\varphi)u_\varphi)_\varphi = 0.$$

Since the PDE and all but one of the BCs are homogeneous, we try a solution of the form

$$u(r, \theta, \varphi) = R(r)\Theta(\theta)\Phi(\varphi), \quad R, \Theta, \Phi \neq 0.$$

As before, the nonhomogeneous BC implies that none of R, Θ, Φ can be the zero function. Replacing in the PDE and multiplying every term by $(r^2 \sin^2 \varphi)/(R\Theta\Phi)$, we find that

$$\frac{\Theta''}{\Theta} = -(\sin^2 \varphi)\frac{(r^2 R')'}{R} - (\sin \varphi)\frac{((\sin \varphi)\Phi')'}{\Phi},$$

where the left-hand side is a function of θ and the right-hand side is a function of r and φ. Consequently, both sides must be equal to one and the same constant, which, for convenience, we denote by $-\mu$. This leads to the pair of equations

$$\Theta''(\theta) + \mu\Theta(\theta) = 0, \quad 0 < \theta < 2\pi, \tag{5.65}$$

$$\frac{(r^2 R')'}{R} + \frac{1}{\sin \varphi}\frac{((\sin \varphi)\Phi')'}{\Phi} - \frac{\mu}{\sin^2 \varphi} = 0,$$

$$0 < r < \alpha, \ 0 < \varphi < \pi. \tag{5.66}$$

(In (5.66), we have also divided by $\sin^2 \varphi$.)

Since $R, \Phi \neq 0$, from the two periodicity conditions in the BVP we deduce in the usual way that

$$\Theta(0) = \Theta(2\pi), \quad \Theta'(0) = \Theta'(2\pi). \tag{5.67}$$

The periodic S–L problem (5.65), (5.67) for Θ is similar to that discussed in Example 3.33. According to Remark 3.34, it has the eigenvalues $\mu = m^2$, $m = 0, 1, 2, \ldots$, and corresponding eigenfunctions (expressed in their most general form)

$$\Theta_m(\theta) = C_{1m} \cos(m\theta) + C_{2m} \sin(m\theta),$$

which, using Euler's formula, we can rewrite as

$$\Theta_m(\theta) = C'_{1m} e^{im\theta} + C'_{2m} e^{-im\theta}. \tag{5.68}$$

We now operate a second separation of variables, this time in (5.66):

$$\frac{1}{\sin \varphi}\frac{((\sin \varphi)\Phi')'}{\Phi} - \frac{m^2}{\sin^2 \varphi} = \frac{(r^2 R')'}{R} = -\lambda = \text{const};$$

this yields the equations

$$\frac{1}{\sin \varphi}((\sin \varphi)\Phi'(\varphi))' + \left(\lambda - \frac{m^2}{\sin^2 \varphi}\right)\Phi(\varphi) = 0, \quad 0 < \varphi < \pi, \tag{5.69}$$

$$(r^2 R')' - \lambda R = 0, \quad 0 < r < \alpha. \tag{5.70}$$

In (5.69), we substitute

$$\xi = \cos\varphi, \quad d\xi = -\sin\varphi\, d\varphi, \quad \frac{d}{d\xi} = -\frac{1}{\sin\varphi}\frac{d}{d\varphi},$$

$$\sin^2\varphi = 1 - \cos^2\varphi = 1 - \xi^2, \quad \Phi(\varphi) = \Psi(\xi)$$

and bring the equation to the form

$$\left((1-\xi^2)\Psi'(\xi)\right)' + \left(\lambda - \frac{m^2}{1-\xi^2}\right)\Psi(\xi) = 0, \quad -1 < \xi < 1. \tag{5.71}$$

The boundary conditions for Ψ are obtained from those at $\varphi = 0$ and $\varphi = \pi$ in the original problem, by recalling that Φ cannot be the zero function:

$$\Psi(\xi), \ \Psi'(\xi) \text{ bounded as } \xi \to -1+ \text{ and as } \xi \to 1-. \tag{5.72}$$

We remark that (5.71), (5.72) is the singular S–L problem for the associated Legendre equation mentioned in Section 3.5, with eigenvalues and eigenfunctions

$$\lambda_n = n(n+1), \quad \Psi(\xi) = P_n^m(\xi), \quad n = m, m+1, \ldots,$$

where P_n^m are the associated Legendre functions; consequently,

$$\Phi_n(\varphi) = P_n^m(\cos\varphi). \tag{5.73}$$

Turning to the Cauchy–Euler equation (5.70) with

$$\lambda = n(n+1),$$

we find that its general solution (see Section 1.4) is

$$R_n(r) = C_1 r^n + C_2 r^{-(n+1)}, \quad C_1, C_2 = \text{const.}$$

When the condition that $R(r)$ and $R'(r)$ be bounded as $r \to 0+$, which follows from the second BC in the given BVP, is applied, we find that $C_2 = 0$, so

$$R_n(r) = r^n, \quad n = m, m+1, \ldots. \tag{5.74}$$

Here, as usual, we have taken $C_1 = 1$.

Finally, we combine (5.68), (5.73), (5.74), (3.33), and (3.34) and write the solution of the original problem in the form

$$u(r, \theta, \varphi) = \sum_{\substack{m=0,1,2,\ldots \\ n=m,m+1,\ldots}} R_n(r)\Theta_m(\theta)\Phi_{nm}(\varphi)$$

$$= \sum_{\substack{m=0,1,2,\ldots \\ n=m,m+1,\ldots}} r^n\left(C'_{1m}e^{im\theta} + C'_{2m}e^{-im\theta}\right)P_n^m(\cos\varphi),$$

or, equivalently,

$$u(r, \theta, \varphi) = \sum_{n=0}^{\infty} \sum_{m=-n}^{m=n} C_{nm} r^n e^{im\theta} P_n^m(\cos \varphi)$$

$$= \sum_{n=0}^{\infty} \sum_{m=-n}^{m=n} c_{n,m} r^n Y_{n,m}(\theta, \varphi), \qquad (5.75)$$

where $Y_{n,m}$ are the orthonormalized spherical harmonics defined by (3.34). According to the nonhomogeneous BC in the given BVP, we must have

$$u(\alpha, \theta, \varphi) = \sum_{n=0}^{\infty} \sum_{m=-n}^{m=n} c_{n,m} \alpha^n Y_{n,m}(\theta, \varphi) = f(\theta, \varphi).$$

The coefficients $c_{n,m}$ are now determined in the standard way, by means of the orthonormality relations (3.35):

$$c_{n,m} = \frac{1}{\alpha^n} \int_0^{2\pi} \int_0^{\pi} f(\theta, \varphi) \bar{Y}_{n,m}(\theta, \varphi) \sin \varphi \, d\varphi \, d\theta. \qquad (5.76)$$

5.29 Example. If in the preceding problem the upper and lower hemispheres of the sphere with the center at the origin and radius $\alpha = 1$ are kept at constant temperatures of 2 and -1, respectively, then

$$f(\theta, \varphi) = \begin{cases} 2, & 0 < \theta < 2\pi, \ 0 < \varphi < \pi/2, \\ -1, & 0 < \theta < 2\pi, \ \pi/2 < \varphi < \pi, \end{cases}$$

so, using formula (5.76), we find that the first two nonzero coefficients $c_{n,m}$ are $c_{0,0} = \sqrt{\pi}$ and $c_{1,0} = 3\sqrt{3\pi}/2$. By (5.75) and the expressions of the spherical harmonics given in Example 3.44, the solution of the problem is

$$u(r, \theta, \varphi) = \sqrt{\pi} Y_{0,0}(\theta, \varphi) + \tfrac{3}{2}\sqrt{3\pi} \, r Y_{1,0}(\theta, \varphi) + \cdots = \tfrac{1}{2} + \tfrac{9}{4} r \cos \varphi + \cdots.$$

Exercises

Use separation of variables to compute, up to r^2-terms, the series solution of the BVP

$$\frac{1}{r^2} (r^2 u_r)_r + \frac{1}{r^2 \sin^2 \varphi} u_{\theta\theta} + \frac{1}{r^2 \sin \varphi} \big((\sin \varphi) u_\varphi\big)_\varphi = 0,$$

$$0 < r < 1, \ 0 < \theta < 2\pi, \ 0 < \varphi < \pi,$$

$$u(1, \theta, \varphi) = f(\theta, \varphi), \quad u(r, \theta, \varphi), \ u_r(r, \theta, \varphi) \text{ bounded as } r \to 0+,$$

$$0 < \theta < 2\pi, \ 0 < \varphi < \pi,$$

$$u(r,0,\varphi) = u(r,2\pi,\varphi), \quad u_\theta(r,0,\varphi) = u_\theta(r,2\pi,\varphi),$$
$$0 < r < 1, \ 0 < \varphi < \pi,$$
$$u(r,\theta,\varphi), \ u_\varphi(r,\theta,\varphi) \text{ bounded as } \varphi \to 0+ \text{ and as } \varphi \to \pi-,$$
$$0 < r < 1, \ 0 < \theta < 2\pi$$

with function f as indicated.

1 $f(\theta,\varphi) = 2\cos^2\theta \sin^2\varphi - \sin\theta \sin\varphi \cos\varphi - \cos^2\varphi.$

2 $f(\theta,\varphi) = \begin{cases} 1, & 0 < \varphi \le \pi/2, \\ \sin\theta \sin\varphi \cos\varphi, & \pi/2 < \varphi < \pi. \end{cases}$

Answers to Odd-Numbered Exercises

1 $u(r,\theta,\varphi) = 1/3 + r^2 \left[2/3 - 2\cos^2\varphi - \sin\theta \sin\varphi \cos\varphi + \cos(2\theta)\sin^2\varphi \right] + \cdots.$

CHAPTER 6

Linear Nonhomogeneous Problems

The method of separation of variables can be applied only if the PDE and BCs are homogeneous. However, a mathematical model may include non-homogeneous terms (that is, terms not containing the unknown function or its derivatives) in the equation, and/or may have nonhomogeneous (nonzero) data prescribed at its boundary points. In this chapter we show how, in certain cases, such problems can be reduced—at least in part—to their corresponding homogeneous versions.

6.1 Equilibrium Solutions

We have already discussed equilibrium (or steady-state, or time-independent) solutions for the higher-dimensional heat equation (see Section 4.2). Below, we consider equilibrium temperature distributions for a uniform rod under a variety of BCs.

For a time-dependent problem, it seems physically reasonable to expect that if an equilibrium solution exists, it should be the result of the temperature settling down, so to speak, after a very long wait. This is why we denote such a solution by u_∞ and write

$$u_\infty(x) = \lim_{t \to \infty} u(x, t).$$

In a problem with time-independent data, the issue of the existence of a steady-state solution is much simpler, as the next few examples illustrate.

6.1 Example. The equilibrium temperature for the IBVP

$$u_t(x,t) = 4u_{xx}(x,t) + 2x - 1, \quad 0 < x < 2, \ t > 0,$$
$$u(0,t) = 1, \quad u(2,t) = -2, \quad t > 0,$$
$$u(x,0) = f(x), \quad 0 < x < 2$$

is obtained by solving the BVP

$$4u_\infty''(x) + 2x - 1 = 0, \quad 0 < x < 2,$$
$$u_\infty(0) = 1, \quad u_\infty(2) = -2.$$

Clearly, since the time variable is not involved, the IC plays no role in this process.

Integrating the ODE twice, we find that

$$u_\infty(x) = -\tfrac{1}{12} x^3 + \tfrac{1}{8} x^2 + C_1 x + C_2, \quad C_1, C_2 = \text{const.}$$

The constants C_1 and C_2 are now easily found from the BCs. The equilibrium solution is

$$u_\infty(x) = -\tfrac{1}{12} x^3 + \tfrac{1}{8} x^2 - \tfrac{17}{12} x + 1.$$

6.2 Example. As above, the steady-state temperature for the IBVP

$$u_t(x,t) = 4u_{xx}(x,t) + \gamma x + 24, \quad 0 < x < 2, \ t > 0,$$
$$u_x(0,t) = 2, \quad u_x(2,t) = 14, \quad t > 0,$$
$$u(x,0) = \pi \sin \frac{\pi x}{2} + 1, \quad 0 < x < 2,$$

where $\gamma = \text{const}$, satisfies the BVP

$$4u_\infty''(x) + \gamma x + 24 = 0, \quad 0 < x < 2,$$
$$u_\infty'(0) = 2, \quad u_\infty'(2) = 14.$$

The ODE yields

$$u_\infty'(x) = -\tfrac{1}{8} \gamma x^2 - 6x + C_1,$$

and the BCs imply that $C_1 = 2$ and $\gamma = -48$. This is the only value of γ for which an equilibrium temperature exists. From here, we find that

$$u_\infty(x) = 2x^3 - 3x^2 + 2x + C_2.$$

To compute C_2, we integrate the PDE with $\gamma = -48$ over $0 \le x \le 2$ to obtain

$$\int_0^2 u_t(x,t)\,dx = \int_0^2 4u_{xx}(x,t)\,dx + \int_0^2 (-48x + 24)\,dx,$$

or

$$\frac{d}{dt} \int_0^2 u(x,t)\,dx = 4u_x(x,t)\big|_{x=0}^{x=2} + 24\big[-x^2 + x\big]_0^2 = 0,$$

which means that

$$\int_0^2 u(x,t)\,dt = \text{const}, \quad 0 \le t < \infty.$$

In this case, we have

$$\int_0^2 u(x,0)\,dx = \int_0^2 \lim_{t\to\infty} u(x,t)\,dx = \int_0^2 u_\infty(x)\,dx,$$

so

$$\int_0^2 \left(\pi \sin\frac{\pi x}{2} + 1\right) dx = \int_0^2 \left(2x^3 - 3x^2 + 2x + C_2\right) dx.$$

Evaluation of the integrals on both sides now yields the value $C_2 = 1$; hence,

$$u_\infty(x) = 2x^3 - 3x^2 + 2x + 1.$$

It is easily verified that, as expected, the total contribution of the source term $\gamma x + 24 = -48x + 24$ over the length of the rod, and of the heat flux at the endpoints, is zero.

6.3 Example. To find the equilibrium temperature for the IBVP

$$u_t(x,t) = 4u_{xx}(x,t) - 16, \quad 0 < x < 2,\ t > 0,$$
$$u(0,t) = -5, \quad u_x(2,t) = 4, \quad t > 0,$$
$$u(x,0) = f(x), \quad 0 < x < 2,$$

we need to solve the BVP

$$u_\infty''(x) - 4 = 0, \quad 0 < x < 2,$$
$$u_\infty(0) = -5, \quad u_\infty'(2) = 4.$$

Direct integration shows that $u_\infty(x) = 2x^2 - 4x - 5$.

6.4 Example. The time-independent temperature for the IBVP

$$u_t(x,t) = 4u_{xx}(x,t) + 32, \quad 0 < x < 2,\ t > 0,$$
$$u_x(0,t) = 2\big[u(0,t) + \tfrac{1}{2}\big], \quad u(2,t) = 1, \quad t > 0,$$
$$u(x,0) = f(x), \quad 0 < x < 2$$

satisfies

$$4u_\infty''(x) + 32 = 0, \quad 0 < x < 2,$$
$$u_\infty'(0) = 2\left[u_\infty(0) + \tfrac{1}{2}\right], \quad u_\infty(2) = 1,$$

with solution $u_\infty(x) = -4x^2 + 7x + 3$.

Equilibrium temperatures can sometimes also be computed explicitly for heat-conduction problems in higher dimensions.

6.5 Example. The IBVP (see Remark 4.12(ii))

$$u_t(r, \theta, t) = k(\Delta u)(r, \theta, t) = k\left[r^{-1}(ru_r(r, \theta, t))_r + r^{-2}u_{\theta\theta}(r, \theta, t)\right],$$
$$1 < r < 2, \quad -\pi < \theta < \pi, \quad t > 0,$$
$$u(1, \theta, t) = 3, \quad u(2, \theta, t) = -1, \quad -\pi < \theta < \pi, \quad t > 0,$$
$$u(r, \theta, 0) = f(r), \quad 1 < r < 2, \quad -\pi < \theta < \pi$$

models heat conduction in a uniform circular annulus with no sources when the inner and outer boundary circles $r = 1$ and $r = 2$ are held at two different constant temperatures and the IC function depends only on r. Since the domain and the prescribed data have circular symmetry, we may assume that the solution u is independent of the polar angle θ. Then the equilibrium temperature $u_\infty(r)$ satisfies the BVP

$$(ru_\infty'(r))' = 0, \quad 1 < r < 2,$$
$$u_\infty(1) = 3, \quad u_\infty(2) = -1.$$

Integrating the ODE once, we find that

$$ru_\infty'(r) = C_1 = \text{const},$$

while a second integration yields the general solution

$$u_\infty(r) = C_1 \ln r + C_2, \quad C_2 = \text{const}.$$

Using the BC at $r = 1$, we see that $C_2 = 3$; then the condition at $r = 2$ leads to $C_1 = -4/(\ln 2)$, so

$$u_\infty(r) = -\frac{4}{\ln 2} \ln r + 3.$$

VERIFICATION WITH MATHEMATICA®. The input

```
u = - (4/Log[2]) * Log[r] + 3;
FullSimplify[{r^(-1) * (D[r * D[u,r],r]), (u /. r->1) - 3,
    (u /. r->2) + 1}]
```

generates the output $\{0, 0, 0\}$.

Exercises

In **1–6**, find the equilibrium solution for the IBVP consisting of the PDE

$$u_t(x,t) = u_{xx}(x,t) + q(x), \quad 0 < x < 1, \ t > 0$$

and function q, BCs, and IC as indicated. (In **5** and **6**, also find the value of γ for which an equilibrium solution exists.)

1 $q(x) = 6 - 12x, \quad u(0,t) = -1, \quad u(1,t) = 2, \quad u(x,0) = f(x).$

2 $q(x) = 4 - 6x, \quad u(0,t) = -3, \quad u_x(1,t) = -1, \quad u(x,0) = f(x).$

3 $q(x) = -4 - 12x^2, \quad u_x(0,t) = 3, \quad -u_x(1,t) = 3[u(1,t) - 23/3],$

$u(x,0) = f(x).$

4 $q(x) = 24x - 12x^2, \quad u_x(0,t) = 3[u(0,t) + 1/3], \quad u(1,t) = -2,$

$u(x,0) = f(x).$

5 $q(x) = \gamma x - 4, \quad u_x(0,t) = -3, \quad u_x(1,t) = 4, \quad u(x,0) = -3x - 1/12.$

6 $q(x) = \gamma + 6x - 12x^2, \quad u_x(0,t) = -2, \quad u_x(1,t) = 3,$

$u(x,0) = (157\pi/120)\sin(\pi x).$

In **7–12**, find the equilibrium solution of the IBVP consisting of the PDE and IC

$$u_t(x,t) = ku_{xx}(x,t) - au_x(x,t) + bu(x,t) + q(x), \quad 0 < x < 1, \ t > 0,$$
$$u(x,0) = f(x), \quad 0 < x < 1$$

with numbers k, a, b, function q, and BCs as indicated.

7 $k = 1, \quad a = 3, \quad b = 2, \quad q(x) = -2x^2 + 2x + 6, \quad u(0,t) = -1, \quad u(1,t) = 2.$

8 $k = 1, \quad a = 1, \quad b = -2, \quad q(x) = 10x + 5, \quad u(0,t) = 2, \quad u_x(1,t) = 5 - 2/e.$

9 $k = 4, \quad a = 4, \quad b = 1, \quad q(x) = -x^2 + 11x - 20, \quad u_x(0,t) = -1,$

$u_x(1,t) = -1 + (7/2)\sqrt{e}.$

10 $k = 4, \quad a = 4, \quad b = 5, \quad q(x) = 9 - 5x, \quad u_x(0,t) = 3/2, \quad u(1,t) = \sqrt{e}\cos 1.$

11 $k = 1, \quad a = 2, \quad b = 1, \quad q(x) = -2\pi\cos(\pi x) + (1 - \pi^2)\sin(\pi x),$

$u(0,t) = 2, \quad -u_x(1,t) = 2[u(1,t) - e - \pi/2].$

12 $k = 1, \quad a = 1, \quad b = -6, \quad q(x) = 6x - 17, \quad u_x(0,t) = u(0,t) + 8,$

$u_x(1,t) = 1 + 6e^3.$

Answers to Odd-Numbered Exercises

1 $u_\infty(x) = 2x^3 - 3x^2 + 4x - 1.$

3 $u_\infty(x) = x^4 + 2x^2 + 3x - 2.$

5 $\gamma = -6, \quad u_\infty(x) = x^3 + 2x^2 - 3x - 1.$

7 $u_\infty(x) = x^2 + 2x - 1.$

9 $u_\infty(x) = (3x - 2)e^{x/2} + x^2 - 3x.$

11 $u_\infty(x) = (2 - x)e^x - \sin(\pi x).$

6.2 Nonhomogeneous Problems

Fully nonhomogeneous linear IBVPs are much easier to solve if we can bring them to a form where the PDE and BCs, or at least the latter, are homogeneous. Sometimes, this can be accomplished by means of a substitution involving the equilibrium solution, if one exists, or some other suitably chosen function.

6.2.1 Time-Independent Sources and Boundary Conditions

We illustrate the procedure with a couple of specific problems.

6.6 Example. The equilibrium solution u_∞ of the IBVP

$$u_t(x,t) = u_{xx}(x,t) + \pi^2 \sin(\pi x), \quad 0 < x < 1, \ t > 0,$$

$$u(0,t) = 1, \quad u(1,t) = -3, \quad t > 0,$$

$$u(x,0) = x^2, \quad 0 < x < 1$$

satisfies

$$u_\infty''(x) + \pi^2 \sin(\pi x) = 0, \quad 0 < x < 1,$$

$$u_\infty(0) = 1, \quad u_\infty(1) = -3.$$

The solution of this BVP is $u_\infty(x) = \sin(\pi x) - 4x + 1$. Then the substitution

$$u(x,t) = v(x,t) + u_\infty(x) = v(x,t) + \sin(\pi x) - 4x + 1$$

reduces the given IBVP to the problem

$$v_t(x,t) = v_{xx}(x,t), \quad 0 < x < 1, \ t > 0,$$

$$v(0,t) = 0, \quad v(1,t) = 0, \quad t > 0,$$

$$v(x,0) = x^2 + 4x - 1 - \sin(\pi x), \quad 0 < x < 1,$$

which can be solved by the method of separation of variables, discussed in the next chapter.

6.7 Example. If the IBVP

$$u_t(x,t) = u_{xx}(x,t) + x - \gamma, \quad 0 < x < 1, \ t > 0,$$
$$u_x(0,t) = 0, \quad u_x(1,t) = 0, \quad t > 0,$$
$$u(x,0) = -\tfrac{1}{2} + x + \tfrac{1}{4}x^2 - \tfrac{1}{6}x^3, \quad 0 < x < 1,$$

where $\gamma = \text{const}$, has a steady-state solution u_∞, then this solution must satisfy

$$u_\infty''(x) + x - \gamma = 0, \quad 0 < x < 1,$$
$$u_\infty'(0) = 0, \quad u_\infty'(1) = 0.$$

Integrating the equation once, we find that

$$u_\infty'(x) = -\tfrac{1}{2}x^2 + \gamma x + C_1, \quad C_1 = \text{const},$$

which satisfies both boundary conditions only if $\gamma = 1/2$ and $C_1 = 0$. Another integration yields

$$u_\infty(x) = -\tfrac{1}{6}x^3 + \tfrac{1}{4}x^2 + C_2, \quad C_2 = \text{const}.$$

Following the procedure in Example 6.2, we deduce that $C_2 = 0$. Then the substitution

$$u(x,t) = v(x,t) + u_\infty(x) = v(x,t) - \tfrac{1}{6}x^3 + \tfrac{1}{4}x^2$$

reduces the original problem to the IBVP

$$v_t(x,t) = v_{xx}(x,t), \quad 0 < x < 1, \ t > 0,$$
$$v_x(0,t) = 0, \quad v_x(1,t) = 0, \quad t > 0,$$
$$v(x,0) = x - \tfrac{1}{2}, \quad 0 < x < 1,$$

to which we can apply the method of separation of variables.

6.2.2 The General Case

When at least one of the source and BC functions is explicitly time-dependent, we may not have an equilibrium solution to perform the substitution used in the preceding cases, so we need to modify the solution technique.

6.8 Example. The IBVP

$$u_t(x,t) = u_{xx}(x,t) + xt, \quad 0 < x < 1, \ t > 0,$$
$$u(0,t) = 1 - t, \quad u(1,t) = t^2, \quad t > 0,$$
$$u(x,0) = x, \quad 0 < x < 1$$

is of the above form. Here, we look for a polynomial $p(x,t) = A(t)x + B(t)$ that satisfies both BCs; in other words,

$$p(0,t) = B(t) = 1 - t, \quad p(1,t) = A(t) + B(t) = t^2,$$

which means that
$$p(x,t) = (t^2 + t - 1)x + 1 - t.$$

The substitution
$$u(x,t) = v(x,t) + p(x,t) = v(x,t) + (t^2 + t - 1)x + 1 - t$$

now reduces the given IBVP to the problem

$$v_t(x,t) = v_{xx}(x,t) + 1 - xt - x, \quad 0 < x < 1, \ t > 0,$$
$$v(0,t) = 0, \quad v(1,t) = 0, \quad t > 0,$$
$$v(x,0) = 2x - 1, \quad 0 < x < 1$$

with a nonhomogeneous PDE but homogeneous BCs. Such problems, and similar ones with other types of BCs, are solved by the method of eigenfunction expansion, discussed, as already mentioned, in Chapter 7.

6.9 Example. The IBVP

$$u_t(x,t) = u_{xx}(x,t) + xt, \quad 0 < x < 1, \ t > 0,$$
$$u_x(0,t) = t, \quad u_x(1,t) = t^2, \quad t > 0,$$
$$u(x,0) = x + 1, \quad 0 < x < 1$$

needs slightly different handling. Here, we take p_x instead of p to be a linear polynomial in x satisfying the BCs; that is, $p_x(x,t) = A(t)x + B(t)$, so

$$p_x(0,t) = B(t) = t, \quad p_x(1,t) = A(t) + B(t) = t^2.$$

Then $p_x(x,t) = (t^2 - t)x + t$, from which, by integration, we obtain

$$p(x,t) = \tfrac{1}{2}x^2(t^2 - t) + xt.$$

(Since we need only one such function p, we have taken the constant of integration to be zero.) The substitution

$$u(x,t) = v(x,t) + \tfrac{1}{2}x^2(t^2 - t) + xt$$

now reduces the original problem to the IBVP

$$v_t(x,t) = v_{xx}(x,t) + xt + t^2 - t - \tfrac{1}{2}x^2(2t - 1) - x,$$
$$0 < x < 1, \ t > 0,$$
$$v_x(0,t) = 0, \quad v_x(1,t) = 0, \quad t > 0,$$
$$v(x,0) = x + 1, \quad 0 < x < 1.$$

6.10 Remark. This technique can also be applied to reduce other types of IBVPs to simpler versions where the PDE is nonhomogeneous but the BCs are homogeneous. In the case of BVPs, we can similarly make two of the BCs homogeneous by means of a suitably chosen linear polynomial.

6.11 Example. Consider the IBVP

$$u_{tt}(x,t) - 2u_t(x,t) + u(x,t) = u_{xx}(x,t) + x^2 - 3t, \quad 0 < x < 1, \ t > 0,$$
$$u(0,t) = 1 - t, \quad u_x(1,t) = 2t + 1, \quad t > 0,$$
$$u(x,0) = 2x, \quad u_t(x,0) = x + 4, \quad 0 < x < 1.$$

We construct $p(x,t) = A(t)x + B(t)$ such that

$$p(0,t) = B(t) = 1 - t, \quad p_x(1,t) = A(t) = 2t + 1,$$

so $p(x,t) = (2t+1)x + 1 - t$. Then the substitution

$$u(x,t) = v(x,t) + p(x,t) = v(x,t) + (2t+1)x + 1 - t$$

brings the given IBVP to the form

$$v_{tt}(x,t) - 2v_t(x,t) + v(x,t) = v_{xx}(x,t) + x^2 - 2xt + 3x - 2t - 3,$$
$$0 < x < 1, \ t > 0,$$
$$v(0,t) = 0, \quad v_x(1,t) = 0, \quad t > 0,$$
$$v(x,0) = x - 1, \quad v_t(x,0) = 5 - x, \quad 0 < x < 1.$$

6.12 Example. The BVP

$$u_{xx}(x,y) + u_{yy}(x,y) - 3u_x(x,y) = x - y^2, \quad 0 < x < 1, \ 0 < y < 2,$$
$$u(0,y) = 2y^2, \quad u(1,y) = 3y - 1, \quad 0 < y < 2,$$
$$u_y(x,0) = x + 2, \quad u(x,2) = 4x - 3, \quad 0 < x < 1$$

can be transformed into a new one where, say, the first two BCs are homogeneous. To do so, we look for $p(x,y) = A(y)x + B(y)$ satisfying

$$p(0,y) = B(y) = 2y^2, \quad p(1,y) = A(y) + B(y) = 3y - 1;$$

that is, $p(x,y) = (-2y^2 + 3y - 1)x + 2y^2$. Then the substitution

$$u(x,y) = v(x,y) + p(x,y) = v(x,y) + (-2y^2 + 3y - 1)x + 2y^2$$

reduces the given BVP to

$$v_{xx}(x,y) + v_{yy}(x,y) - 3v_x(x,y) = 5x - 7y^2 + 9y - 7,$$
$$0 < x < 1, \ 0 < y < 2,$$
$$v(0,y) = 0, \quad v(1,y) = 0, \quad 0 < y < 2,$$
$$v_y(x,0) = 2 - 2x, \quad v(x,2) = 7x - 11, \quad 0 < x < 1.$$

Exercises

In **1–6**, find a substitution that reduces the IBVP consisting of the PDE

$$u_t(x,t) = u_{xx}(x,t) + q(x,t), \quad 0 < x < 1, \ t > 0,$$

function q, BCs, and IC as indicated, to an equivalent IBVP with homogeneous BCs. In each case, compute the nonhomogeneous term $\tilde{q}$ in the transformed PDE, and the transformed IC.

1 $q(x,t) = x + t - 2, \quad u(0,t) = t + 1, \quad u(1,t) = 2t - t^2, \quad u(x,0) = 2x.$

2 $q(x,t) = 2x - t, \quad u_x(0,t) = 2 - t^2, \quad u_x(1,t) = t - 3, \quad u(x,0) = x - 1.$

3 $q(x,t) = 3 - 2xt, \quad u_x(0,t) = 3t, \quad u_x(1,t) = t^2 + t, \quad u(x,0) = 2 - x.$

4 $q(x,t) = x^2 - t + 2, \quad u(0,t) = t + 2, \quad u_x(1,t) = 1 - 2t, \quad u(x,0) = x^2 - x.$

5 $q(x,t) = 3x + 2t - 4, \quad u_x(0,t) = 2[u(0,t) - t], \quad u_x(1,t) = 2t + 3,$
$u(x,0) = 2x - 3.$

6 $q(x,t) = 2xt - 3t, \quad u(0,t) = t^2, \quad -u_x(1,t) = 4[u(1,t) - 3], \quad u(x,0) = x + 3.$

In **7–10**, find a substitution that reduces the IBVP consisting of the PDE

$$u_{tt}(x,t) = u_{xx}(x,t) + q(x,t), \quad 0 < x < 1, \ t > 0,$$

function q, BCs, and ICs as indicated, to an equivalent IBVP with homogeneous BCs. In each case, compute the nonhomogeneous term $\tilde{q}$ in the transformed PDE, and the transformed ICs.

7 $q(x,t) = 2x - xt, \quad u(0,t) = 2t + 3, \quad u(1,t) = t^2 - 4,$
$u(x,0) = x + 1, \quad u_t(x,0) = x.$

8 $q(x,t) = x + 2t - 1, \quad u(0,t) = 3t - 1, \quad u(1,t) = t - t^2,$
$u(x,0) = -x, \quad u_t(x,0) = 2x + 1.$

9 $q(x,t) = 2 - t + xt, \quad u_x(0,t) = t - 1, \quad u_x(1,t) = t^2 + 2,$
$u(x,0) = 4x, \quad u_t(x,0) = 2x - 1.$

10 $q(x,t) = t^2 + 2x, \quad u_x(0,t) = t^2 - t, \quad u_x(1,t) = 2t + 3,$
$u(x,0) = 1 - x, \quad u_t(x,0) = x + 2.$

In **11–14**, find a substitution that reduces the BVP consisting of the PDE

$$u_{xx}(x,y) + u_{yy}(x,y) = q(x,y), \quad 0 < x < L, \ 0 < y < K,$$

numbers L, K, function q, and BCs as indicated, to an equivalent BVP with homogeneous BCs at $x = 0$ and $x = L$. In each case, compute the nonhomogeneous term $\tilde{q}$ in the transformed PDE, and the transformed BCs at $y = 0$ and $y = K$.

11 $L = 1$, $K = 1$, $q(x, y) = -2x + y$, $u(0, y) = 2y^2 - 1$, $u(1, y) = y - y^2$,
$u(x, 0) = 3x$, $u(x, 1) = 1 - 2x$.

12 $L = 1$, $K = 2$, $q(x, y) = x - x^2 y - 1$, $u(0, y) = y^2 + 2$, $u(1, y) = 3y - 1$,
$u(x, 0) = 1 - x$, $u(x, 2) = x^2$.

13 $L = 2$, $K = 1$, $q(x, y) = -2x + 3y$, $u_x(0, y) = y - y^2$, $u_x(2, y) = y^2 + 1$,
$u(x, 0) = 2x$, $u(x, 1) = x + 3$.

14 $L = 2$, $K = 1$, $q(x, y) = x - 4y - 1$, $u_x(0, y) = 2y^2$, $u_x(2, y) = 1 - y^2$,
$u(x, 0) = 2x + 1$, $u(x, 1) = 3 - 2x$.

In **15–18**, find a substitution that reduces the IBVP consisting of the PDE

$$u_t(x, t) = ku_{xx}(x, t) - au_x(x, t) + bu(x, t) + q(x, t), \quad 0 < x < 1, \ t > 0,$$

numbers k, a, b, function q, BCs, and IC as indicated, to an equivalent IBVP
with homogeneous BCs. In each case, compute the nonhomogeneous term $\tilde{q}$
in the transformed PDE, and the transformed IC.

15 $k = 1$, $a = 2$, $b = 0$, $q(x, t) = xt - x + 1$,
$u(0, t) = 2t + 3$, $u(1, t) = 4 - 3t$, $u(x, 0) = x + 2$.

16 $k = 2$, $a = 1$, $b = 1$, $q(x, t) = 3x - 2t$,
$u(0, t) = t + 4$, $u(1, t) = 2t - 1$, $u(x, 0) = 1 - 2x$.

17 $k = 2$, $a = 1$, $b = 2$, $q(x, t) = x^2 + 2xt$,
$u_x(0, t) = 1 - 2t$, $u_x(1, t) = t + 3$, $u(x, 0) = x^2$.

18 $k = 1$, $a = 2$, $b = -3$, $q(x, t) = xt + 3t^2$,
$u_x(0, t) = t^2$, $u_x(1, t) = 3t + 2$, $u(x, 0) = 2x$.

Answers to Odd-Numbered Exercises

1 $v(x, t) = u(x, t) - t - 1 + (t^2 - t + 1)x$, $\tilde{q}(x, t) = 2xt + t - 3$, $v(x, 0) = 3x - 1$.

3 $v(x, t) = u(x, t) - 3xt - (t^2/2 - t)x^2$,
$\tilde{q}(x, t) = -x^2 t + x^2 + t^2 - 2xt - 3x - 2t + 3$, $v(x, 0) = 2 - x$.

5 $v(x, t) = u(x, t) - 2t - 3/2 - (2t + 3)x$, $\tilde{q}(x, t) = x + 2t - 6$,
$v(x, 0) = -x - 9/2$.

7 $v(x, t) = u(x, t) - 2t - 3 - (t^2 - 2t - 7)x$,
$\tilde{q}(x, t) = -xt$, $v(x, 0) = 8x - 2$, $v_t(x, 0) = 3x - 2$.

9 $v(x, t) = u(x, t) - (t - 1)x - (1/2)(t^2 - t + 3)x^2$,
$\tilde{q}(x, t) = -x^2 + xt + t^2 - 2t + 5$, $v(x, 0) = 5x - 3x^2/2$,
$v_t(x, 0) = x^2/2 + x - 1$.

11 $v(x, y) = u(x, y) - 2y^2 + 1 - x(1 + y - 3y^2)$,

$\tilde{q}(x, y) = 4x + y - 4$, $v(x, 0) = 2x + 1$, $v_t(x, 0) = -x$.

13 $v(x, y) = u(x, y) - x(y - y^2) - (x^2/4)(2y^2 - y + 1)$,

$\tilde{q}(x, y) = -x^2 - y^2 + 7y/2 - 1/2$, $v(x, 0) = 2x - x^2/4$,

$v_t(x, 0) = 3 + x - x^2/2$.

15 $v(x, t) = u(x, t) - 2t - 3 - (1 - 5t)x$,

$\tilde{q}(x, t) = xt + 4x + 10t - 3$, $v(x, 0) = -1$.

17 $v(x, t) = u(x, t) + (2t - 1)x - (3t/2 + 1)x^2$,

$\tilde{q}(x, t) = 3x^2t + 3x^2/2 - 5xt + 2x + 8t + 3$, $v(x, 0) = -x$.

CHAPTER 7

The Method of Eigenfunction Expansion

Separation of variables cannot be performed if the PDE and/or BCs are not homogeneous. As we saw in Chapter 6, there are particular situations where we can reduce the problem to an equivalent one with homogeneous PDE and BCs, but this is not always possible. In the general case, the best we can do is make the boundary conditions homogeneous. The eigenfunction expansion technique is designed for IBVPs with a nonhomogeneous equation and homogeneous BCs.

7.1 The Nonhomogeneous Heat Equation

Below, we consider problems with several types of BCs.

7.1.1 Rod with Zero Temperature at the Endpoints

We start with the IBVP

$$u_t(x,t) = ku_{xx}(x,t) + q(x,t), \quad 0 < x < L, \ t > 0, \qquad \text{(PDE)}$$

$$u(0,t) = 0, \quad u(L,t) = 0, \quad t > 0, \qquad \text{(BCs)}$$

$$u(x,0) = f(x), \quad 0 < x < L. \qquad \text{(IC)}$$

The eigenvalues and eigenfunctions of the corresponding homogeneous problem ($q = 0$) are, respectively (see Section 5.1),

$$\lambda_n = \left(\frac{n\pi}{L}\right)^2, \quad X_n(x) = \sin\frac{n\pi x}{L}, \quad n = 1, 2, \ldots.$$

Since $\{X_n\}_{n=1}^{\infty}$ is a complete set, we may consider for the solution an expansion of the form

$$u(x,t) = \sum_{n=1}^{\infty} c_n(t) X_n(x). \tag{7.1}$$

Differentiating series (7.1) term by term and recalling that $X_n'' + \lambda_n X_n = 0$, from the PDE we obtain

$$\sum_{n=1}^{\infty} c_n'(t) X_n(x) = k \sum_{n=1}^{\infty} c_n(t) X_n''(x) + q(x,t)$$

$$= -k \sum_{n=1}^{\infty} c_n(t) \lambda_n X_n(x) + q(x,t),$$

or

$$\sum_{n=1}^{\infty} \left[c_n'(t) + k \lambda_n c_n(t) \right] X_n(x) = q(x,t). \tag{7.2}$$

Multiplying this equality by $X_m(x)$, integrating from 0 to L, and taking the orthogonality of the X_n on $[0, L]$ (see Theorem 3.8(iii)) into account, we find that

$$\left[c_m'(t) + k \lambda_m c_m(t) \right] \int_0^L X_m^2(x)\, dx = \int_0^L q(x,t) X_m(x)\, dx,$$

which (with m replaced by n) yields the equations

$$c_n'(t) + k \lambda_n c_n(t) = \frac{\int\limits_0^L q(x,t) X_n(x)\, dx}{\int\limits_0^L X_n^2(x)\, dx}, \qquad t > 0, \quad n = 1, 2, \ldots. \tag{7.3}$$

The BCs are automatically satisfied since each of the X_n in (7.1) satisfies them.

From (7.1) and the IC we see that

$$u(x,0) = f(x) = \sum_{n=1}^{\infty} c_n(0) X_n(x), \tag{7.4}$$

so, proceeding as above, we arrive at the initial conditions

$$c_n(0) = \frac{\int\limits_0^L f(x) X_n(x)\, dx}{\int\limits_0^L X_n^2(x)\, dx}, \qquad n = 1, 2, \ldots. \tag{7.5}$$

Clearly, this is the same as finding formal expansions

$$q(x,t) = \sum_{n=1}^{\infty} q_n(t) X_n(x), \quad f(x) = \sum_{n=1}^{\infty} f_n X_n(x),$$

where the $q_n(t)$ and f_n are given by the right-hand sides in (7.3) and (7.5), respectively. Replacing these expressions in (7.2) and (7.4), we arrive at the equalities

$$\sum_{n=1}^{\infty} [c_n'(t) + k\lambda_n c_n(t)] X_n(x) = \sum_{n=1}^{\infty} q_n(t) X_n(x),$$

$$\sum_{n=1}^{\infty} c_n(0) X_n(x) = \sum_{n=1}^{\infty} f_n X_n(x).$$

Since the X_n are linearly independent, we deduce that for $n = 1, 2, \ldots$,

$$c_n'(t) + k\lambda_n c_n(t) = q_n(t),$$
$$c_n(0) = f_n.$$

In our case, these equations take the specific form

$$c_n'(t) + k\left(\frac{n\pi}{L}\right)^2 c_n(t) = q_n(t) = \frac{2}{L} \int_0^L q(x,t) \sin \frac{n\pi x}{L} \, dx,$$

$$\tag{7.6}$$

$$c_n(0) = f_n = \frac{2}{L} \int_0^L f(x) \sin \frac{n\pi x}{L} \, dx.$$

In conclusion, the solution of the given IBVP is of the form (7.1) with the coefficients $c_n(t)$ computed by means of (7.6).

IBVPs with other types of BCs are treated similarly.

7.1 Example. The eigenvalues and eigenfunctions for the IBVP

$$u_t(x,t) = u_{xx}(x,t) + \pi^2 e^{-24\pi^2 t} \sin(5\pi x), \quad 0 < x < 1, \; t > 0,$$
$$u(0,t) = 0, \quad u(1,t) = 0, \quad t > 0,$$
$$u(x,0) = 3\sin(4\pi x), \quad 0 < x < 1$$

are (here $k = 1$ and $L = 1$)

$$\lambda_n = n^2\pi^2, \quad X_n(x) = \sin(n\pi x), \quad n = 1, 2, \ldots.$$

Since

$$q(x,t) = \pi^2 e^{-24\pi^2 t} \sin(5\pi x), \quad f(x) = 3\sin(4\pi x)$$

coincide with their eigenfunction expansions, the above formulas yield the IVPs

$$c_n'(t) + n^2\pi^2 c_n(t) = \begin{cases} \pi^2 e^{-24\pi^2 t}, & n = 5, \\ 0, & n \neq 5, \end{cases} \quad t > 0,$$

$$c_n(0) = \begin{cases} 3, & n = 4, \\ 0, & n \neq 4. \end{cases}$$

Hence, for $n = 4$ we have

$$c_4'(t) + 16\pi^2 c_4(t) = 0, \quad t > 0,$$
$$c_4(0) = 3,$$

with solution $c_4(t) = 3e^{-16\pi^2 t}$; for $n = 5$,

$$c_5'(t) + 25\pi^2 c_5(t) = \pi^2 e^{-24\pi^2 t}, \quad t > 0,$$
$$c_5(0) = 0,$$

with solution $c_5(t) = e^{-24\pi^2 t} - e^{-25\pi^2 t}$; and for $n \neq 4, 5$,

$$c_n'(t) + n^2\pi^2 c_n(t) = 0, \quad t > 0,$$
$$c_n(0) = 0,$$

with solution $c_n(t) = 0$. Consequently, by (7.1), the solution of the IBVP is

$$u(x,t) = c_4(t)X_4(x) + c_5(t)X_5(x)$$
$$= 3e^{-16\pi^2 t}\sin(4\pi x) + (e^{-24\pi^2 t} - e^{-25\pi^2 t})\sin(5\pi x).$$

VERIFICATION WITH MATHEMATICA®. The input

```
u = 3 * E^( - 16 * Pi^2 * t) * Sin[4 * Pi * x] + (E^( - 24 * Pi^2 * t)
  - E^( - 25 * Pi^2 * t)) * Sin[5 * Pi * x];
{q, f} = {Pi^2 * E^( - 24 * Pi^2 * t) * Sin[5 * Pi * x], 3 * Sin[4 * Pi * x]};
Simplify[{D[u,t] - D[u,x,x] - q, u /. x -> 0, u /. x -> 1,
  (u /. t -> 0) - f}]
```

generates the output $\{0, 0, 0, 0\}$.

7.2 Example. The IBVP

$$u_t(x,t) = u_{xx}(x,t) + xe^{-t}, \quad 0 < x < 1, \ t > 0,$$
$$u(0,t) = 0, \quad u(1,t) = 0, \quad t > 0,$$
$$u(x,0) = x - 1, \quad 0 < x < 1$$

has the same eigenvalues and eigenfunctions as that in Example 7.1. Here, $q(x,t) = xe^{-t}$ and $f(x) = x - 1$. Hence, by (7.3), (7.5), and (2.5), and using integration by parts, we find that

$$c_n'(t) + n^2\pi^2 c_n(t) = 2e^{-t}\int_0^1 x\sin(n\pi x)\,dx$$

$$= (-1)^{n+1}\frac{2}{n\pi}e^{-t}, \quad t > 0, \tag{7.7}$$

and

$$c_n(0) = 2\int_0^1 (x-1)\sin(n\pi x)\,dx = -\frac{2}{n\pi}. \tag{7.8}$$

The solution of the IVP (7.7), (7.8) is

$$c_n(t) = \frac{2}{n\pi}\left\{\frac{(-1)^{n+1}}{n^2\pi^2 - 1}e^{-t} - \left[1 + \frac{(-1)^{n+1}}{n^2\pi^2 - 1}\right]e^{-n^2\pi^2 t}\right\},$$

which means that, by (7.1), the solution of the given IBVP is

$$u(x,t) = \sum_{n=1}^{\infty}\frac{2}{n\pi}\left\{\frac{(-1)^{n+1}}{n^2\pi^2 - 1}e^{-t}\right.$$

$$\left. - \left[1 + \frac{(-1)^{n+1}}{n^2\pi^2 - 1}\right]e^{-n^2\pi^2 t}\right\}\sin(n\pi x). \tag{7.9}$$

VERIFICATION WITH MATHEMATICA®. As remarked in Chapter 5, it is impractical to use the infinite series (7.9) in the PDE, BCs, and ICs to show that the function u defined by its sum is the solution of our IBVP. Instead, we cut off the series after, say, 7 terms and show —graphically—that this truncation is a good approximation of the exact solution. First, we write the PDE in the operator form $Lu = q$; then the input

```
un = (2/(n*Pi)) * (((-1)^(n+1)/(n^2*Pi^2-1)) * E^(-t)
    - (1+(-1)^(n+1)/(n^2*Pi^2-1)) * E^(-n^2*Pi^2*t))
    * Sin[n*Pi*x];
Approx=Sum[un,{n,1,7}];
EqApprox = D[Approx,t] - D[Approx,x,x];
p1 = Plot3D[EqApprox, {t,0,10}, {x,0,1}, PlotRange -> {0,1},
    Ticks -> {{10},{1},{1}}];
p2 = Plot3D[x*E^(-t), {t,0,10}, {x,0,1}, PlotRange -> {0,1},
    Ticks -> {{10},{1},{1}}];
GraphicsRow[{p1, p2}]
```

generates the 3D graphs (for $0 < x < 1$ and $0 < t < 10$) of the function obtained by applying the partial differential operator L to the truncation, and of the nonhomogeneous term $q(x,t) = xe^{-t}$:

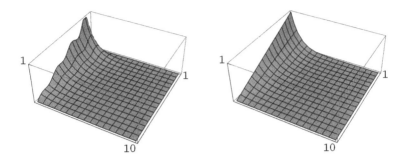

As can be seen, the two graphs show good agreement, the observable discrepancy between them occurring near $t = 0$. Next, the input

```
{Approx /. x -> 0, Approx /. x -> 1}
```

generates the output $\{0, 0\}$, confirming that the truncation satisfies the BCs. Finally, the input

```
Plot [{Approx /. t -> 0, x - 1}, {x,0,1}, PlotRange -> {{-0.18,1.1},
    {-1.1,0.1}}, PlotStyle -> Black, ImageSize -> Scaled[0.2],
    Ticks -> {{1},{-1}}, AspectRatio -> 0.9]
```

outputs the graphs of the chosen truncation at $t = 0$ and of the IC function $f(x) = x - 1$:

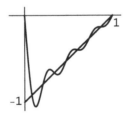

These show good agreement away from the point $x = 0$, where the infinite series (7.9) with $t = 0$ converges to $0 \neq f(0) = -1$.

Exercises

Use the method of eigenfunction expansion to find the solution of the IBVP consisting of the PDE and BCs

$$u_t(x,t) = u_{xx}(x,t) + q(x,t), \quad 0 < x < 1, \ t > 0,$$
$$u(0,t) = 0, \quad u(1,t) = 0, \quad t > 0$$

and IC and function q as indicated.

1 $u(x,0) = \sin(2\pi x) - 5\sin(4\pi x)$, $\quad q(x,t) = 2t\sin(2\pi x)$.

2 $u(x,0) = -2\sin(\pi x)$,

$q(x,t) = \left[3 + \pi^2(3t-2)\right]\sin(\pi x) + (9\pi^2 t^2 + 2t)\sin(3\pi x)$.

3 $u(x,0) = \sin(\pi x) + 2\sin(3\pi x)$, $\quad q(x,t) = e^{-t}\sin(3\pi x) - \sin(5\pi x)$.

4 $u(x,0) = \sin(2\pi x) + 3\sin(4\pi x)$,

$q(x,t) = \left[(\pi^2 - 1)e^{-t} - \pi^2\right]\sin(\pi x) + (4\pi^2 t + 4\pi^2 + 1)\sin(2\pi x)$

$$+ 48\pi^2\sin(4\pi x).$$

5 $u(x,0) = \sin(\pi x) + 2\sin(2\pi x)$, $\quad q(x,t) = (t-1)\sin(\pi x)$.

6 $u(x,0) = 3\sin(\pi x) - 2\sin(2\pi x)$,

$q(x,t) = \pi^2 e^{-4\pi^2 t}\sin(2\pi x) + (2t+3)\sin(3\pi x)$.

7 $u(x,0) = \begin{cases} 1, & 0 < x \le 1/2, \\ 0, & 1/2 < x < 1, \end{cases}$ $\quad q(x,t) = xt$.

8 $u(x,0) = x$, $\quad q(x,t) = (x-1)t/2$.

Answers to Odd-Numbered Exercises

1 $u(x,t) = \left[((8\pi^4 + 1)/(8\pi^4))e^{-4\pi^2 t} + (1/(2\pi^2))t - 1/(8\pi^4)\right]\sin(2\pi x)$

$$- 5e^{-16\pi^2 t}\sin(4\pi x).$$

3 $u(x,t) = e^{-\pi^2 t}\sin(\pi x)$

$$+ \left[((18\pi^2 - 3)/(9\pi^2 - 1))e^{-9\pi^2 t} + (1/(9\pi^2 - 1))e^{-t}\right]\sin(3\pi x)$$

$$+ (1/(25\pi^2))(e^{-25\pi^2 t} - 1)\sin(5\pi x).$$

5 $u(x,t) = \left[(1/\pi^2)(t-1) - 1/\pi^4 + (1 + 1/\pi^2 + 1/\pi^4)e^{-\pi^2 t}\right]\sin(\pi x)$

$$+ 2e^{-4\pi^2 t}\sin(2\pi x).$$

7 $u(x,t) = \sum_{n=1}^{\infty} (2/(n^5\pi^5))\{[(-1)^{n+1} + n^4\pi^4(1 - \cos(n\pi/2))]e^{-n^2\pi^2 t}$

$$+ (-1)^n(1 - n^2\pi^2 t)\}\sin(n\pi x).$$

7.1.2 Rod with Insulated Endpoints

Since the general discussion here is similar to that in Subsection 7.1.1, we proceed directly to an illustration.

7.3 Example. The eigenvalues and eigenfunctions associated with the IBVP

$$u_t(x,t) = u_{xx}(x,t) + 2t + \cos(2\pi x), \quad 0 < x < 1, \ t > 0,$$
$$u_x(0,t) = 0, \quad u_x(1,t) = 0, \quad t > 0,$$
$$u(x,0) = \frac{1}{2\pi^2} \cos(2\pi x), \quad 0 < x < 1$$

are (see Subsection 5.1.2)

$$\lambda_n = n^2\pi^2, \quad X_n(x) = \cos(n\pi x), \quad n = 0,1,2,\ldots$$

(for convenience, this time we have taken $X_0(x) = 1$ instead of $1/2$ in (5.11)). Reasoning just as in the case of the sine eigenfunctions, we deduce that (7.1)–(7.5) still hold except that now $n = 0,1,2,\ldots$ and the X_n are cosines. Since

$$q(x,t) = 2t + \cos(2\pi x), \quad f(x) = \frac{1}{2\pi^2} \cos(2\pi x)$$

are linear combinations of the eigenfunctions, (7.3) (with $k = 1$) and (7.5) lead to the IVPs

$$c'_n(t) + n^2\pi^2 c_n(t) = \begin{cases} 2t, & n = 0, \\ 1, & n = 2, \\ 0, & n \neq 0,2, \end{cases} \quad t > 0,$$

$$c_n(0) = \begin{cases} \dfrac{1}{2\pi^2}, & n = 2, \\ 0, & n \neq 2. \end{cases}$$

Thus, for $n = 0$ we have

$$c'_0(t) = 2t, \quad t > 0,$$
$$c_0(0) = 0,$$

with solution $c_0(t) = t^2$; for $n = 2$,

$$c'_2(t) + 4\pi^2 c_2(t) = 1, \quad t > 0,$$
$$c_2(0) = \frac{1}{2\pi^2},$$

with solution $c_2(t) = (4\pi^2)^{-1}(1 + e^{-4\pi^2 t})$; and for $n \neq 0,2$,

$$c'_n(t) + n^2\pi^2 c_n(t) = 0, \quad t > 0,$$
$$c_n(0) = 0,$$

with solution $c_n(t) = 0$. Hence, the solution of the given IBVP is

$$u(x,t) = c_0(t)X_0(x) + c_2(t)X_2(x) = t^2 + \frac{1}{4\pi^2}(1 + e^{-4\pi^2 t})\cos(2\pi x).$$

VERIFICATION WITH MATHEMATICA®. The input

```
u = t^2 + (1/(4*Pi^2)) * (1 + E^(-4*Pi^2*t)) * Cos[2*Pi*x];
{q, f} = 2*t + Cos[2*Pi*x], (1/(2*Pi^2)) * Cos[2*Pi*x];
Simplify[{D[u,t] - D[u,x,x] - q, D[u,x] /. x -> 0, D[u,x] /. x -> 1,
   (u /. t -> 0) - f}]
```

generates the output $\{0, 0, 0, 0\}$.

Exercises

Use the method of eigenfunction expansion to find the solution of the IBVP consisting of the PDE and BCs

$$u_t(x, t) = u_{xx}(x, t) + q(x, t), \quad 0 < x < 1, \ t > 0,$$
$$u_x(0, t) = 0, \quad u_x(1, t) = 0, \quad t > 0$$

and IC and function q as indicated.

1 $u(x, 0) = 2\cos(\pi x) - \cos(2\pi x), \quad q(x, t) = 2 + \cos(2\pi x).$

2 $u(x, 0) = 4\cos(2\pi x) + 2\cos(3\pi x),$
   $q(x, t) = 2 + 16\pi^2 \cos(2\pi x) + \left[(9\pi^2 - 1)e^{-t} + 9\pi^2\right]\cos(3\pi x).$

3 $u(x, 0) = 1 + 3\cos(4\pi x), \quad q(x, t) = t - t\cos(\pi x).$

4 $u(x, 0) = 2\cos(2\pi x), \quad q(x, t) = \pi^2 e^{-\pi^2 t}\cos(\pi x) + (1 - 2t)\cos(2\pi x).$

5 $u(x, 0) = 2 - \cos(3\pi x), \quad q(x, t) = e^{-t}\cos(\pi x).$

6 $u(x, 0) = \begin{cases} -1, & 0 < x \le 1/2, \\ 1, & 1/2 < x < 1, \end{cases} \quad q(x, t) = 3t - 4.$

7 $u(x, 0) = 1 - 3\cos(2\pi x), \quad q(x, t) = -2xt.$

8 $u(x, 0) = x, \quad q(x, t) = (1 - x)t.$

Answers to Odd-Numbered Exercises

1 $u(x, t) = 2t + 2e^{-\pi^2 t}\cos(\pi x) + [1/(4\pi^2) - ((4\pi^2 + 1)/(4\pi^2))e^{-4\pi^2 t}]$
   $$\times \cos(2\pi x).$$

3 $u(x, t) = (1/2)t^2 + 1 + [1/\pi^4 - (1/\pi^2)t - (1/\pi^4)e^{-\pi^2 t}]\cos(\pi x)$
   $$+ 3e^{-16\pi^2 t}\cos(4\pi x).$$

5 $u(x, t) = 2 + (1/(\pi^2 - 1))(e^{-t} - e^{-\pi^2 t})\cos(\pi x) - e^{-9\pi^2 t}\cos(3\pi x).$

7 $u(x,t) = 1 - t^2/2 - 3e^{-4\pi^2 t}\cos(2\pi x)$

$$+ \sum_{n=1}^{\infty} [1 - (-1)^n](4/(n^6 \pi^6))(e^{-n^2\pi^2 t} - 1 + n^2\pi^2 t)\cos(n\pi x).$$

7.1.3 Rod with Mixed Boundary Conditions

The method of eigenfunction expansion works equally well for IBVPs with other combinations of BCs.

7.4 Example. The eigenvalues and eigenfunctions for the IBVP

$$u_t(x,t) = u_{xx}(x,t) - \tfrac{1}{4}(16 + 81\pi^4)\sin t \cos\frac{3\pi x}{2}, \quad 0 < x < 1,\ t > 0,$$

$$u_x(0,t) = 0, \quad u(1,t) = 0, \quad t > 0,$$

$$u(x,0) = 4\cos\frac{3\pi x}{2}, \quad 0 < x < 1$$

are (see Section 5.1)

$$\lambda_n = \frac{(2n-1)^2\pi^2}{4}, \quad X_n(x) = \cos\frac{(2n-1)\pi x}{2}, \quad n = 1, 2, \ldots.$$

Following the general scheme (with $k = 1$ and $L = 1$), we deduce that

$$c_n'(t) + \frac{(2n-1)^2\pi^2}{4}c_n(t) = \begin{cases} -\tfrac{1}{4}(16 + 81\pi^4)\sin t, & n = 2, \\ 0, & n \neq 2, \end{cases} \quad t > 0,$$

$$c_n(0) = \begin{cases} 4, & n = 2, \\ 0, & n \neq 2. \end{cases}$$

Thus, for $n = 2$ we have

$$c_2'(t) + \frac{9\pi^2}{4}c_2(t) = -\tfrac{1}{4}(16 + 81\pi^4)\sin t, \quad t > 0,$$

$$c_2(0) = 4,$$

with solution $c_2(t) = 4\cos t - 9\pi^2\sin t$, and for $n \neq 2$,

$$c_n(t) + \frac{(2n-1)^2\pi^2}{4}c_n(t) = 0, \quad t > 0,$$

$$c_n(0) = 0,$$

with solution $c_n(t) = 0$. Hence, by (7.1), the solution of the given IBVP is

$$u(x,t) = (4\cos t - 9\pi^2\sin t)\cos\frac{3\pi x}{2}.$$

VERIFICATION WITH MATHEMATICA®. The input

```
u = (4 * Cos[t] - 9 * Pi^2 * Sin[t]) * Cos[3 * Pi * x/2];
{q, f} = { - (1/4) * (16 + 81 * Pi^4) * Sin[t] * Cos[3 * Pi * x/2],
   4 * Cos[3 * Pi * x/2]};
Simplify[{D[u,t] - D[u,x,x] - q, D[u,x] /. x -> 0, u /. x -> 1,
   (u /. t -> 0) - f}]
```

generates the output $\{0, 0, 0, 0\}$.

Exercises

Use the method of eigenfunction expansion to find the solution of the IBVP consisting of the PDE

$$u_t(x, t) = u_{xx}(x, t) + q(x, t), \quad 0 < x < 1, \ t > 0,$$

BCs, IC, and function q as indicated.

1 $u_x(0, t) = 0, \quad u(1, t) = 0, \quad u(x, 0) = \cos(\pi x/2) + 2\cos(5\pi x/2),$

$q(x, t) = t\cos(\pi x/2).$

2 $u_x(0, t) = 0, \quad u(1, t) = 0, \quad u(x, 0) = -\cos(\pi x/2),$

$q(x, t) = (1/4)(2\pi^2 t + 8 - \pi^2)\cos(\pi x/2) + (1/4)(9\pi^2 t^2 + 8t)\cos(3\pi x/2).$

3 $u(0, t) = 0, \quad u_x(1, t) = 0, \quad u(x, 0) = \sin(3\pi x/2),$

$q(x, t) = \sin(3\pi x/2) - 2\sin(5\pi x/2).$

4 $u(0, t) = 0, \quad u_x(1, t) = 0, \quad u(x, 0) = 2\sin(5\pi x/2),$

$q(x, t) = (1/4)[9\pi^2 t^2 + 2(4 - 9\pi^2)t - 8]\sin(3\pi x/2)$

$$+ (1/2)(25\pi^2 - 4)e^{-t}\sin(5\pi x/2).$$

Answers to Odd-Numbered Exercises

1 $u(x, t) = [(1 + 16/\pi^4)e^{-\pi^2 t/4} + 4t/\pi^2 - 16/\pi^4]\cos(\pi x/2)$

$$+ 2e^{-25\pi^2 t/4}\cos(5\pi x/2).$$

3 $u(x, t) = [((9\pi^2 - 4)/(9\pi^2))e^{-9\pi^2 t/4} + 4/(9\pi^2)]\sin(3\pi x/2)$

$$+ (8/(25\pi^2))(e^{-25\pi^2 t/4} - 1)\sin(5\pi x/2).$$

7.2 The Nonhomogeneous Wave Equation

As an illustration, we examine three different sets of BCs.

7.2.1 Vibrating String with Fixed Endpoints

Consider the IBVP

$$u_{tt}(x,t) = c^2 u_{xx}(x,t) + q(x,t), \quad 0 < x < L, \ t > 0, \qquad \text{(PDE)}$$
$$u(0,t) = 0, \quad u(L,t) = 0, \quad t > 0, \qquad \text{(BCs)}$$
$$u(x,0) = f(x), \quad u_t(x,0) = g(x), \quad 0 < x < L. \qquad \text{(ICs)}$$

The eigenvalues and eigenfunctions associated with this problem are (see Section 5.2)

$$\lambda_n = \left(\frac{n\pi}{L}\right)^2, \quad X_n(x) = \sin\frac{n\pi x}{L}, \quad n = 1, 2, \dots.$$

Starting with an expansion of the form (7.1) and following the same general procedure as for the heat equation in Section 7.1.1, this time also including the second IC, we find that, for $n = 1, 2, \dots,$

$$c_n''(t) + c^2 \lambda_n c_n(t) = \frac{\int_0^L q(x,t)X_n(x)\,dx}{\int_0^L X_n^2(x)\,dx}, \quad t > 0,$$

$$c_n(0) = \frac{\int_0^L f(x)X_n(x)\,dx}{\int_0^L X_n^2(x)\,dx}, \qquad (7.10)$$

$$c_n'(0) = \frac{\int_0^L g(x)X_n(x)\,dx}{\int_0^L X_n^2(x)\,dx}.$$

The solution of the IBVP is then given by (7.1), with the c_n determined from formulas (7.10), which in this case take the specific form

$$c_n''(t) + c^2\left(\frac{n\pi}{L}\right)^2 c_n(t) = \frac{2}{L}\int_0^L q(x,t)\sin\frac{n\pi x}{L}\,dx, \quad t > 0,$$

$$c_n(0) - \frac{2}{L}\int_0^L f(x)\sin\frac{n\pi x}{L}\,dx, \quad c_n'(0) = \frac{2}{L}\int_0^L g(x)\sin\frac{n\pi x}{L}\,dx.$$

7.5 Example. In the IBVP

$$u_{tt}(x,t) = u_{xx}(x,t) + (4 + 9\pi^2)e^{-2t}\sin(3\pi x), \quad 0 < x < 1, \ t > 0,$$

$$u(0,t) = 0, \quad u(1,t) = 0, \quad t > 0,$$

$$u(x,0) = 3\sin(3\pi x), \quad u_t(x,0) = -2\sin(3\pi x), \quad 0 < x < 1,$$

we have $L = 1$ and notice that q, f, and g are already expanded in the eigenfunctions—specifically, they are multiples of X_3. This leads to

$$c_n''(t) + n^2\pi^2 c_n(t) = \begin{cases} (4 + 9\pi^2)e^{-2t}, & n = 3, \\ 0, & n \neq 3, \end{cases} \quad t > 0,$$

$$c_n(0) = \begin{cases} 1, & n = 3, \\ 0, & n \neq 3, \end{cases} \quad c_n'(0) = \begin{cases} -2, & n = 3, \\ 0, & n \neq 3. \end{cases}$$

Hence, for $n = 3$,

$$c_3''(t) + 9\pi^2 c_3(t) = (4 + 9\pi^2)e^{-2t}, \quad t > 0,$$

$$c_3(0) = 3, \quad c_3'(0) = -2,$$

with solution $c_3(t) = 2\cos(3\pi t) + e^{-2t}$, and for $n \neq 3$,

$$c_n''(t) + n^2\pi^2 c_n(t) = 0, \quad t > 0,$$

$$c_n(0) = 0, \quad c_n'(0) = 0,$$

with solution $c_n(t) = 0$. Consequently, by (7.1),

$$u(x,t) = \left[2\cos(3\pi t) + e^{-2t}\right]\sin(3\pi x).$$

VERIFICATION WITH MATHEMATICA®. The input

```
u = (2 * Cos[3 * Pi * t] + E^(-2 * t)) * Sin[3 * Pi * x];
{q,f,g}={(4 + 9 * Pi^2) * E^(-2 * t) * Sin(3 * Pi * x),
  3 * Sin[3 * Pi * x], -2 * Sin[3 * Pi * x]};
Simplify[{D[u,t,t]-D[u,x,x] - q, {u /. x -> 0, u /. x -> 1},
  {(u /. t -> 0) - f, (D[u,t] /. t -> 0) - g}}
```

generates the output $\{0, \{0, 0\}, \{0, 0\}\}$, which shows that our computed solution u satisfies the PDE, the BCs, and the ICs.

7.6 Example. For the IBVP

$$u_{tt}(x,t) = u_{xx}(x,t) + \pi^2\sin(\pi x), \quad 0 < x < 1, \ t > 0,$$

$$u(0,t) = 0, \quad u(1,t) = 0, \quad t > 0,$$

$$u(x,0) = \pi, \quad u_t(x,0) = 2\pi\sin(2\pi x), \quad 0 < x < 1,$$

the above formulas with $c^2 = 1$, $L = 1$, and functions

$$q(x,t) = \pi^2 \sin(\pi x), \quad f(x) = \pi, \quad g(x) = 2\pi \sin(2\pi x)$$

yield

$$c_n''(t) + n^2\pi^2 c_n(t) = \begin{cases} \pi^2, & n = 1, \\ 0, & n \neq 1, \end{cases} \quad t > 0,$$

$$c_n(0) = 2 \int_0^1 \pi \sin(n\pi x)\, dx = \left[1 - (-1)^n\right]\frac{2}{n}, \quad n = 1, 2, \ldots,$$

$$c_n'(0) = \begin{cases} 2\pi, & n = 2, \\ 0, & n \neq 2. \end{cases}$$

Hence, for $n = 1$,

$$c_1''(t) + \pi^2 c_1(t) = \pi^2, \quad t > 0,$$
$$c_1(0) = 4, \quad c_1'(0) = 0,$$

with solution $c_1(t) = 3\cos(\pi t) + 1$; for $n = 2$,

$$c_2''(t) + 4\pi^2 c_2(t) = 0, \quad t > 0,$$
$$c_2(0) = 0, \quad c_2'(0) = 2\pi,$$

with solution $c_2(t) = \sin(2\pi t)$; and for $n \neq 1, 2$,

$$c_n''(t) + n^2\pi^2 c_n(t) = 0, \quad t > 0,$$
$$c_n(0) = \left[1 - (-1)^n\right]\frac{2}{n}, \quad c_n'(0) = 0,$$

with solution

$$c_n(t) = \left[1 - (-1)^n\right]\frac{2}{n}\cos(n\pi t).$$

Consequently, by (7.1),

$$u(x,t) = \left[3\cos(\pi t) + 1\right]\sin(\pi x) + \sin(2\pi t)\sin(2\pi x)$$

$$+ \sum_{n=3}^{\infty} \left[1 - (-1)^n\right]\frac{2}{n}\cos(n\pi t)\sin(n\pi x). \qquad (7.11)$$

VERIFICATION WITH MATHEMATICA®. The input

```
un = (1 - ( - 1)^n) * (2/n) * Cos [n * Pi * t] * Sin [n * Pi * x] ;
Approx = (3 * Cos [Pi * t] + 1) * Sin [Pi * x] + Sin [2 * Pi * t]
   * Sin [2 * Pi * x] + Sum [un, {n,3,7}] ;
```

```
{D[Approx,t,t] -D[Approx,x,x] -Pi^2*Sin[Pi*x]//Simplify,
  {Approx/. x->0, Approx/. x->1}, (D[Approx,t]/. t->0)
  -2*Pi*Sin[2*Pi*x]}
Plot[{Approx/. t->0, Pi}, {x,0,1}, PlotRange->{{-0.2,1.1},
  {-0.7,3.8}}, PlotStyle->Black, ImageSize->Scaled[0.2],
  Ticks->{{1},{Pi}}, AspectRatio->[0.7]
```

generates the numerical output $\{0, \{0, 0\}, 0\}$, which shows that the truncation after 7 terms of u given by (7.11) satisfies the PDE, the BCs, and the second IC exactly, and the graph

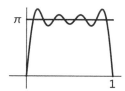

which indicates that, at $t = 0$, this truncation is also a good approximation of the first IC function.

Exercises

Use the method of eigenfunction expansion to find the solution of the IBVP consisting of the PDE and BCs

$$u_{tt}(x, t) = u_{xx}(x, t) + q(x, t), \quad 0 < x < 1, \ t > 0,$$
$$u(0, t) = 0, \quad u(1, t) = 0, \quad t > 0,$$

with ICs and function q as indicated.

1 $u(x, 0) = \sin(\pi x), \quad u_t(x, 0) = -3\sin(2\pi x), \quad q(x, t) = 2\sin(2\pi x).$

2 $u(x, 0) = 3\sin(\pi x) + \sin(3\pi x), \quad u_t(x, 0) = \sin(3\pi x),$

$q(x, t) = (3\pi^2 - 4 - 2\pi^2 t^2)\sin(\pi x) + 9\pi^2(1 + t)\sin(3\pi x).$

3 $u(x, 0) = \sin(\pi x), \quad u_t(x, 0) = 2\sin(\pi x) + 4\sin(3\pi x), \quad q(x, t) = t\sin(\pi x).$

4 $u(x, 0) = 0, \quad u_t(x, 0) = 2\sin(2\pi x) + 2\sin(4\pi x),$

$q(x, t) = 4\left[\pi^2 - (\pi^2 + 1)e^{-2t}\right]\sin(2\pi x) + 2(16\pi^2 - 1)\sin t \sin(4\pi x).$

5 $u(x, 0) = 0, \quad u_t(x, 0) = 2x - 1, \quad q(x, t) = (t + 1)\sin(2\pi x).$

6 $u(x, 0) = \begin{cases} 0, & 0 < x \le 1/2, \\ 2, & 1/2 < x < 1, \end{cases} \quad u_t(x, 0) = -1, \quad q(x, t) = 2xt.$

Answers to Odd-Numbered Exercises

1 $u(x,t) = \cos(\pi t) \sin(\pi x)$
$$+ [1/(2\pi^2) - (1/(2\pi^2)) \cos(2\pi t) - (3/(2\pi)) \sin(2\pi t)] \sin(2\pi x).$$

3 $u(x,t) = [\cos(\pi t) + ((2\pi^2 - 1)/\pi^3) \sin(\pi t) + (1/\pi^2)t] \sin(\pi x)$
$$+ (4/(3\pi)) \sin(3\pi t) \sin(3\pi x).$$

5 $u(x,t) = (1/(8\pi^3))[2\pi(1 + t) - 2\pi \cos(2\pi t) - \sin(2\pi t)] \sin(2\pi x)$
$$+ \sum_{n=1}^{\infty} [(-1)^{n+1} - 1](2/(n^2\pi^2)) \sin(n\pi t) \sin(n\pi x).$$

7.2.2 Vibrating String with Free Endpoints

The solution procedure here is similar to that in Subsection 7.2.1, with the obvious changes.

7.7 Example. The IBVP

$$u_{tt}(x,t) = u_{xx}(x,t) + 1 + t\cos(\pi x), \quad 0 < x < 1, \ t > 0,$$
$$u_x(0,t) = 0, \quad u_x(1,t) = 0, \quad t > 0,$$
$$u(x,0) = 2, \quad u_t(x,0) = -2\cos(2\pi x), \quad 0 < x < 1$$

gives rise to the eigenvalues and eigenfunctions (see Subsection 5.2.2)

$$\lambda_n = n^2\pi^2, \quad X_n(x) = \cos(n\pi x), \quad n = 0, 1, 2, \ldots,$$

where, once again, we have chosen $X_0(x) = 1$ instead of $1/2$. Taking $c^2 = 1$ and $L = 1$ and seeing that

$$q(x,t) = 1 + t\cos(\pi x), \quad f(x) = 2, \quad g(x) = -2\cos(2\pi x)$$

coincide with their eigenfunction expansions, we find that equations (7.10) lead to the IVPs

$$c_n''(t) + n^2\pi^2 c_n(t) = \begin{cases} 1, & n = 0, \\ t, & n = 1, \\ 0, & n \neq 0, 1, \end{cases} \quad t > 0,$$

$$c_n(0) = \begin{cases} 2, & n = 0, \\ 0, & n \neq 0, \end{cases} \quad c_n'(0) = \begin{cases} -2, & n = 2, \\ 0, & n \neq 2. \end{cases}$$

So for $n = 0$ we have

$$c_0''(t) = 1, \quad t > 0,$$
$$c_0(0) = 2, \quad c_0'(0) = 0,$$

with solution $c_0(t) = t^2/2 + 2$; for $n = 1$,

$$c_1''(t) + \pi^2 c_1(t) = t, \quad t > 0,$$
$$c_1(0) = 0, \quad c_1'(0) = 0,$$

with solution $c_1(t) = -\pi^{-3} \sin(\pi t) + \pi^{-2} t$; for $n = 2$,

$$c_2''(t) + 4\pi^2 c_2(t) = 0, \quad t > 0,$$
$$c_2(0) = 0, \quad c_2'(0) = -2,$$

with solution $c_2(t) = -\pi^{-1} \sin(2\pi t)$; and for $n \neq 0, 1, 2$,

$$c_n''(t) + n^2 \pi^2 c_n(t) = 0, \quad t > 0,$$
$$c_n(0) = 0, \quad c_n'(0) = 0,$$

with solution $c_n(t) = 0$. Consequently, by (7.1), the solution of the IBVP is

$$u(x,t) = \tfrac{1}{2} t^2 + 2 + \left[\frac{1}{\pi^2} t - \frac{1}{\pi^3} \sin(\pi t) \right] \cos(\pi x) - \frac{1}{\pi} \sin(2\pi t) \cos(2\pi x).$$

VERIFICATION WITH MATHEMATICA®. The input

```
u = (1/2) * t^2 + 2 + ((1/(Pi^2)) * t - (1/(Pi^3)) * Sin[Pi * t])
    * Cos[Pi * x] - (1/Pi) * Sin[2 * Pi * t] * Cos[2 * Pi * x];
{q, f, g} = {1 + t * Cos[Pi * x], 2, -2 * Cos[2 * Pi * x]};
Simplify[{D[u,t,t] - D[u,x,x] - q, D[u,x] /. x -> 0, D[u,x] /. x -> 1,
    (u /. t -> 0) - f, (D[u,t] /. t -> 0) - g}]
```

generates the output $\{0, 0, 0, 0, 0\}$.

Exercises

Use the method of eigenfunction expansion to find the solution of the IBVP consisting of the PDE and BCs

$$u_{tt}(x,t) = u_{xx}(x,t) + q(x,t), \quad 0 < x < 1, \ t > 0,$$
$$u_x(0,t) = 0, \quad u_x(1,t) = 0, \quad t > 0,$$

with ICs and function q as indicated.

1 $u(x,0) = 1 + 2\cos(2\pi x), \quad u_t(x,0) = \cos(3\pi x), \quad q(x,t) = 3.$

2 $u(x,0) = 3\cos(\pi x), \quad u_t(x,0) = -\cos(\pi x) + 6\pi \cos(3\pi x),$
   $q(x,t) = \pi^2(3 - t)\cos(\pi x).$

3 $u(x,0) = 2 + \cos(\pi x)$, $u_t(x,0) = -3 - \cos(\pi x) + \cos(2\pi x)$,

   $q(x,t) = 2 + (1 + \pi^2)e^{-t}\cos(\pi x) + 4\pi^2 t \cos(2\pi x)$.

4 $u(x,0) = -\cos(3\pi x)$, $u_t(x,0) = 1 + \cos(2\pi x)$,

   $q(x,t) = (t-2)e^{-t} + 4\pi^2 t \cos(2\pi x)$.

5 $u(x,0) = x - 1$, $u_t(x,0) = 0$, $q(x,t) = -x$.

6 $u(x,0) = x$, $u_t(x,0) = \begin{cases} 0, & 0 < x \le 1/2, \\ 1, & 1/2 < x < 1, \end{cases}$ $q(x,t) = 2xt$.

Answers to Odd-Numbered Exercises

1 $u(x,t) = 3t^2/2 + 1 + 2\cos(2\pi t)\cos(2\pi x) + (1/(3\pi))\sin(3\pi t)\cos(3\pi x)$.

3 $u(x,t) = t^2 - 3t + 2 + e^{-t}\cos(\pi x) + t\cos(2\pi x)$.

5 $u(x,t) = -1/2 - t^2/4$

$\qquad + \sum_{n=1}^{\infty} [(-1)^n - 1](2/(n^4\pi^4))[(n^2\pi^2 + 1)\cos(n\pi t) - 1]\cos(n\pi x)$.

7.2.3 Vibrating String with Mixed Boundary Conditions

The method of solution for this type of problem follows the general pattern established in the preceding two subsections.

7.8 Example. The standard procedure in the case of the IBVP

$$u_{tt} = u_{xx} + 2\pi^2 \sin\frac{\pi x}{2} + (8 + 25\pi^2 t^2)\sin\frac{5\pi x}{2}, \quad 0 < x < 1, \ t > 0,$$

$$u(0,t) = 0, \quad u_x(1,t) = 0, \quad t > 0,$$

$$u(x,0) = 8\sin\frac{\pi x}{2}, \quad u_t(x,0) = 0, \quad 0 < x < 1$$

starts with the associated eigenvalues and eigenfunctions (see Appendix A.2)

$$\lambda_n = \frac{(2n-1)^2\pi^2}{4}, \quad X_n(x) = \sin\frac{(2n-1)\pi x}{2}, \quad n = 1, 2, \ldots.$$

Once again, we notice that the nonhomogeneous terms are linear combinations of the X_n, which leads us to the IVPs

$$c_n''(t) + \frac{(2n-1)^2\pi^2}{4} c_n(t) = \begin{cases} 2\pi^2, & n = 1, \\ 8 + 25\pi^2 t^2, & n = 3, \\ 0, & n \ne 1, 3, \end{cases} \quad t > 0,$$

$$c_n(0) = \begin{cases} 8, & n = 1, \\ 0, & n \ne 1, \end{cases} \quad c_n'(0) = 0, \quad n = 1, 2, \ldots.$$

For $n = 1$, we have

$$c_1'' + \frac{\pi^2}{4} c_1 = 2\pi^2, \quad t > 0,$$

$$c_1(0) = 8, \quad c_1'(0) = 0,$$

so $c_1(t) = 8$. For $n = 3$,

$$c_3'' + \frac{25\pi^2}{4} c_3 = 8 + 25\pi^2 t^2, \quad t > 0,$$

$$c_3(0) = 0, \quad c_3'(0) = 0,$$

with solution $c_3(t) = 4t^2$. Finally, for $n \neq 1, 3$,

$$c_n'' + \frac{(2n-1)^2 \pi^2}{4} c_n = 0, \quad t > 0,$$

$$c_n(0) = 0, \quad c_n'(0) = 0,$$

from which $c_n(t) = 0$. Thus,

$$u(x, t) = 8 \sin \frac{\pi x}{2} + 4t^2 \sin \frac{5\pi x}{2}.$$

VERIFICATION WITH MATHEMATICA®. The input

```
u = 8 * Sin[Pi * x /2] + 4 * t^2 * Sin[5 * Pi * x /2];
{q,f,g} = {2 * Pi^2 * Sin[Pi * x /2] +
  (8 + 25 * Pi^2 * t^2) * Sin[5 * Pi * x /2], 8 * Sin[Pi * x /2], 0};
Simplify[{D[u,t,t] - D[u,x,x] - q, {u/.x->0, D[u,x] /.x->1},
  {(u/.t->0) - f, (D[u,t] /.t->0) - g}}]
```

generates the output $\{0, \{0, 0\}, \{0, 0\}\}$.

Exercises

Use the method of eigenfunction expansion to find the solution of the IBVP consisting of the PDE and ICs

$$u_{tt}(x, t) = u_{xx}(x, t) + q(x, t), \quad 0 < x < 1, \ t > 0,$$

$$u(x, 0) = f(x), \quad u_t(x, 0) = g(x), \quad 0 < x < 1,$$

with the BCs and functions f, g, and q as indicated.

1 $u(0, t) = 0, \quad u_x(1, t) = 0, \quad f(x) = \sin(\pi x/2) + \sin(3\pi x/2),$
   $g(x) = 2\pi \sin(\pi x/2), \quad q(x, t) = \pi^2 \sin(\pi x/2) + 9\pi^2 \sin(3\pi x/2).$

2 $u(0,t) = 0$, $u_x(1,t) = 0$, $f(x) = 4\sin(\pi x/2) + (8/25)\sin(5\pi x/2)$,

$g(x) = 0$, $q(x,t) = (\pi^2 - 16)\cos(2t)\sin(\pi x/2) + 2\pi^2\sin(5\pi x/2)$.

3 $u(0,t) = 0$, $u_x(1,t) = 0$, $f(x) = 2\sin(\pi x/2)$, $g(x) = x$,

$q(x,t) = \pi^3 t\sin(\pi x/2)$.

4 $u_x(0,t) = 0$, $u(1,t) = 0$, $f(x) = 3\cos(\pi x/2)$, $g(x) = -\pi\cos(3\pi x/2)$,

$q(x,t) = \pi^2\cos(\pi x/2) + 3\pi^3 t\cos(3\pi x/2)$.

5 $u_x(0,t) = 0$, $u(1,t) = 0$, $f(x) = 4\cos(\pi x/2) - 2\cos(5\pi x/2)$,

$g(x) = 4\cos(\pi x/2)$, $q(x,t) = (16 + \pi^2)e^{2t}\cos(\pi x/2) - 25\pi^2\cos(5\pi x/2)$.

6 $u_x(0,t) = 0$, $u(1,t) = 0$, $f(x) = x - 1$, $g(x) = \cos(3\pi x/2)$,

$q(x,t) = (9\pi^2/4)(t+1)\cos(3\pi x/2)$.

Answers to Odd-Numbered Exercises

1 $u(x,t) = \big[4 - 3\cos(\pi t/2) + 4\sin(\pi t/2)\big]\sin(\pi x/2)$
$$+ \big[4 - 3\cos(3\pi t/2)\big]\sin(3\pi x/2).$$

3 $u(x,t) = \big[4\pi t + 2\cos(\pi t/2) - 8\sin(\pi t/2)\big]\sin(\pi x/2)$
$$+ \sum_{n=1}^{\infty}\big[16(-1)^{n+1}/((2n-1)^3\pi^3)\big]\sin\big((2n-1)\pi t/2\big)$$
$$\times \sin\big((2n-1)\pi x/2\big).$$

5 $u(x,t) = 4e^{2t}\cos(\pi x/2) - 4\cos(5\pi x/2)$.

7.3 The Nonhomogeneous Laplace Equation

We solve some typical boundary value problems for this type of equation in both Cartesian and polar coordinates.

7.3.1 Equilibrium Temperature in a Rectangle

Consider the BVP

$$u_{xx}(x,y) + u_{yy}(x,y) = q(x,y), \quad 0 < x < L,\ 0 < y < K,$$
$$u(0,y) = 0, \quad u(L,y) = 0, \quad 0 < y < K,$$
$$u(x,0) = f_1(x), \quad u(x,K) = f_2(x), \quad 0 < x < L,$$

where, for convenience, the source term has been shifted to the right-hand side of the PDE. The eigenvalues and eigenfunctions associated with this problem, found in Section 5.3, are

$$\lambda_n = \left(\frac{n\pi}{L}\right)^2, \quad X_n(x) = \sin\frac{n\pi x}{L}, \quad n = 1, 2, \ldots.$$

Seeking a solution of the form

$$u(x, y) = \sum_{n=1}^{\infty} c_n(y) \sin \frac{n\pi x}{L},$$

we deduce, as in Sections 7.1 and 7.2, that the coefficients c_n, $n = 1, 2, \ldots$, satisfy the boundary value problems

$$c_n''(y) - n^2\pi^2 c_n(y) = \frac{2}{L} \int_0^L q(x, y) \sin \frac{n\pi x}{L} \, dx, \quad 0 < y < K,$$

$$c_n(0) = \frac{2}{L} \int_0^L f_1(x) \sin \frac{n\pi x}{L} \, dx, \tag{7.12}$$

$$c_n(K) = \frac{2}{L} \int_0^L f_2(x) \sin \frac{n\pi x}{L} \, dx.$$

7.9 Example. In the BVP

$$u_{xx}(x, y) + u_{yy}(x, y) = \pi^2 \sin(\pi x), \quad 0 < x < 1, \ 0 < y < 2,$$
$$u(0, y) = 0, \quad u(1, y) = 0, \quad 0 < y < 2,$$
$$u(x, 0) = 2\sin(3\pi x), \quad u(x, 2) = -\sin(\pi x), \quad 0 < x < 1,$$

we have $L = 1$, $K = 2$, and

$$q(x, y) = \pi^2 \sin(\pi x), \quad f_1(x) = 2\sin(3\pi x), \quad f_2(x) = -\sin(\pi x).$$

Since q, f_1, and f_2 are finite linear combinations of the eigenfunctions, we can bypass (7.12) and see directly that c_n satisfies

$$c_n''(y) - n^2\pi^2 c_n(y) = \begin{cases} \pi^2, & n = 1, \\ 0, & n \neq 1, \end{cases} \quad 0 < y < 2,$$

$$c_n(0) = \begin{cases} 2, & n = 3, \\ 0, & n \neq 3, \end{cases}$$

$$c_n(2) = \begin{cases} -1, & n = 1, \\ 0, & n \neq 1. \end{cases}$$

Then for $n = 1$,

$$c_1''(y) - \pi^2 c_1(y) = \pi^2, \quad 0 < y < 2,$$
$$c_1(0) = 0, \quad c_1(2) = -1,$$

with solution $c_1(y) = -\operatorname{csch}(2\pi)\sinh\left(\pi(y-2)\right) - 1$; for $n = 3$,

$$c_3''(y) - 9\pi^2 c_3(y) = 0, \quad 0 < y < 2,$$
$$c_3(0) = 2, \quad c_3(2) = 0,$$

with solution $c_3(y) = -2\operatorname{csch}(6\pi)\sinh\left(3\pi(y-2)\right)$; and for $n \neq 1, 3$,

$$c_n''(y) - n^2\pi^2 c_n(y) = 0, \quad 0 < y < 2,$$
$$c_n(0) = 0, \quad c_n(2) = 0,$$

with solution $c_n(y) = 0$. Hence, the solution of the BVP is

$$u(x,y) = -\left[\operatorname{csch}(2\pi)\sinh\left(\pi(y-2)\right) + 1\right]\sin(\pi x)$$
$$- 2\operatorname{csch}(6\pi)\sinh\left(3\pi(y-2)\right)\sin(3\pi x).$$

VERIFICATION WITH MATHEMATICA®. The input

```
u = - (Csch[2*Pi] * Sinh[Pi * (y-2)] + 1) * Sin[Pi*x]
  - 2*Csch[6*Pi] * Sinh[3*Pi * (y-2)] * Sin[3*Pi*x];
{q, f, g} = {Pi^2*Sin[Pi*x], 2*Sin[3*Pi*x], -Sin[Pi*x]};
Simplify[{D[u,x,x] +D[u,y,y] -q, {u/. x->0, u/. x->1},
  {(u/. y->0) -f, (u/. y->2) -g}}]
```

generates the output $\{0, \{0,0\}, \{0,0\}\}$.

7.10 Example. A similar solution shortcut may be implemented for the BVP

$$u_{xx}(x,y) + u_{yy}(x,y) = \frac{\pi^2}{2}(1 - 2x)\cos\frac{\pi y}{2}, \quad 0 < x < 1, \ 0 < y < 2,$$

$$u_x(0,y) = 1 + \left(4 - \pi\cosh\frac{\pi}{2}\right)\cos\frac{\pi y}{2}, \quad 0 < y < 2,$$

$$u(1,y) = 2 + \left(2 + \cosh\frac{\pi}{2}\right)\cos\frac{\pi y}{2}, \quad 0 < y < 2,$$

$$u_y(x,0) = 0, \quad u_y(x,2) = 0, \quad 0 < x < 1,$$

where the associated eigenfunctions are $Y_n(y) = \cos(n\pi y/2)$, $n = 0, 1, 2, \ldots$, so the solution may be represented as a series of the form

$$u(x,y) = \sum_{n=0}^{\infty} c_n(x)\cos\frac{n\pi y}{2}.$$

As in Example 7.9, we notice that the nonhomogeneous term in the PDE and the functions prescribed at $x = 0$ and $x = 1$ are finite linear combinations of the eigenfunctions, which leads to the equalities

$$c_n''(x) - \frac{n^2\pi^2}{4} c_n(x) = \begin{cases} \frac{\pi^2}{2}(1 - 2x), & n = 1, \\ 0, & n \neq 1, \end{cases} \quad 0 < x < 1,$$

$$c_n'(0) = \begin{cases} 1, & n = 0, \\ 4 - \pi \cosh\frac{\pi}{2}, & n = 1, \\ 0, & n \neq 0, 1, \end{cases}$$

$$c_n(1) = \begin{cases} 2, & n = 0, \\ 2 + \cosh\frac{\pi}{2}, & n = 1, \\ 0, & n \neq 0, 1. \end{cases}$$

Consequently, the individual cases are

$$c_0''(x) = 0, \quad 0 < x < 1,$$
$$c_0'(0) = 1, \quad c_0(1) = 2,$$

$$c_1''(x) - \frac{\pi^2}{4} c_1(x) = \frac{\pi^2}{2}(1 - 2x), \quad 0 < x < 1,$$
$$c_1'(0) = 4 - \pi \cosh\frac{\pi}{2}, \quad c_1(1) = 2 + \cosh\frac{\pi}{2},$$

$$c_n''(x) - \frac{n^2\pi^2}{4} c_n(x) = 0, \quad 0 < x < 1, \ n \neq 0, 1,$$
$$c_n'(0) = 0, \quad c_n(1) = 0,$$

with solutions

$$c_0(x) = x + 1, \quad 0 < x < 1,$$
$$c_1(x) = \cosh\frac{\pi x}{2} - 2\sinh\frac{\pi(x - 1)}{2} + 2(2x - 1), \quad 0 < x < 1,$$
$$c_n(x) = 0, \quad 0 < x < 1, \ n \neq 0, 1,$$

respectively, which means that

$$u(x, y) = x + 1 + \left[\cosh\frac{\pi x}{2} - 2\sinh\frac{\pi(x - 1)}{2} + 2(2x - 1) \right] \cos\frac{\pi y}{2}.$$

VERIFICATION WITH MATHEMATICA®. The input

```
u = x + 1 + (Cosh[Pi*x/2] - 2*Sinh[Pi*(x-1)/2]
    + 2*(2*x-1))*Cos[Pi*y/2];
```

```
{q,f,g} = {(Pi^2/2) * (1-2*x) * Cos[Pi*y/2],
   1 + (4 - Pi * Cosh[Pi/2]) * Cos[Pi*y/2],
   2 + (2 + Cosh[Pi/2]) * Cos[Pi*y/2]};
Simplify[{D[u,x,x] + D[u,y,y] - q, {(D[u,x] /.x->0) - f,
   (u/.x->1)-g}, {D[u,y] /.y->0, D[u,y] /.y->2}}]
```

generates the output $\{0, \{0, 0\}, \{0, 0\}\}$.

7.11 Example. The same solution procedure carries over to the BVP

$$u_{xx}(x, y) + u_{yy}(x, y) = \frac{3\pi^2}{4} \sin \frac{\pi x}{2}, \quad 0 < x < 1, \ 0 < y < 2,$$

$$u(0, y) = 0, \quad u_x(1, y) = 0, \quad 0 < y < 2,$$

$$u(x, 0) = -3 \sin \frac{\pi x}{2} + 2 \sinh(3\pi) \sin \frac{3\pi x}{2}, \quad 0 < x < 1,$$

$$u(x, 2) = (\sinh \pi - 3) \sin \frac{\pi x}{2}, \quad 0 < x < 1,$$

with associated eigenfunctions $X_n(x) = \sin\left((2n - 1)\pi x/2\right)$, $n = 1, 2, \ldots$. This allows us to seek the solution as

$$u(x, y) = \sum_{n=1}^{\infty} c_n(y) \sin \frac{(2n - 1)\pi x}{2}.$$

Once again, we see that the data functions in the problem are finite linear combinations of the eigenfunctions, which, after appropriate manipulation, generate for the c_n the equations

$$c_n''(y) - \frac{(2n - 1)^2 \pi^2}{4} c_n(y) = \begin{cases} \dfrac{3\pi^2}{4}, & n = 1, \\ 0, & n \neq 1, \end{cases} \quad 0 < y < 2,$$

$$c_n(0) = \begin{cases} -3, & n = 1, \\ 2 \sinh(3\pi), & n = 2, \\ 0, & n \neq 1, 2, \end{cases}$$

$$c_n(2) = \begin{cases} \sinh \pi - 3, & n = 1, \\ 0, & n \neq 1. \end{cases}$$

The three distinctive cases here are

$$c_1''(y) - \frac{\pi^2}{4} c_1(y) = \frac{3\pi^2}{4}, \quad 0 < y < 2,$$

$$c_1(0) = -3, \quad c_1(2) = \sinh \pi - 3,$$

$$c_2''(y) - \frac{9\pi^2}{4} c_2(y) = 0, \quad 0 < y < 2,$$

$$c_2(0) = 2\sinh(3\pi), \quad c_2(2) = 0,$$

$$c_n''(y) - \frac{(2n-1)^2\pi^2}{4} c_n(y) = 0, \quad 0 < y < 2, \ n \neq 1, 2,$$

$$c_n(0) = 0, \quad c_n(2) = 0,$$

with solutions

$$c_1(y) = \sinh\frac{\pi y}{2} - 3, \quad 0 < y < 2,$$

$$c_2(y) = -2\sinh\frac{3\pi(y-2)}{2}, \quad 0 < y < 2,$$

$$c_n(y) = 0, \quad 0 < y < 2, \ n \neq 1, 2,$$

leading to

$$u(x, y) = \left(\sinh\frac{\pi y}{2} - 3\right)\sin\frac{\pi x}{2} - 2\sinh\frac{3\pi(y-2)}{2}\sin\frac{3\pi x}{2}.$$

VERIFICATION WITH MATHEMATICA®. The input

```
u= (Sinh[Pi*y/2]-3)*Sin[Pi*x/2]-2*Sinh[3*Pi(y-2)/2]*
    Sin[3*Pi*x/2];
{q,f,g}={(3*Pi^2/4)*Sin[Pi*x/2], -3*Sin[Pi*x/2]+
    2*Sinh[3*Pi]*Sin[3*Pi*x/2], (Sinh[Pi]-3)*Sin[Pi*x/2];
Simplify[{D[u,x,x]+D[u,y,y]-q, {u/.x->0, D[u,x]/.x->1},
{(u/.y->0)-f, (u/.y->2)-g}}]
```

generates the output $\{0, \{0, 0\}, \{0, 0\}\}$.

7.12 Example. The eigenfunctions associated with the BVP

$$u_{xx}(x, y) + u_{yy}(x, y) = (4 - 9\pi^2)e^{-x}\cos\frac{3\pi y}{2}, \quad 0 < x < 2, \ 0 < y < 1,$$

$$u(0, y) = \left[4 + 2\cosh(3\pi)\right]\cos\frac{3\pi y}{2}, \quad 0 < y < 1,$$

$$u_x(2, y) = \left[3\pi\cosh(3\pi) - 4e^{-2}\right]\cos\frac{3\pi y}{2}, \quad 0 < y < 1,$$

$$u_y(x, 0) = 0, \quad u(x, 1) = 0, \quad 0 < x < 2$$

are $Y_n(y) = \cos\left((2n-1)\pi y/2\right)$, $n = 1, 2, \ldots$. Consequently, we may seek the solution in the form

$$u(x, y) = \sum_{n=1}^{\infty} c_n(x)\cos\frac{(2n-1)\pi y}{2}.$$

Since the data functions are finite linear combinations of the Y_n, a simple computation yields for the c_n the BVPs

$$c_n''(x) - \frac{(2n-1)^2\pi^2}{4} c_n(x) = \begin{cases} (4-9\pi^2)e^{-x}, & n=2, \\ 0, & n \neq 2, \end{cases} \quad 0 < x < 2,$$

$$c_n(0) = \begin{cases} 4 + 2\cosh(3\pi), & n=2, \\ 0, & n \neq 2, \end{cases}$$

$$c_n'(2) = \begin{cases} 3\pi\cosh(3\pi) - 4e^{-2}, & n=2, \\ 0, & n \neq 2. \end{cases}$$

There are two distinct cases, namely

$$c_2''(x) - \frac{9\pi^2}{4} c_2(x) = (4-9\pi^2)e^{-x}, \quad 0 < x < 2,$$

$$c_2(0) = 4 + 2\cosh(3\pi), \quad c_2'(2) = 3\pi\cosh(3\pi) - 4e^{-2}$$

and

$$c_n''(x) - \frac{(2n-1)^2\pi^2}{4} c_n(x) = 0, \quad 0 < x < 2, \ n \neq 2,$$

$$c_n(0) = 0, \quad c_n'(2) = 0,$$

which produce, respectively, the solutions

$$c_2(x) = 2\left\{ \sinh\frac{3\pi x}{2} + \cosh\frac{3\pi(x-2)}{2} + 2e^{-x} \right\}, \quad 0 < x < 2,$$

$$c_n(x) = 0, \quad 0 < x < 2, \ n \neq 2.$$

Hence,

$$u(x,y) = 2\left\{ \sinh\frac{3\pi x}{2} + \cosh\frac{3\pi(x-2)}{2} + 2e^{-x} \right\} \cos\frac{3\pi y}{2}.$$

VERIFICATION WITH MATHEMATICA®. The input

```
u = 2 * (Sinh[3 * Pi * x /2] + Cosh[3 * Pi * (x - 2) /2] +
    2 * E^(-x)) * Cos[3 * Pi * y /2];
{q,f1,f2} = {(4 - 9 * Pi^2) * E^(-x) * Cos[3 * Pi * y /2],
    (4 + 2 * Cosh[3 * Pi]) * Cos[3 * Pi * y /2],
    (3 * Pi * Cosh[3 * Pi] - 4 * E^(-2)) * Cos[3 * Pi * y /2]};
Simplify[{D[u,x,x] + D[u,y,y] - q, {(u /.x->0) - f1,
    (D[u,x] /.x->2) - f2}, {D[u,y] /.y->0, u /.y->1}}]
```

generates the output $\{0, \{0,0\}, \{0,0\}\}$.

Exercises

Use the method of eigenfunction expansion to find the solution of the BVP consisting of the PDE

$$u_{xx}(x,y) + u_{yy}(x,y) = q(x,y), \quad 0 < x < 1, \ 0 < y < 2,$$

function q, and BCs as indicated.

1 $q(x,y) = \sin(2\pi x), \quad u(0,y) = 0, \quad u(1,y) = 0,$
$u(x,0) = \sin(\pi x) - 2\sin(3\pi x), \quad u(x,2) = -\sin(2\pi x).$

2 $q(x,y) = -\pi^2(2y+3)\sin(\pi x) + (1 - 4\pi^2)e^{-y}\sin(2\pi x),$
$u(0,y) = 0, \quad u(1,y) = 0,$
$u(x,0) = 3\sin(\pi x) + \sin(2\pi x), \quad u_y(x,2) = 2\sin(\pi x) - e^{-2}\sin(2\pi x).$

3 $q(x,y) = 2y\sin(\pi x), \quad u(0,y) = 0, \quad u(1,y) = 0,$
$u_y(x,0) = x - 1, \quad u_y(x,2) = 0.$

4 $q(x,y) = -\pi^2\sin(\pi y), \quad u(0,y) = 2\sin(\pi y), \quad u(1,y) = \sin(\pi y/2),$
$u(x,0) = 0, \quad u(x,2) = 0.$

5 $q(x,y) = \left[2(1-\pi^2)e^x + \pi^2\right]\sin(\pi y) - (9\pi^2/4)(x+3)\sin(3\pi y/2),$
$u(0,y) = \sin(\pi y) + 3\sin(3\pi y/2), \quad u_x(1,y) = 2e\sin(\pi y) + \sin(3\pi y/2),$
$u(x,0) = 0, \quad u(x,2) = 0.$

6 $q(x,y) = 3x\sin(\pi y/2), \quad u(0,y) = y, \quad u(1,y) = 0,$
$u(x,0) = 0, \quad u(x,2) = 0.$

7 $q(x,y) = -2 - 4\pi^2(2y-1)\cos(2\pi x), \quad u_x(0,y) = 0, \quad u_x(1,y) = 0,$
$u(x,0) = 1 - \cos(2\pi x), \quad u(x,2) = -3 + 3\cos(2\pi x).$

8 $q(x,y) = y^2\cos(2\pi x), \quad u_x(0,y) = 0, \quad u_x(1,y) = 0,$
$u(x,0) = 2\cos(\pi x), \quad u_y(x,2) = -\cos(2\pi x).$

9 $q(x,y) = 6y - 2 + (1 - 4\pi^2)e^y\cos(2\pi x), \quad u_x(0,y) = 0, \quad u_x(1,y) = 0,$
$u(x,0) = \cos(2\pi x), \quad u(x,2) = x.$

10 $q(x,y) = (1-x^2)\cos(\pi y), \quad u(0,y) = -\cos(\pi y), \quad u(1,y) = 2\cos(\pi y/2),$
$u_y(x,0) = 0, \quad u_y(x,2) = 0.$

11 $q(x,y) = 27\pi^3 x\cos(3\pi y/2),$
$u_x(0,y) = -12\pi, \quad u(1,y) = -17\pi\cos(3\pi y/2),$
$u_y(x,0) = 0, \quad u_y(x,2) = 0.$

12 $q(x, y) = 4y - 2$, $u(0, y) = -\cos(\pi y)$, $u(1, y) = 0$,

   $u_y(x, 0) = 0$, $u_y(x, 2) = 0$.

13 $q(x, y) = (\pi^2/4)\big[9\sin(3\pi x/2) + 25y\sin(5\pi x/2)\big]$,

   $u(0, y) = 0$, $u_x(1, y) = 0$,

   $u(x, 0) = -\sin(3\pi x/2) - \sinh(5\pi)\sin(5\pi x/2)$,

   $u(x, 2) = \big[3\sinh(3\pi) - 1\big]\sin(3\pi x/2) - 2\sin(5\pi x/2)$.

14 $q(x, y) = (4 - \pi^2/16)e^{2x}\sin(\pi y/4)$, $u(x, 0) = 0$, $u_y(x, 2) = 0$,

   $u_x(0, y) = \big[2 + (\pi/2)\cosh(\pi/4)\big]\sin(\pi y/4)$,

   $u(1, y) = \big[e^2 - \cosh(\pi/4)\big]\sin(\pi y/4) + 2\cosh(3\pi/4)\sin(3\pi y/4)$.

15 $q(x, y) = (\pi^2/16)(1 - 2x)\cos(\pi y/4)$, $u_y(x, 0) = 0$, $u(x, 2) = 0$,

   $u(0, y) = -\big[1 + \cosh(\pi/4)\big]\cos(\pi y/4) - 3\cosh(3\pi/4)\cos(3\pi y/4)$,

   $u_x(1, y) = \big[2 + (\pi/2)\cosh(\pi/4)\big]\cos(\pi y/4)$.

16 $q(x, y) = \big[(4 - \pi^2)e^{-y} + 2\pi^2\big]\cos(\pi x/2)$, $u_x(0, y) = 0$, $u(1, y) = 0$,

   $u(x, 0) = -4\cos(\pi x/2)$, $u(x, 2) = 4(e^{-2} - 2 + 2\sinh\pi)\cos(\pi x/2)$. `

Answers to Odd-Numbered Exercises

1 $u(x, y) = -\operatorname{cosech}(2\pi)\sinh\big(\pi(y - 2)\big)\sin(\pi x)$

   $+ \big(1/(4\pi^2)\big)\big[(1 - 4\pi^2)\operatorname{cosech}(4\pi)\sinh(2\pi y)$

   $- \operatorname{cosech}(4\pi)\sinh\big(2\pi(y - 2)\big) - 1\big]\sin(2\pi x)$

   $+ 2\operatorname{cosech}(6\pi)\sinh\big(3\pi(y - 2)\big)\sin(3\pi x)$.

3 $u(x, y) = (2/\pi^3)\big\{\operatorname{csch}(2\pi)\big[\cosh(\pi y) - \cosh(\pi(y - 2))\big] - \pi y\big\}\sin(\pi x)$

   $+ \sum_{n=1}^{\infty} \big(2/(n^2\pi^2)\big)\operatorname{csch}(2n\pi)\sin(n\pi x)\cosh\big(n\pi(y - 2)\big)$.

5 $u(x, y) = (2e^x - 1)\sin(\pi y) + (x + 3)\sin(3\pi y/2)$.

7 $u(x, y) = 1 - y^2 + (2y - 1)\cos(2\pi x)$.

9 $u(x, y) = y^3 - y^2 - 2y + \big[e^y - e^2\operatorname{csch}(4\pi)\sinh(2\pi y)\big]\cos(2\pi x)$

   $+ \tfrac{1}{4}y + \sum_{n=1}^{\infty} \big(2/(n^2\pi^2)\big)\big[(-1)^n - 1\big]\operatorname{csch}(2n\pi)$

   $\times \cos(n\pi x)\sinh(n\pi y)$.

11 $u(x, y) = 12\pi(1 - x) - \pi\big[5\operatorname{sech}(3\pi/2)\cosh(3\pi x/2) + 12x\big]\cos(3\pi y/2)$.

13 $u(x, y) = \left[3 \sinh(3\pi y/2) - 1\right] \sin(3\pi x/2)$
$$+ \left[\sinh\left(5\pi(y-2)/2\right) - y\right] \sin(5\pi x/2).$$

15 $u(x, y) = \left[2 \sinh(\pi x/4) - \cosh\left(\pi(x-1)/4\right) + 2x - 1\right] \cos(\pi y/4)$
$$- 3 \cosh\left(3\pi(x-1)/4\right) \cos(3\pi y/4).$$

7.3.2 Equilibrium Temperature in a Circular Disc

The same method can be applied to the nonhomogeneous Laplace equation in polar coordinates.

7.13 Example. In Subsection 5.3.2, we established that the eigenvalues and eigenfunctions associated with a BVP such as

$$u_{rr}(r, \theta) + r^{-1} u_r(r, \theta) + r^{-2} u_{\theta\theta}(r, t) = 4,$$
$$0 < r < 1, \quad -\pi < \theta < \pi,$$
$$u(1, \theta) = 2 \cos \theta - \sin(2\theta), \quad -\pi < \theta < \pi$$

are

$$\lambda_0 = 0, \quad \Theta_0(\theta) = 1,$$
$$\lambda_n = n^2, \quad \Theta_{1n}(\theta) = \cos(n\theta), \quad \Theta_{2n}(\theta) = \sin(n\theta), \quad n = 1, 2, \ldots,$$

where, as in previous cases, we have taken the constant eigenfunction Θ_0 to be 1 instead of 1/2. Seeking a solution of the form

$$u(r, \theta) = c_0(r) + \sum_{n=1}^{\infty} \left[c_{1n}(r)\Theta_{1n}(\theta) + c_{2n}(r)\Theta_{2n}(\theta)\right]$$

and reasoning exactly as in Example 7.9, we deduce that the coefficients c_0, c_{1n}, and c_{2n}, $n = 1, 2, \ldots$, satisfy, respectively,

$$c_0''(r) + r^{-1}c_0'(r) = 4, \quad 0 < r < 1,$$
$$c_0(1) = 0,$$

$$c_{1n}''(r) + r^{-1}c_{1n}'(r) - n^2 r^{-2} c_{1n}(r) = 0, \quad 0 < r < 1,$$
$$c_{1n}(1) = \begin{cases} 2, & n = 1, \\ 0, & n \neq 1, \end{cases}$$

$$c_{2n}''(r) + r^{-1}c_{2n}'(r) - n^2 r^{-2} c_{2n}(r) = 0, \quad 0 < r < 1,$$
$$c_{2n}(1) = \begin{cases} -1, & n = 2, \\ 0, & n \neq 2. \end{cases}$$

Since the BC function is smooth, it follows that, according to the explanation given in Subsection 5.3.2, $u(r, \theta)$, $u_r(r, \theta)$, and $u_\theta(r, \theta)$ must be continuous (hence, bounded) for $0 \leq r \leq 1$, $-\pi < \theta \leq \pi$; consequently, we look for solutions $c_0(r)$, $c_{1n}(r)$, and $c_{2n}(r)$ that remain bounded together with their derivatives as $r \to 0+$.

The ODE for $n = 0$ is integrated by first noting that, after multiplication by r, the left-hand side can be written as $(rc_0')'$; in the end, we find that the desired bounded solution is

$$c_0(r) = r^2 - 1.$$

Multiplying the ODEs for $n \geq 1$ by r^2, we arrive at BVPs for Cauchy–Euler equations. Thus, for $n = 1$,

$$r^2 c_{11}''(r) + r c_{11}'(r) - c_{11}(r) = 0, \quad 0 < r < 1,$$
$$c_{11}(1) = 2,$$

and

$$r^2 c_{21}''(r) + r c_{21}'(r) - c_{21}(r) = 0, \quad 0 < r < 1,$$
$$c_{21}(1) = 0,$$

with bounded solutions $c_{11}(r) = 2r$ and $c_{21}(r) = 0$; for $n = 2$,

$$r^2 c_{12}''(r) + r c_{12}'(r) - 4c_{12}(r) = 0, \quad 0 < r < 1,$$
$$c_{12}(1) = 0,$$

and

$$r^2 c_{22}''(r) + r c_{22}'(r) - 4c_{22}(r) = 0, \quad 0 < r < 1,$$
$$c_{22}(1) = -1,$$

with bounded solutions $c_{12}(r) = 0$ and $c_{22}(r) = -r^2$; and for $n = 3, 4, \ldots$,

$$r^2 c_{1n}''(r) + r c_{1n}'(r) - n^2 c_{1n}(r) = 0, \quad 0 < r < 1,$$
$$c_{1n}(1) = 0,$$

and

$$r^2 c_{2n}''(r) + r c_{2n}'(r) - n^2 c_{2n}(r) = 0, \quad 0 < r < 1,$$
$$c_{2n}(1) = 0,$$

with bounded solutions $c_{1n}(r) = 0$ and $c_{2n}(r) = 0$. Therefore, the solution of the BVP is

$$u(r, \theta) = c_0(r) + c_{11}(r)\Theta_{11}(\theta) + c_{22}(r)\Theta_{22}(\theta)$$
$$= r^2 - 1 + 2r \cos \theta - r^2 \sin(2\theta).$$

VERIFICATION WITH MATHEMATICA®. The input

```
u = r^2 - 1 + 2 * r * Cos[θ] - r^2 * Sin[2 * θ];
{q, f} = {4, 2 * Cos[θ] - Sin[2 * θ]};
Simplify[{D[u,r,r] + r^(-1) * D[u,r] + r^(-2) * D[u,θ,θ] - q,
  (u /. r -> 1) - f}]
```

generates the output $\{0, 0\}$.

7.14 Example. In the case of the BVP

$$u_{rr}(r, \theta) + r^{-1}u_r(r, \theta) + r^{-2}u_{\theta\theta}(r, t)$$

$$= 3r^{-1} + 3\cos\theta - 2r^{-2}\sin\theta + (3r^{-1} - 4r^{-2})\cos(2\theta),$$

$$0 < r < 1, \quad -\pi < \theta < \pi,$$

$$u(1, \theta) = 3 + 3\cos\theta + 7\sin\theta, \quad -\pi < \theta < \pi$$

we have

$$\lambda_0 = 0, \quad \Theta_0(\theta) = 1,$$

$$\lambda_n = n^2, \quad \Theta_{1n}(\theta) = \cos(n\theta), \quad \Theta_{2n}(\theta) = \sin(n\theta), \quad n = 1, 2, \ldots,$$

so we expect a solution of the form

$$u(r, \theta) = c_0(r) + \sum_{n=1}^{\infty} \left[c_{1n}(r)\Theta_{1n}(\theta) + c_{2n}(r)\Theta_{2n}(\theta) \right].$$

The standard procedure now yields the BVPs

$$rc_0''(r) + c_0'(r) = 3, \quad 0 < r < 1,$$
$$c_0(1) = 3, \quad c_0(r), c_0'(r) \text{ bounded as } r \to 0+,$$

$$r^2c_{11}''(r) + rc_{11}'(r) - c_{11}(r) = 3r^2, \quad 0 < r < 1,$$
$$c_{11}(1) = 3, \quad c_{11}(r), c_{11}'(r) \text{ bounded as } r \to 0+,$$

$$r^2c_{12}''(r) + rc_{12}'(r) - 4c_{12}(r) = 3r - 4, \quad 0 < r < 1,$$
$$c_{12}(1) = 0, \quad c_{12}(r), c_{12}'(r) \text{ bounded as } r \to 0+,$$

$$r^2c_{1n}''(r) + rc_{1n}'(r) - n^2c_{1n}(r) = 0, \quad 0 < r < 1,$$
$$c_{1n}(1) = 0, \quad c_{1n}(r), c_{1n}'(r) \text{ bounded as } r \to 0+ \quad (n \neq 1, 2),$$

$$r^2c_{21}''(r) + rc_{21}'(r) - c_{21}(r) = -2, \quad 0 < r < 1,$$
$$c_{21}(1) = 7, \quad c_{21}(r), c_{21}'(r) \text{ bounded as } r \to 0+,$$

$$r^2c_{2n}''(r) + rc_{2n}'(r) - n^2c_{2n}(r) = 0, \quad 0 < r < 1,$$
$$c_{2n}(1) = 0, \quad c_{2n}(r), c_{2n}'(r) \text{ bounded as } r \to 0+ \quad (n \neq 1),$$

with solutions

$$c_0(r) = 3r, \quad c_{11}(r) = r^2 + 2r, \quad c_{21}(r) = 5r + 2, \quad c_{12}(r) = 1 - r,$$
$$c_{1n}(r) = 0 \ (n \neq 1, 2), \quad c_{2n}(r) = 0 \ (n \neq 1).$$

Consequently,

$$u(r, \theta) = 3r + (r^2 + 2r) \cos \theta + (1 - r) \cos(2\theta) + (5r + 2) \sin \theta.$$

VERIFICATION WITH MATHEMATICA®. The input

```
u = 3 * r + (r^2 + 2 * r) * Cos[θ] + (1 - r) * Cos[2 * θ] + (2 + 5 * r) * Sin[θ];
{q, f} = {3 * r^(-1) + 3 * Cos[θ] + (3 * r^(-1) -
   4 * r^(-2)) * Cos[2 * θ] - 2 * r^(-2) * Sin[θ],
   3 + 3 * Cos[θ] + 7 * Sin[θ]};
Simplify[{D[u,r,r] + r^(-1) * D[u,r] + r^(-2) * D[u,θ,θ] - q,
   (u/.r -> 1) - f}]
```

generates the output $\{0, 0\}$.

Exercises

Use the method of eigenfunction expansion to find the solution of the BVP consisting of the PDE

$$u_{rr}(r, \theta) + r^{-1}u_r(r, \theta) + r^{-2}u_{\theta\theta}(r, \theta) = q(r, \theta),$$
$$0 < x < 1, \ -\pi < \theta < \pi,$$
$$u(r, \theta), \ u_r(r, \theta) \text{ bounded as } r \to 0+, \quad u(1, \theta) = f(\theta), \quad -\pi < \theta < \pi,$$
$$u(r, -\pi) = u(r, \pi), \quad u_\theta(r, -\pi) = u_\theta(r, \pi), \quad 0 < r < 1$$

and functions q and f as indicated.

1 $q(r, \theta) = -8, \quad f(\theta) = -1 + 2\sin\theta + 2\cos(3\theta).$

2 $q(r, \theta) = 9r + 6\cos\theta + 3\sin\theta, \quad f(\theta) = 1 - 3\sin\theta.$

3 $q(r, \theta) = 18r - 4 + 8r\cos\theta + (7r^2 - 5)\sin(3\theta),$

   $f(\theta) = 1 + \cos\theta + 2\sin(3\theta).$

4 $q(r, \theta) = 3\cos\theta + 10\cos(3\theta), \quad f(\theta) = \cos\theta - \cos(3\theta) - \sin(4\theta).$

5 $q(r, \theta) = (1 - r)\sin(2\theta), \quad f(\theta) = \begin{cases} 0, & -\pi < \theta \leq 0, \\ 1, & 0 < \theta < \pi. \end{cases}$

6 $q(r, \theta) = r \cos \theta - 2 \sin(3\theta), \quad f(\theta) = \begin{cases} \theta, & -\pi < \theta \le 0, \\ -1, & 0 < \theta < \pi. \end{cases}$

Answers to Odd-Numbered Exercises

1 $u(r, \theta) = 1 - 2r^2 + 2r \sin \theta + 2r^3 \cos(3\theta)$.

3 $u(r, \theta) = 2r^3 - r^2 + r^3 \cos \theta + (r^4 + r^2) \sin(3\theta)$.

5 $u(r, \theta) = (1/20)(-4r^3 + 4r^2 + 5r^2 \ln r) \sin(2\theta) + 1/2$
$$+ \sum_{n=1}^{\infty} [1 - (-1)^n](1/(n\pi))r^n \sin(n\theta).$$

7.4 Other Nonhomogeneous Equations

The method of eigenfunction expansion can be extended to some IBVPs or BVPs for more general partial differential equations. Below, we make a selection from among the linear ones mentioned in Section 4.4.

7.15 Example. Consider the IBVP

$$u_t(x, t) = u_{xx}(x, t) - 2u_x(x, t) + u(x, t) + 2te^x \sin(2\pi x),$$
$$0 < x < 1, \ t > 0,$$
$$u(0, t) = 0, \quad u(1, t) = 0, \quad t > 0,$$
$$u(x, 0) = 0, \quad 0 < x < 1.$$

Separating the variables in the associated homogeneous PDE, we arrive at the regular S–L problem

$$X''(x) - 2X'(x) + \lambda X(x) = 0, \quad 0 < x < 1,$$
$$X(0) = 0, \quad X(1) = 0,$$

so, by the formulas in Remark 3.27(iii) with $a = -2$, $b = 0$, $c = 1$, and $L = 1$,

$$\lambda_n = n^2\pi^2 + 1, \quad X_n(x) = e^x \sin(n\pi x), \quad n = 1, 2, \ldots.$$

This suggests that we should seek the solution of our IBVP in the form

$$u(x, t) = \sum_{n=1}^{\infty} c_n(t)e^x \sin(n\pi x).$$

The standard procedure now leads to the IVPs

$$c_n'(t) + (\lambda_n - 1)c_n(t) = \begin{cases} 2t, & n = 2, \\ 0, & n \ne 2, \end{cases} \quad t > 0,$$
$$c_n(0) = 0.$$

Specifically, for $n = 2$ we have

$$c_2'(t) + 4\pi^2 c_2(t) = 2t, \quad t > 0,$$
$$c_2(0) = 0,$$

from which

$$c_2(t) = \frac{1}{8\pi^4} \left(e^{-4\pi^2 t} + 4\pi^2 t - 1 \right).$$

For $n \neq 2$, we find that $c_n(t) = 0$. Replaced in the above expansion, these coefficients give rise to the solution

$$u(x, t) = \frac{1}{8\pi^4} \left(e^{-4\pi^2 t} + 4\pi^2 t - 1 \right) e^x \sin(2\pi x).$$

VERIFICATION WITH MATHEMATICA®. The input

```
u = (1/(8*Pi^4)) * (E^(-4*Pi^2*t) +4*Pi^2*t-1) *E^x
    * Sin[2*Pi*x];
{q, f} = {2*t*E^x*Sin[2*Pi*x] , 0};
Simplify[{D[u,t] - D[u,x,x] +2*D[u,x] -u-q, {u/.x->0, u/.x->1},
    (u/.t->0) - f}]
```

generates the output $\{0, \{0, 0\}, 0\}$.

7.16 Example. The same technique applied to the IBVP

$$u_{tt}(x, t) + 2u_t(x, t) = u_{xx}(x, t) - u_x(x, t)$$
$$+ \left[2 + \left(9\pi^2 + \tfrac{1}{4} \right) t \right] e^{x/2} \sin(3\pi x), \quad 0 < x < 1, \ t > 0,$$
$$u(0, t) = 0, \quad u(1, t) = 0, \quad t > 0,$$
$$u(x, 0) = 0, \quad u_t(x, 0) = e^{x/2} \sin(3\pi x), \quad 0 < x < 1$$

yields

$$\lambda_n = n^2 \pi^2 + \tfrac{1}{4}, \quad X_n(x) = e^{x/2} \sin(n\pi x), \quad n = 1, 2, \ldots,$$

so we seek the solution in the form

$$u(x, t) = \sum_{n=1}^{\infty} c_n(t) e^{x/2} \sin(n\pi x),$$

with the coefficients $c_n(t)$ satisfying

$$c_n''(t) + 2c_n'(t) + \lambda_n c_n(t) = \begin{cases} 2 + \left(9\pi^2 + \tfrac{1}{4} \right) t, & n = 3, \\ 0, & n \neq 3, \end{cases} \quad t > 0,$$

$$c_n(0) = 0, \quad c_n'(0) = \begin{cases} 1, & n = 3, \\ 0, & n \neq 3. \end{cases}$$

The IVP for $n = 3$ is

$$c_3''(t) + 2c_3'(t) + \left(9\pi^2 + \tfrac{1}{4}\right)c_3(t) = 2 + \left(9\pi^2 + \tfrac{1}{4}\right)t, \quad t > 0,$$
$$c_3(0) = 0, \quad c_3'(0) = 1,$$

with solution

$$c_3(t) = t;$$

the fully homogeneous IVP for $n \neq 3$ produces the coefficients $c_n(t) = 0$. Hence, the solution of our problem is

$$u(x, t) = te^{x/2}\sin(3\pi x).$$

VERIFICATION WITH MATHEMATICA®. The input

```
u = t * E^(x/2) * Sin[3 * Pi * x];
{q, f, g} = {(2 + (9 * Pi^2 + 1/4) * t) * E^(x/2) * Sin[3 * Pi * x], 0,
  E^(x/2) * Sin[3 * Pi * x]};
Simplify[{D[u,t,t] + 2 * D[u,t] - D[u,x,x] + D[u,x] - q, {u/. x ->0,
  u/. x ->1}, {(u/. t ->0) - f, (D[u,t] /. t ->0) - g}}]
```

generates the output $\{0, \{0,0\}, \{0,0\}\}$.

7.17 Example. In the application of the method of eigenfunction expansion to the BVP

$$u_{xx}(x, y) + u_{yy}(x, y) - 4u_x(x, y) = (y^2 - y)e^{2x}\sin(\pi x),$$
$$0 < x < 1, \ 0 < y < 2,$$
$$u(0, y) = 0, \quad u(1, y) = 0, \quad 0 < y < 2,$$
$$u(x, 0) = 2e^{2x}\sin(2\pi x), \quad u(x, 2) = 0, \quad 0 < x < 1,$$

the eigenvalues and eigenfunctions turn out to be

$$\lambda_n = n^2\pi^2 + 4, \quad X_n(x) = e^{2x}\sin(n\pi x), \quad n = 1, 2, \dots,$$

so we seek the solution in the form

$$u(x, y) = \sum_{n=1}^{\infty} c_n(y)e^{2x}\sin(n\pi x),$$

where the coefficients $c_n(y)$ are shown to satisfy

$$c_n''(y) - \lambda_n c_n(y) = \begin{cases} y^2 - y, & n = 1, \\ 0, & n \neq 1, \end{cases} \quad 0 < y < 2,$$

$$c_n(0) = \begin{cases} 2, & n = 2, \\ 0, & n \neq 2, \end{cases} \quad c_n(2) = 0.$$

The individual coefficients are now easily computed. For $n = 1$,

$$c_1''(y) - (\pi^2 + 4)c_1(y) = y^2 - y, \quad 0 < y < 2,$$
$$c_1(0) = 0, \quad c_1(2) = 0,$$

with solution

$$c_1(y) = \frac{1}{(\pi^2 + 4)^2} \left\{ 2\operatorname{csch}\left(2\sqrt{\pi^2 + 4}\right)\left[(\pi^2 + 5)\sinh\left(\sqrt{\pi^2 + 4}\,y\right)\right.\right.$$
$$\left.\left. - \sinh\left(\sqrt{\pi^2 + 4}\,(y - 2)\right)\right] + (\pi^2 + 4)(y - y^2) - 2\right\};$$

for $n = 2$,

$$c_2''(y) - 4(\pi^2 + 1)c_2(y) = 0, \quad 0 < y < 2,$$
$$c_2(0) = 2, \quad c_2(2) = 0,$$

with solution

$$c_2(y) = -2\operatorname{csch}\left(4\sqrt{\pi^2 + 1}\right)\sinh\left(2\sqrt{\pi^2 + 1}\,(y - 2)\right);$$

and for $n \neq 1, 2$,

$$c_n''(y) + (n^2\pi^2 + 4)c_n(y) = 0, \quad 0 < y < 2,$$
$$c_n(0) = 0, \quad c_n(2) = 0,$$

with solution $c_n(y) = 0$. Consequently, the solution of the given BVP is

$$u(x, y) = c_1(y)e^{2x}\sin(\pi x) + c_2(y)e^{2x}\sin(2\pi x),$$

with $c_1(y)$ and $c_2(y)$ as determined above.

VERIFICATION WITH MATHEMATICA®. The input

```
c1 = (1/(Pi^2+4)^2) * (2*Csch[2*(Pi^2+4)^(1/2)] * ((Pi^2+5)
    * Sinh[((Pi^2+4)^(1/2)) * y] - Sinh[((Pi^2+4)^(1/2)) * (y-2)])
    + (Pi^2+4) * (y-y^2) - 2);
c2 = -2*Csch[4*(Pi^2+1)^(1/2)] * Sinh[2*((Pi^2+1)^(1/2))
    * (y-2)];
u = c1 * E^(2*x) * Sin[Pi*x] + c2 * E^(2*x) * Sin[2*Pi*x];
{q, f, g} = {(y^2-y) * E^(2*x) * Sin[Pi*x], 2*E^(2*x)
    * Sin[2*Pi*x], 0};
Simplify[{D[u,x,x] + D[u,y,y] - 4*D[u,x] - q, {u/. x->0, u/. x->1},
    {(u/. y->0) - f, (u/. y->2) - g}}]
```

generates the output $\{0, \{0, 0\}, \{0, 0\}\}$.

7.18 Example. Consider the BVP

$$u_{xx}(x,y) + u_{yy}(x,y) - u_y(x,y) = (15 - 16\pi^2)e^{2x+y/2}\sin(2\pi y),$$
$$0 < x < 2, \ 0 < y < 1,$$
$$u(0,y) = 8e^{y/2}\sin(2\pi y), \quad 0 < y < 1,$$
$$u_x(2,y) = \left(8e^4 + 2\sqrt{16\pi^2+1}\,\cosh\sqrt{16\pi^2+1}\,\right)e^{y/2}\sin(2\pi y),$$
$$0 < y < 1,$$
$$u(x,0) = 0, \quad u(x,1) = 0, \quad 0 < x < 2,$$

with associated eigenvalues and eigenfunctions

$$\lambda_n = n^2\pi^2 + \tfrac{1}{4}, \quad Y_n(y) = e^{y/2}\sin(n\pi y), \quad n = 1,2,\ldots.$$

Looking for a solution of the form

$$u(x,y) = \sum_{n=1}^{\infty} c_n(x)e^{y/2}\sin(n\pi y),$$

we find that the coefficients $c_n(x)$ are determined from the BVPs

$$c_n''(x) - \lambda_n c_n(x) = \begin{cases} (15 - 16\pi^2)e^{2x}, & n = 2, \\ 0, & n \neq 2, \end{cases} \quad 0 < x < 2,$$

$$c_n(0) = \begin{cases} 8, & n = 2, \\ 0, & n \neq 2, \end{cases}$$

$$c_n'(2) = \begin{cases} 8e^4 + 2\sqrt{16\pi^2+1}\,\cosh\sqrt{16\pi^2+1}, & n = 2, \\ 0, & n \neq 2. \end{cases}$$

For $n = 2$, we have

$$c_2''(x) - \tfrac{1}{4}(16\pi^2 + 1)c_2(x) = (15 - 16\pi^2)e^{2x}, \quad 0 < x < 2,$$
$$c_1(0) = 8, \quad c_2'(2) = 8e^4 + 2\sqrt{16\pi^2+1}\,\cosh\sqrt{16\pi^2+1},$$

which yields

$$c_2(x) = 4\left[\sinh\left(\tfrac{1}{2}\sqrt{16\pi^2+1}\,x\right) + e^{2x}\right].$$

For $n \neq 2$,

$$c_n''(x) + \lambda_n c_n(x) = 0, \quad 0 < x < 2,$$
$$c_n(0) = 0, \quad c_n'(2) = 0,$$

so $c_n(x) = 0$. This means that

$$u(x,y) = 4\left[\sinh\left(\tfrac{1}{2}\sqrt{16\pi^2+1}\,x\right) + e^{2x}\right]e^{y/2}\sin(2\pi y).$$

VERIFICATION WITH MATHEMATICA®. The input

```
u = 4 * (Sinh[(1/2) * (16 * Pi^2 + 1)^(1/2) * x] +
    E^(2 * x)) * E^(y/2) * Sin[2 * Pi * y];
{q, f, g} = {(15 - 16 * Pi^2) * E^(2 * x + y/2) * Sin[2 * Pi * y],
    4 * E^(y/2) * Sin[2 * Pi * y], (8 * E^4 + 2 * (16 * Pi^2 + 1)^(1/2) *
    Cosh[(16 * Pi^2 + 1)^(1/2)] * E^(y/2) * Sin[2 * Pi * y]};
Simplify[{D[u,x,x] + D[u,y,y] - D[u,y] - q,
    {u/.y -> 0, u/.y -> 1}, {(u/.x -> 0) - f, (D[u,x] /.x -> 2) - g}}]
```

generates the output $\{0, \{0, 0\}, \{0, 0\}\}$.

Exercises

In **1–4**, use the method of eigenfunction expansion to find the solution of the IBVP

$$u_t(x, t) = ku_{xx}(x, t) - au_x(x, t) + bu(x, t) + q(x, t),$$
$$0 < x < 1, \ t > 0,$$

$$u(0, t) = 0, \quad u(1, t) = 0, \quad t > 0,$$
$$u(x, 0) = f(x), \quad 0 < x < 1$$

with numbers k, a, b and functions q, f as indicated.

1 $k = 1, \ a = 1, \ b = 2, \ q(x, t) = (1/4)(16\pi^2 - 11)e^{-t + x/2} \sin(2\pi x),$
$f(x) = e^{x/2} \sin(2\pi x).$

2 $k = 2, \ a = 1, \ b = 1, \ q(x, t) = [2 + (18\pi^2 - 7/8)(2t - 3)]e^{x/4} \sin(3\pi x),$
$f(x) = -3e^{x/4} \sin(3\pi x).$

3 $k = 1, \ a = 2, \ b = 2, \ q(x, t) = -2[1 + (\pi^2 - 1)t]e^x \sin(\pi x),$
$f(x) = e^x \sin(\pi x).$

4 $k = 1, \ a = 1, \ b = 0, \ q(x, t) = [4 + (1 + 16\pi^2)t]e^{x/2} \sin(2\pi x),$
$f(x) = 1.$

In **5–8**, use the method of eigenfunction expansion to find the solution of the IBVP

$$u_{tt}(x, t) + au_t(x, t) + bu(x, t) = c^2 u_{xx}(x, t) - du_x(x, t) + q(x, t),$$
$$0 < x < 1, \ t > 0,$$

$$u(0, t) = 0, \quad u(1, t) = 0, \quad t > 0,$$
$$u(x, 0) = f(x), \quad u_t(x, 0) = g(x), \quad 0 < x < 1$$

with numbers a, b, c, d and functions q, f, g as indicated.

5 $a = 1$, $b = 1$, $c = 1$, $d = 2$, $q(x,t) = \left[(\pi^2 + 2)t + 1\right]e^x \sin(\pi x)$,

  $f(x) = 0$, $g(x) = e^x \sin(\pi x)$.

6 $a = 1$, $b = 2$, $c = 1$, $d = 1$, $q(x,t) = -4(4\pi^2 + 3)e^{t/2 + x/2} \sin(2\pi x)$,

  $f(x) = -4e^{x/2} \sin(2\pi x)$, $g(x) = -2e^{x/2} \sin(2\pi x)$.

7 $a = 2$, $b = 1$, $c = 1/2$, $d = 1$, $q(x,t) = -(\pi^2 + 2)e^{2x} \sin(2\pi x)$,

  $f(x) = 0$, $g(x) = -e^{2x} \sin(2\pi x)$.

8 $a = 2$, $b = 0$, $c = 1$, $d = 4$,

  $q(x,t) = \left[(9\pi^2 + 4)t^2 + 4t + 2\right]e^{2x} \sin(3\pi x)$, $f(x) = 0$, $g(x) = 1$.

In **9–12**, use the method of eigenfunction expansion to find the solution of the BVP

$$u_{xx}(x,y) + u_{yy}(x,y) - au_x(x,y) - bu_y(x,y) + cu(x,y) = q(x,y),$$
$$0 < x < L, \quad 0 < y < K,$$
$$u(0,y) = 0, \quad u(L,y) = 0, \quad 0 < y < K,$$
$$u(x,0) = f(x), \quad u(x,K) = g(x), \quad 0 < x < L$$

with numbers a, b, c, L, K and functions q, f, g as indicated.

9 $a = 2$, $b = 1$, $c = 1$, $L = 1$, $K = 2$, $q(x,y) = (6 - \pi^2)e^{x-2y} \sin(\pi x)$,

  $f(x) = e^x \sin(\pi x)$, $g(x) = e^{x-4} \sin(\pi x)$.

10 $a = 1$, $b = 1$, $c = 2$, $L = 2$, $K = 1$,

  $q(x,y) = \left[9/2 - 8\pi^2 + (4\pi^2 - 7/4)y\right]e^{x/2} \sin(2\pi x)$,

  $f(x) = 2e^{x/2} \sin(2\pi x)$, $g(x) = e^{x/2} \sin(2\pi x)$.

11 $a = 2$, $b = 1$, $c = 1$, $L = 2$, $K = 1$,

  $q(x,y) = -(5/4)(4\pi^2 + 1)e^{x+y/2} \sin(\pi x) \sin\left((4\pi^2 + 1)^{1/2}y\right)$,

  $f(x) = 0$, $g(x) = e^{x+1/2} \sin\left((4\pi^2 + 1)^{1/2}\right) \sin(\pi x)$.

12 $a = 0$, $b = 0$, $c = -1$, $L = 1$, $K = 2$,

  $q(x,y) = (\pi^2 + 1) \sinh\left(2(\pi^2 + 1)^{1/2}\right)y \sin(\pi x)$, $f(x) = 1$, $g(x) = 0$.

Answers to Odd-Numbered Exercises

1 $u(x,t) = e^{-t+x/2} \sin(2\pi x)$.

3 $u(x,t) = \left[e^{(1-\pi^2)t} - 2t\right]e^x \sin(\pi x)$.

5 $u(x,t) = te^x \sin(\pi x)$.

7 $u(x,t) = [e^{-t} \cos((\pi^2 + 1)^{1/2}t) - 1]e^{2x} \sin(2\pi x)$.

9 $u(x,y) = e^{x-2y} \sin(\pi x)$.

11 $u(x,y) = e^{x+y/2} \sin(\pi x) \sin((4\pi^2 + 1)^{1/2}y)$.

CHAPTER 8

The Fourier Transformations

Some problems of practical importance are beyond the reach of the method of eigenfunction expansion. This is the case, for example, when the space variable is defined on the entire real line and where, as a consequence, there are no boundary points. Such problems may have a continuum of eigenvalues instead of a countable set, which means that they require other techniques of solution. The Fourier transformations—developed, in fact, from the Fourier series representations of functions—are particularly useful tools when dealing with infinite or semi-infinite spatial regions because they are designed for exactly this type of setup and have the added advantage that they reduce the number of 'active' variables in the given PDE problem.

8.1 The Full Fourier Transformation

Consider, for simplicity, a function f that is continuous and periodic with period $2L$ on $\mathbb{R}$. Then we have the representation (see Chapter 2)

$$f(x) = \tfrac{1}{2} a_0 + \sum_{n=1}^{\infty} \left(a_n \cos \frac{n\pi x}{L} + b_n \sin \frac{n\pi x}{L} \right), \tag{8.1}$$

where

$$a_n = \frac{1}{L} \int_{-L}^{L} f(x) \cos \frac{n\pi x}{L} \, dx, \quad n = 0, 1, 2, \ldots,$$

$$b_n = \frac{1}{L} \int_{-L}^{L} f(x) \sin \frac{n\pi x}{L} \, dx, \quad n = 1, 2, \ldots. \tag{8.2}$$

Using Euler's formula (see Section 14.1)

$$e^{i\theta} = \cos\theta + i\sin\theta, \quad i^2 = -1,$$

and its version with θ replaced by $-\theta$, we deduce that

$$\cos\theta = \tfrac{1}{2}\left(e^{i\theta} + e^{-i\theta}\right), \quad \sin\theta = -\tfrac{1}{2}i\left(e^{i\theta} - e^{-i\theta}\right);$$

consequently, (8.1) with $\theta = n\pi x/L$ becomes

$$f(x) = \tfrac{1}{2}a_0 + \sum_{n=1}^{\infty} \tfrac{1}{2}(a_n - ib_n)e^{in\pi x/L} + \sum_{n=1}^{\infty} \tfrac{1}{2}(a_n + ib_n)e^{-in\pi x/L}.$$

Replacing n by $-n$ in the first sum above and noticing from (8.2) that $a_{-n} = a_n$ and $b_{-n} = -b_n$, from the last equality we obtain

$$f(x) = \tfrac{1}{2}a_0 + \sum_{n=-1}^{-\infty} \tfrac{1}{2}(a_n + ib_n)e^{-in\pi x/L} + \sum_{n=1}^{\infty} \tfrac{1}{2}(a_n + ib_n)e^{-in\pi x/L},$$

or

$$f(x) = \sum_{n=-\infty}^{\infty} c_n e^{-in\pi x/L},$$

where, by (8.2),

$$c_n = \tfrac{1}{2}(a_n + ib_n) = \frac{1}{2L} \int_{-L}^{L} f(x)\left(\cos\frac{n\pi x}{L} + i\sin\frac{n\pi x}{L}\right) dx$$

$$= \frac{1}{2L} \int_{-L}^{L} f(x)e^{in\pi x/L}\,dx \quad \text{(with } b_0 = 0\text{)};$$

hence,

$$f(x) = \sum_{n=-\infty}^{\infty} \left[\frac{1}{2L} \int_{-L}^{L} f(\xi)e^{in\pi\xi/L}\,d\xi\right] e^{-in\pi x/L}. \qquad (8.3)$$

If f is not periodic in the proper sense of the word, then we may regard it as 'periodic' with an 'infinite period'. Using some advanced calculus arguments, we find that, as $L \to \infty$, representation (8.3) for such a function takes the form

$$f(x) = \int_{-\infty}^{\infty} \left[\frac{1}{2\pi} \int_{-\infty}^{\infty} f(\xi)\,e^{i\omega\xi}d\xi\right] e^{-i\omega x}d\omega. \qquad (8.4)$$

We define the (full) *Fourier transform* of f by

$$\mathcal{F}[f](\omega) = F(\omega) = \frac{1}{\sqrt{2\pi}} \int_{-\infty}^{\infty} f(x)e^{i\omega x}dx. \qquad (8.5)$$

From (8.4) it is clear that the *inverse Fourier transform* of F is

$$\mathcal{F}^{-1}[F](x) = f(x) = \frac{1}{\sqrt{2\pi}} \int_{-\infty}^{\infty} F(\omega)e^{-i\omega x}\,d\omega. \tag{8.6}$$

The integral operators $\mathcal{F}$ and $\mathcal{F}^{-1}$ are called the *Fourier transformation* and *inverse Fourier transformation*, respectively. The variable ω is called the *transformation parameter*.

8.1 Remarks. (i) Formulas (8.5) and (8.6) are valid, respectively, if

$$\int_{-\infty}^{\infty} |f(x)|\,dx < \infty, \quad \int_{-\infty}^{\infty} |F(\omega)|\,d\omega < \infty;$$

that is, if f (in (8.5)) and F (in (8.6)) are *absolutely integrable* on $\mathbb{R}$. However, the Fourier transform (inverse Fourier transform) may also be defined for some functions that do not have the above property. This is done in a generalized sense through a limiting process involving transforms (inverse transforms) of absolutely integrable functions. For simplicity, we refer to a function that has a Fourier transform as a *transformable function*.

(ii) The construction of the full Fourier transform can be extended to piecewise continuous functions f. In this case, $f(x)$ must be replaced by $\frac{1}{2}\left[f(x-) + f(x+)\right]$ in (8.4) and (8.6).

(iii) The pair $\mathcal{F}$, $\mathcal{F}^{-1}$ is defined slightly differently in some branches of science and engineering. Thus, in classical physics, systems engineering, and signal processing the preferred definitions are, respectively,

$$\mathcal{F}[f](\omega) = \frac{1}{2\pi} \int_{-\infty}^{\infty} f(x)e^{i\omega x}\,dx, \quad \mathcal{F}^{-1}[F](x) = \int_{-\infty}^{\infty} F(\omega)e^{-i\omega x}\,d\omega,$$

$$\mathcal{F}[f](\omega) = \int_{-\infty}^{\infty} f(x)e^{-i\omega x}\,dx, \quad \mathcal{F}^{-1}[F](x) = \frac{1}{2\pi} \int_{-\infty}^{\infty} F(\omega)e^{i\omega x}\,d\omega,$$

$$\mathcal{F}[f](\omega) = \int_{-\infty}^{\infty} f(x)e^{-2\pi i\omega x}\,dx, \quad \mathcal{F}^{-1}[F](x) = \int_{-\infty}^{\infty} F(\omega)e^{2\pi i\omega x}\,d\omega.$$

Mathematicians can work with any of these alternatives. Here, we selected formulas (8.5) and (8.6) since they are the default in the *Mathematica*® package used in this book.

8.2 Example. The Fourier transform of the function

$$f(x) = \begin{cases} 1, & -a < x \le a, \\ 0 & \text{otherwise,} \end{cases}$$

where $a > 0$ is a constant, is

$$F(\omega) = \frac{1}{\sqrt{2\pi}} \int\limits_{-\infty}^{\infty} f(x)e^{i\omega x} \, dx = \frac{1}{\sqrt{2\pi}} \int\limits_{-a}^{a} e^{i\omega x} \, dx$$

$$= \frac{1}{\sqrt{2\pi}} \frac{1}{i\omega}(e^{i\omega a} - e^{-i\omega a}) = \sqrt{\frac{2}{\pi}} \frac{\sin(a\omega)}{\omega}.$$

8.3 Definition. Let f and g be absolutely integrable on $\mathbb{R}$. By the *convolution* of f and g we understand the function $f * g$ defined by

$$(f * g)(x) = \frac{1}{\sqrt{2\pi}} \int\limits_{-\infty}^{\infty} f(x - \xi)g(\xi) \, d\xi, \quad -\infty < x < \infty.$$

8.4 Remark. Making the change of variable $x - \xi = \eta$ and then replacing η by ξ, we easily convince ourselves that

$$(f * g)(x) = \frac{1}{\sqrt{2\pi}} \int\limits_{-\infty}^{\infty} f(\xi)g(x - \xi) \, d\xi = (g * f)(x).$$

This means that the operation of convolution is commutative.

8.5 Theorem. (i) $\mathcal{F}$ *is linear; that is,*

$$\mathcal{F}[c_1 f_1 + c_2 f_2] = c_1 \mathcal{F}[f_1] + c_2 \mathcal{F}[f_2]$$

for any functions f_1 and f_2 (to which $\mathcal{F}$ can be applied) and any numbers c_1 and c_2.

(ii) *If $u = u(x, t)$, $u(x, t) \to 0$ as $x \to \pm\infty$, and $\mathcal{F}[u](\omega, t) = U(\omega, t)$, then*

$$\mathcal{F}[u_x](\omega, t) = -i\omega U(\omega, t).$$

(iii) *If, in addition, $u_x(x, t) \to 0$ as $x \to \pm\infty$, then*

$$\mathcal{F}[u_{xx}](\omega, t) = -\omega^2 U(\omega, t).$$

(iv) *Time differentiation and the Fourier transformation with respect to x commute:*

$$\mathcal{F}[u_t](\omega, t) = \big(\mathcal{F}[u]\big)_t(\omega, t) = U'(\omega, t).$$

(v) *The Fourier transform of a convolution is*

$$\mathcal{F}[f * g] = \mathcal{F}[f]\mathcal{F}[g].$$

8.6 Remarks. (i) In general, $\mathcal{F}[fg] \neq \mathcal{F}[f]\mathcal{F}[g]$.

(ii) It is obvious from its definition that $\mathcal{F}^{-1}$ is also linear.

(iii) The Fourier transforms of a few elementary functions are given in Appendix A.3.

The properties of the Fourier transformation listed in Theorem 8.5 play an essential role in the solution of certain types of PDE problems. The solution strategy in such situations is best illustrated by examples.

8.1.1 Cauchy Problem for an Infinite Rod

Heat conduction in a very long, uniform rod where the diffusion activity diminishes towards the end-points is modeled mathematically by the IVP (called a *Cauchy problem*)

$$u_t(x,t) = ku_{xx}(x,t), \quad -\infty < x < \infty, \ t > 0,$$
$$u(x,t), \ u_x(x,t) \to 0 \quad \text{as } x \to \pm\infty, \ t > 0,$$
$$u(x,0) = f(x), \quad -\infty < x < \infty.$$

Adopting the notation

$$\mathcal{F}[u](\omega,t) = U(\omega,t), \quad \mathcal{F}[f](\omega) = F(\omega),$$

we apply $\mathcal{F}$ to the PDE and IC and use the properties of $\mathcal{F}$ in Theorem 8.5 to reduce the given IVP to an initial value problem for an ordinary differential equation in the so-called transform domain. This new IVP (in which ω is an 'inert' parameter) is

$$U'(\omega,t) + k\omega^2 U(\omega,t) = 0, \quad t > 0,$$
$$U(\omega,0) = F(\omega),$$

with solution

$$U(\omega,t) = F(\omega)e^{-k\omega^2 t}. \tag{8.7}$$

To find the solution of the original IVP, we now need to compute the inverse Fourier transform of U. Let $P(\omega,t) = e^{-k\omega^2 t}$. Using formula 7 (with $a = (4kt)^{-1/2}$) in Appendix A.3, we easily see that the inverse transform of P is

$$\mathcal{F}^{-1}[P](x,t) = p(x,t) = \frac{1}{\sqrt{2kt}} e^{-x^2/(4kt)}.$$

By Theorem 8.5(v), we can now write (8.7) as

$$\mathcal{F}[u](\omega,t) = \mathcal{F}[f](\omega)\mathcal{F}[p](\omega,t) = \mathcal{F}[f * p](\omega,t),$$

so $u = f * p$, or, by Definition 8.3 and Remark 8.4,

$$u(x,t) = \frac{1}{\sqrt{2\pi}} \int_{-\infty}^{\infty} f(\xi)\frac{1}{\sqrt{2kt}} e^{-(x-\xi)^2/(4kt)} d\xi.$$

An alternative form for this is

$$u(x,t) = \int_{-\infty}^{\infty} G(x,t;\xi,0)u(\xi,0)\, d\xi, \tag{8.8}$$

where

$$G(x,t;\xi,0) = \frac{1}{2\sqrt{\pi kt}} e^{-(x-\xi)^2/(4kt)}$$

is called the *Gauss–Weierstrass kernel*, or *influence function*. Formula (8.8) shows how the initial temperature distribution $u(x, 0)$ influences the subsequent evolution of the temperature in the rod. This type of representation formulas will be discussed in more detail in Chapter 10.

8.7 Example. In the IVP

$$u_t(x, t) = 2u_{xx}(x, t), \quad -\infty < x < \infty, \ t > 0,$$
$$u(x, t), \ u_x(x, t) \to 0 \quad \text{as } x \to \pm\infty, \ t > 0,$$
$$u(x, 0) = f(x) = \begin{cases} -3, & |x| \le 1, \\ 0, & |x| > 1, \end{cases}$$

we have $k = 2$, so

$$G(x, t; \xi, 0) = \frac{1}{2\sqrt{2\pi t}} e^{-(x-\xi)^2/(8t)},$$

which, replaced together with $f(x)$ in (8.8), yields

$$u(x, t) = -\frac{3}{2\sqrt{2\pi t}} \int_{-1}^{1} e^{-(x-\xi)^2/(8t)} \, d\xi.$$

If instead of (8.8) we use formula 12 (with $a = 1$) in Appendix A.3 to compute $F(\omega) = -3\sqrt{2/\pi} \, (\sin\omega)/\omega$ and then solve the transformed problem

$$U'(\omega, t) + 2\omega^2 U(\omega, t) = 0, \quad t > 0,$$
$$U(\omega, 0) = -3\sqrt{\frac{2}{\pi}} \frac{\sin\omega}{\omega},$$

we find that

$$U(\omega, t) = -3\sqrt{\frac{2}{\pi}} \frac{\sin\omega}{\omega} e^{-2\omega^2 t}.$$

From Euler's formula for $e^{i\omega}$ and $e^{-i\omega}$ it follows that

$$\sin\omega = -\tfrac{1}{2} i(e^{i\omega} - e^{-i\omega}),$$

so

$$U(\omega, t) = \tfrac{3}{2} i \sqrt{\frac{2}{\pi}} \frac{1}{\omega} (e^{i\omega} - e^{-i\omega}) e^{-2\omega^2 t}.$$

Using formula 16 with $a = 1/(2\sqrt{2t})$ applied to

$$V(\omega) = i\sqrt{\frac{2}{\pi}} \frac{1}{\omega} e^{-2\omega^2 t},$$

we deduce that

$$v(x) = \mathcal{F}^{-1}[V](x) = \operatorname{erf} \frac{x}{2\sqrt{2t}};$$

hence, by formula 3 with $a = 1$, $b = -1$ and then $a = 1$, $b = 1$,

$$\mathcal{F}^{-1}[e^{i\omega}V(\omega)] = v(x-1) = \text{erf}\,\frac{x-1}{2\sqrt{2t}},$$

$$\mathcal{F}^{-1}[e^{-i\omega}V(\omega)] = v(x+1) = \text{erf}\,\frac{x+1}{2\sqrt{2t}},$$

which yields

$$u(x,t) = \mathcal{F}^{-1}[U](x,t) = \frac{3}{2}\left(\text{erf}\,\frac{x-1}{2\sqrt{2t}} - \text{erf}\,\frac{x+1}{2\sqrt{2t}}\right).$$

The function $y = \text{erf}(x)$ is called the *error function*. The graphs of the error function and of the *complementary error function* defined on $(-\infty, \infty)$, respectively, by

$$y = \text{erf}(x) = \frac{2}{\sqrt{\pi}}\int_0^x e^{-\xi^2}\,d\xi,$$

$$y = \text{erfc}(x) = \frac{2}{\sqrt{\pi}}\int_x^\infty e^{-\xi^2}\,d\xi = 1 - \text{erf}(x),$$

are shown in Fig. 8.1.

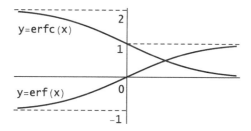

Figure 8.1: The graphs of $\text{erf}(x)$ and $\text{erfc}(x)$.

VERIFICATION WITH MATHEMATICA®. Because of the piecewise definition of the IC function, we use the input

```
u = (3/2) * (Erf [(x - 1)/(2 * (2 * t)^(1/2))]
    - Erf [(x + 1)/(2 * (2 * t)^(1/2))]);
{f1, f2, f3} = {0, -3, 0};
Simplify[{D[u,t] - 2 * D[u,x,x] , Assuming [x< -1, Limit [u,t -> 0]
    - f1] , Assuming [ -1<x<1, Limit [u,t -> 0] - f2] , Assuming [x>1,
    Limit [u,t -> 0] - f3] }]
```

which generates the output $\{0, 0, 0, 0\}$.

8.8 Example. To solve the IBVP

$$u_t(x,t) = 2u_{xx}(x,t) + (9 - 32x^2)e^{t-2x^2}, \quad -\infty < x < \infty, \ t > 0,$$

$$u(x,t), \ u_x(x,t) \to 0 \quad \text{as } x \to \pm\infty, \ t > 0,$$

$$u(x,0) = e^{-2x^2}, \quad -\infty < x < \infty,$$

we set $\mathcal{F}[u](x,t) = U(\omega,t)$ and transform the nonhomogeneous terms by means of formulas 7 and 9 (with $a = \sqrt{2}$) in Appendix A.3:

$$\mathcal{F}\big[(9 - 32x^2)e^{t-2x^2}\big] = \big\{9\mathcal{F}[e^{-2x^2}] - 32\mathcal{F}[x^2 e^{-2x^2}]\big\}e^t$$

$$= \big\{9 \cdot \tfrac{1}{2} e^{-\omega^2/8} - 32 \cdot \tfrac{1}{32}(4 - \omega^2)e^{-\omega^2/8}\big\}e^t$$

$$= \tfrac{1}{2}(1 + 2\omega^2)e^{-\omega^2/8}e^t,$$

$$\mathcal{F}\big[e^{-2x^2}\big] = \tfrac{1}{2} e^{-\omega^2/8}.$$

Then, using formula 2, we convert the IBVP to the ODE problem

$$U'(\omega,t) + 2\omega^2 U(\omega,t) = \tfrac{1}{2}(1 + 2\omega^2)e^{-\omega^2/8}e^t, \quad t > 0,$$

$$U(\omega,0) = \tfrac{1}{2} e^{-\omega^2/8},$$

where the general solution of the equation is of the form

$$U(\omega,t) = C(\omega)e^{-2\omega^2 t} + A(\omega)e^t.$$

After we find that $A(\omega) = \tfrac{1}{2} e^{-\omega^2/8}$, we apply the IC and see that $C(\omega) = 0$; consequently,

$$U(\omega,t) = \tfrac{1}{2} e^{-\omega^2/8}e^t,$$

so, by formula 7,

$$u(x,t) = \mathcal{F}^{-1}[U](x,t) = e^{-2x^2}e^t = e^{t-2x^2}.$$

VERIFICATION WITH MATHEMATICA®. The input

```
u = E^(t - 2 * x^2);
{q, f} = {(9 - 32 * x^2) * E^(t - 2 * x^2), E^(-2 * x^2)};
Simplify[{D[u,t] - 2 * D[u,x,x] - q, (u/. t -> 0) - f}]
```

generates the output $\{0, 0\}$.

Exercises

In **1–6**, use the full Fourier transformation to solve the IVP consisting of the PDE

$$u_t(x,t) = ku_{xx}(x,t) + q(x,t), \quad -\infty < x < \infty, \ t > 0,$$

where
$$u(x,t),\ u_x(x,t) \to 0 \text{ as } x \to \pm\infty,$$

and constant k, function q, and IC as indicated.

1 $k = 1$, $q(x,t) = 0$, $u(x,0) = -3e^{-x^2}$.

2 $k = 2$, $q(x,t) = 0$, $u(x,0) = 5e^{-x^2/4}$.

3 $k = 1$, $q(x,t) = (4x^2 - 5)e^{3t-x^2}$, $u(x,0) = -e^{-x^2}$.

4 $k = 1$, $q(x,t) = (7 - 64x^2)e^{-t-4x^2}$, $u(x,0) = e^{-4x^2}$.

5 $k = 1$, $q(x,t) = (1 + 4t - 16x^2t)e^{-2x^2}$, $u(x,0) = 0$.

6 $k = 2$, $q(x,t) = 2[(8t - 4)x^2 - 4t + 1]e^{-x^2}$, $u(x,0) = e^{-x^2}$.

Answers to Odd-Numbered Exercises

1 $u(x,t) = -3(1 + 4t)^{-1/2}e^{-x^2/(1+4t)}$.

3 $u(x,t) = -e^{3t-x^2}$.

5 $u(x,t) = te^{-2x^2}$.

8.1.2   Vibrations of an Infinite String

The vibrations of a very long string initially at rest, with negligible body force and where the effects of mechanical activity at the endpoints are insignificant, are modeled by the IVP

$$u_{tt}(x,t) = c^2 u_{xx}(x,t), \quad -\infty < x < \infty,\ t > 0,$$
$$u(x,t),\ u_x(x,t) \to 0 \quad \text{as } x \to \pm\infty,\ t > 0,$$
$$u(x,0) = f(x), \quad u_t(x,0) = 0, \quad -\infty < x < \infty.$$

As above, let $\mathcal{F}[u](\omega,t) = U(\omega,t)$ and $\mathcal{F}[f](\omega) = F(\omega)$. Applying $\mathcal{F}$ to the PDE and ICs, we arrive at the ODE problem

$$U''(\omega,t) + c^2\omega^2 U(\omega,t) = 0, \quad t > 0,$$
$$U(\omega,0) = F(\omega), \quad U'(\omega,0) = 0,$$

with general solution

$$U(\omega,t) = C_1(\omega)\cos(c\omega t) + C_2(\omega)\sin(c\omega t),$$

where $C_1(\omega)$ and $C_2(\omega)$ are arbitrary functions of the transformation parameter. Hence, using the ICs, we find that

$$U(\omega,t) = F(\omega)\cos(c\omega t).$$

By Euler's formula, we have $\cos(c\omega t) = \frac{1}{2}(e^{ic\omega t} + e^{-ic\omega t})$, so from (8.6) it follows that if f is transformable, then

$$u(x,t) = \mathcal{F}^{-1}[U](x,t) = \frac{1}{\sqrt{2\pi}} \int_{-\infty}^{\infty} F(\omega) \cos(c\omega t) e^{-i\omega x} d\omega$$

$$= \frac{1}{2\sqrt{2\pi}} \int_{-\infty}^{\infty} F(\omega)[e^{-i\omega(x-ct)} + e^{-i\omega(x+ct)}] d\omega$$

$$= \frac{1}{2}\left[f(x - ct) + f(x + ct)\right]. \tag{8.9}$$

This problem will be approached from a different angle in Chapter 12.

8.9 Example. In the IVP

$$u_{tt}(x,t) = u_{xx}(x,t), \quad -\infty < x < \infty, \ t > 0,$$
$$u(x,t), \ u_x(x,t) \to 0 \quad \text{as } x \to \pm\infty, \ t > 0,$$
$$u(x,0) = e^{-x^2}, \quad u_t(x,0) = 0, \quad -\infty < x < \infty,$$

we have $c = 1$ and $f(x) = e^{-x^2}$, so, by (8.9),

$$u(x,t) = \frac{1}{2}\left[e^{-(x-t)^2} + e^{-(x+t)^2}\right].$$

VERIFICATION WITH MATHEMATICA®. The input

```
u = (1/2) * (^(-(x-t)^2) +E^(-(x+t)^2));
{q, f, g} = {0, E^(-x^2), 0};
Simplify[{D[u,t,t] - D[u,x,x]-q, (u/.t->0)-f,
  (D[u,t]/.t->0)-g}]
```

generates the output $\{0, 0, 0\}$.

8.10 Example. Consider the IVP

$$u_{tt}(x,t) = 4u_{xx}(x,t) + 2(2 + 2t^2 - x^2t^2)e^{-x^2/4},$$
$$-\infty < x < \infty, \ t > 0,$$
$$u(x,t), \ u_x(x,t) \to 0 \quad \text{as } x \to \pm\infty, \ t > 0,$$
$$u(x,0) = 0, \quad u_t(x,0) = 0, \quad -\infty < x < \infty.$$

Setting $\mathcal{F}[u](\omega,t) = U(\omega,t)$ and using formulas 2, 7, and 9 (with $a = 1/2$) in Appendix A.3, we change the IVP into the ODE problem

$$U''(\omega,t) + 4\omega^2 U(\omega,t) = 4\sqrt{2}(1 + 2\omega^2t^2)e^{-\omega^2}, \quad t > 0,$$
$$U(\omega,0) = 0, \quad U'(\omega,0) = 0.$$

Here, the general solution of the equation has the form

$$U(\omega,t) = C_1(\omega)\cos(2\omega t) + C_2(\omega)\sin(2\omega t) + A(\omega)t^2 + B(\omega)t + C(\omega),$$

and the usual technique yields $A(\omega) = 2\sqrt{2}\,e^{-\omega^2}$ and $B(\omega) = C(\omega) = 0$. Applying the ICs now leads to $C_1(\omega) = C_2(\omega) = 0$, so

$$U(\omega,t) = 2\sqrt{2}\,t^2 e^{-\omega^2}.$$

Hence, by formula (7), the solution of the IVP is

$$u(x,t) = \mathcal{F}^{-1}[U](x,t) = 2t^2 e^{-x^2/4}.$$

VERIFICATION WITH MATHEMATICA®. The input

```
u = 2 * t^2 * E^(-x^2/4);
{q, f, g} = {2 * (2 + 2 * t^2 - x^2 * t^2) * E^(-x^2/4), 0, 0};
Simplify[{D[u,t,t] - 4 * D[u,x,x] - q, (u/. t->0) - f,
   (D[u,t] /. t->0) - g}]
```

generates the output $\{0,0,0\}$.

Exercises

Use the full Fourier transformation to solve the IVP consisting of the PDE

$$u_{tt}(x,t) = c^2 u_{xx}(x,t) + q(x,t), \quad -\infty < x < \infty, \ t > 0,$$

where

$$u(x,t), \ u_x(x,t) \to 0 \text{ as } x \to \pm\infty,$$

and constant c, function q, and ICs as indicated.

1 $c = 2$, $q(x,t) = (4t^2 + 2 - 16t^2 x^2)e^{-2x^2}$, $u(x,0) = 0$, $u_t(x,0) = 0$.

2 $c = 1$, $q(x,t) = 2[(2t^2 - 2)x^2 - t^2]e^{-x^2}$, $u(x,0) = e^{-x^2}$, $u_t(x,0) = 0$.

3 $c = 2$, $q(x,t) = [2t^2 + 4t + 10 - (t+1)^2 x^2]e^{-x^2/4}$,

$u(x,0) = 4e^{-x^2/4}$, $u_t(x,0) = 8e^{-x^2/4}$.

4 $c = 1$, $q(x,t) = [8 - 6t + 2t^2 + (3t - t^2)x^2]e^{-x^2/4}$,

$u(x,0) = 0$, $u_t(x,0) = -12e^{-x^2/4}$.

5 $c = 2$, $q(x,t) = 2(9 - 16x^2)e^{-t-x^2}$, $u(x,0) = 2e^{-x^2}$, $u_t(x,0) = -2e^{-x^2}$.

6 $c = 1$, $q(x,t) = [4 - 3t + (4t - 8)x^2]e^{t-x^2}$,

$u(x,0) = 2e^{-x^2}$, $u_t(x,0) = e^{-x^2}$.

Answers to Odd-Numbered Exercises

1 $u(x,t) = t^2 e^{-2x^2}$.

3 $u(x,t) = 4(t+1)^2 e^{-x^2/4}$.

5 $u(x,t) = 2e^{-t-x^2}$.

8.1.3 Equilibrium Temperature in an Infinite Strip

To solve the BVP

$$u_{xx}(x,y) + u_{yy}(x,y) = 0, \quad 0 < x < 1, \quad -\infty < y < \infty,$$
$$u(0,y) = f_1(y), \quad u(1,y) = f_2(y), \quad -\infty < y < \infty,$$
$$u(x,y), u_y(x,y), f_1(y), f_2(y) \to 0 \ \text{ as } y \to \pm\infty, \ 0 < x < 1,$$

we set

$$\mathcal{F}[u](x,\omega) = U(x,\omega), \quad \mathcal{F}[f_1](\omega) = F_1(\omega), \quad \mathcal{F}[f_2](\omega) = F_2(\omega),$$

and reduce the BVP to the ODE problem

$$U''(x,\omega) - \omega^2 U(x,\omega) = 0, \quad 0 < x < 1,$$
$$U(0,\omega) = F_1(\omega), \quad U(1,\omega) = F_2(\omega).$$

Writing the general solution of the equation in the form (see Remark 1.4(i))

$$U(x,\omega) = C_1(\omega)\sinh(\omega x) + C_2(\omega)\sinh\left(\omega(x-1)\right)$$

and applying the BCs, we find that

$$U(x,\omega) = (\operatorname{csch}\omega)\left[F_2(\omega)\sinh(\omega x) - F_1(\omega)\sinh\left(\omega(x-1)\right)\right],$$

from which, if U is invertible,

$$u(x,t) = \mathcal{F}^{-1}[U](x,t)$$

$$= \frac{1}{\sqrt{2\pi}} \int\limits_{-\infty}^{\infty} (\operatorname{csch}\omega)\left[F_2(\omega)\sinh(\omega x)\right.$$
$$\left. - F_1(\omega)\sinh\left(\omega(x-1)\right)\right]e^{-i\omega x}\,d\omega. \qquad (8.10)$$

8.11 Remarks. (i) When f_1 and f_2 have certain specific forms, the inversion formula (8.10) may yield a result in terms of simple, familiar functions. In the great majority of cases, however, the inversion needs to be performed numerically.

(ii) The same procedure also works when the PDE is nonhomogeneous, and for other types of BCs.

8.12 Example. Consider the BVP

$$u_{xx}(x,y) + u_{yy}(x,y) = 2\big[x^2 - 2 + 2(1-x^2)y^2\big]e^{-y^2},$$
$$0 < x < 1, \quad -\infty < y < \infty,$$
$$u(0,y) = e^{-y^2}, \quad u(1,y) = 0, \quad -\infty < y < \infty,$$
$$u(x,y), \ u_y(x,y) \to 0 \ \text{ as } y \to \pm\infty, \ 0 < x < 1.$$

Instead of using the generic approach that leads to (8.10), we apply formulas 7 and 9 in the table in Appendix A.3 and move the problem to the transform domain, where, with $U(x,\omega) = \mathcal{F}[u](x,\omega)$, it becomes

$$U''(x,\omega) - \omega^2 U(x,\omega) = \tfrac{1}{\sqrt{2}}\big[(x^2-1)\omega^2 - 2\big]e^{-\omega^2/4},$$
$$U(0,\omega) = \tfrac{1}{\sqrt{2}}\, e^{-\omega^2/4}, \quad U(1,\omega) = 0.$$

The general solution of the ODE in the new problem is

$$U(x,\omega) = C_1(\omega)\sinh(\omega x) + C_2(\omega)\sinh\big(\omega(x-1)\big) + \tfrac{1}{\sqrt{2}}(1-x^2)e^{-\omega^2/4},$$

with the constants determined from the BCs as $C_1(\omega) = C_2(\omega) = 0$. Hence,

$$U(x,\omega) = \tfrac{1}{\sqrt{2}}(1-x^2)e^{-\omega^2/4},$$

from which, by inversion, we obtain

$$u(x,y) = \mathcal{F}^{-1}[U](x,y) = (1-x^2)e^{-y^2}.$$

VERIFICATION WITH MATHEMATICA®. The input

```
u = (1 - x^2) * E^(-y^2);
{q,f,g} = {2 * (x^2 - 2 + 2 * (1 - x^2) * y^2) * E^(-y^2), E^(-y^2), 0};
Simplify[{D[u,x,x] + D[u,y,y] - q, (u /. x -> 0) - f, (u /. x -> 1) - g}]
```

generates the output $\{0, 0, 0\}$.

8.13 Example. Applying the same method to the BVP

$$u_{xx}(x,y) + u_{yy}(x,y) = 4(x - 2xy^2)e^{-y^2}, \quad 0 < x < 1, \ -\infty < y < \infty,$$
$$u(0,y) = 0, \quad u_x(1,y) = -2e^{-y^2}, \quad -\infty < y < \infty,$$
$$u(x,y), \ u_y(x,y) \to 0 \ \text{ as } y \to \pm\infty, \ 0 < x < 1,$$

we convert it into the ODE problem

$$U''(x,\omega) - \omega^2 U(x,\omega) = \sqrt{2}\,\omega^2 x e^{-\omega^2/4}, \quad 0 < x < 1,$$
$$U(0,\omega) = 0, \quad U'(1,\omega) = -\sqrt{2}\, e^{-\omega^2/4},$$

where the nonzero right-hand sides have been computed with formula 7 in Appendix A.3. The general solution of the equation is (see Remark 1.4(i))

$$U(x,\omega) = C_1(\omega)\sinh(\omega x) + C_2(\omega)\cosh\left(\omega(x-1)\right) - \sqrt{2}\,xe^{-\omega^2/4},$$

from which, using the BCs, we now obtain $C_1(\omega) = C_2(\omega) = 0$. Hence,

$$U(x,\omega) = -\sqrt{2}\,xe^{-\omega^2/4},$$

so, again by formula 7,

$$u(x,y) = \mathcal{F}^{-1}[U](x,y) = -2xe^{-y^2}.$$

VERIFICATION WITH MATHEMATICA®. The input

```
u = -2*x*E^(-y^2);
{q, f1, f2} = 4*(x-2*x*y^2)*E^(-y^2), 0, -2*E^(-y^2);
Simplify[{D[u,x,x] +D[u,y,y] -q, (u/. x->0) -f1,
   (D[u,x] /. x->1) -f2}]
```

generates the output $\{0, 0, 0\}$.

8.14 Example. The above technique can also be applied to the BVP

$$u_{xx}(x,y) + u_{yy}(x,y) = 2\big[y^2 - y - 4 + 2(3 + y - y^2)x^2\big]e^{-x^2},$$
$$-\infty < x < \infty,\ 0 < y < 1,$$
$$u(x,y),\ u_x(x,y) \to 0\ \text{ as } x \to \pm\infty,\ 0 < y < 1,$$
$$u_y(x,0) = e^{-x^2},\quad u(x,1) = 3e^{-x^2},\quad -\infty < x < \infty,$$

which, with $\mathcal{L}[u](\omega, y) = U(\omega, y)$, is reduced to the simpler problem

$$U''(\omega,y) - \omega^2 U(\omega,y) = \frac{1}{\sqrt{2}}\,e^{-\omega^2/4}\big[(y^2 - y - 3)\omega^2 - 2\big],\quad 0 < y < 1,$$
$$U'(\omega,0) = \frac{1}{\sqrt{2}}\,e^{-\omega^2/4},\quad U(\omega,1) = \frac{3}{\sqrt{2}}\,e^{-\omega^2/4}.$$

Given the nature of the BCs, it is convenient to write the GS of the above ODE in the form

$$U(\omega,y) = C_1(\omega)\cosh(\omega y) + C_2(\omega)\sinh\left(\omega(y-1)\right) + \frac{1}{\sqrt{2}}\,e^{-\omega^2/4}(3 + y - y^2).$$

Then, evaluating U and U' at $y = 0$, we obtain $C_1(\omega) = C_2(\omega) = 0$. Hence,

$$U(\omega,y) = \frac{1}{\sqrt{2}}\,e^{-\omega^2/4}(3 + y - y^2),$$

and, by inversion,

$$u(x, y) = \mathcal{F}^{-1}[U](x, y) = (3 + y - y^2)e^{-x^2}.$$

VERIFICATION WITH MATHEMATICA®. The input

```
u = (3 + y - y^2) * E^( - x^2);
{q, f1, f2} = {2 * [y^2 - y - 4 + 2 * (3 + y - y^2) * x^2] * E^( - x^2),
  E^( - x^2), 3 * E^( - x^2)};
Simplify[{D[u,x,x] + D[u,y,y] - q, {(D[u,y] /.y -> 0) - f1,
  (u /.y -> 1) - f2}}]
```

generates the output $\{0, \{0, 0\}\}$.

Exercises

Use the full Fourier transformation to solve the BVP consisting of the PDE

$$u_{xx}(x, y) + u_{yy}(x, y) = q(x, y)$$

and ranges for x, y, function q, BCs, and asymptotic decay of the solution as indicated.

1 $0 < x < 1$, $-\infty < y < \infty$, $q(x, y) = 2x(2y^2 - 1)e^{-y^2}$,

$u(0, y) = 0$, $u(1, y) = e^{-y^2}$, $u(x, y), u_y(x, y) \to 0$ as $y \to \pm\infty$.

2 $-\infty < x < \infty$, $0 < y < 1$, $q(x, y) = 4(y - 1)(4x^2 - 1)e^{-2x^2}$,

$u(x, 0) = -e^{-2x^2}$, $u(x, 1) = 0$, $u(x, y), u_x(x, y) \to 0$ as $x \to \pm\infty$.

3 $-\infty < x < \infty$, $0 < y < 1$, $q(x, y) = (2 - y^2 + x^2y^2)e^{-x^2/2}$,

$u_y(x, 0) = 0$, $u(x, 1) = e^{-x^2/2}$, $u(x, y), u_x(x, y) \to 0$ as $x \to \pm\infty$.

4 $0 < x < 1$, $-\infty < y < \infty$, $q(x, y) = \left[2 + 6x - 2x^2 + (4x^2 - 12x)y^2\right]e^{-y^2}$,

$u(0, y) = 0$, $u_x(1, y) = -e^{-y^2}$, $u(x, y), u_y(x, y) \to 0$ as $y \to \pm\infty$.

5 $0 < x < 1$, $-\infty < y < \infty$, $q(x, y) = \left[3e^x - 12 + (48 - 16e^x)y^2\right]e^{-2y^2}$,

$u_x(0, y) = -e^{-2y^2}$, $u_x(1, y) = -e^{1-2y^2}$,

$u(x, y), u_y(x, y) \to 0$ as $y \to \pm\infty$.

6 $-\infty < x < \infty$, $0 < y < 1$, $q(x, y) = (4x^2y - y - 2)e^{-x^2-y}$,

$u(x, 0) = 0$, $u_y(x, 1) = 0$, $u(x, y), u_x(x, y) \to 0$ as $x \to \pm\infty$.

Answers to Odd-Numbered Exercises

1 $u(x, y) = xe^{-y^2}$. **3** $u(x, y) = y^2 e^{-x^2/2}$. **5** $u(x, y) = (3 - e^x)e^{-2y^2}$.

8.2   The Fourier Sine and Cosine Transformations

The Fourier sine and cosine transforms, constructed as generalizations of the corresponding Fourier sine and cosine series, are defined for functions f that are piecewise continuous on $0 < x < \infty$ by

$$\mathcal{F}_S[f](\omega) = F(\omega) = \sqrt{\frac{2}{\pi}} \int_0^\infty f(x)\sin(\omega x)\, dx,$$

$$\mathcal{F}_C[f](\omega) = F(\omega) = \sqrt{\frac{2}{\pi}} \int_0^\infty f(x)\cos(\omega x)\, dx.$$

It can be shown that these transforms exist if f is absolutely integrable on $(0, \infty)$; that is,

$$\int_0^\infty |f(x)|\, dx < \infty.$$

The corresponding inverse transforms are

$$\mathcal{F}_S^{-1}[F](x) = f(x) = \sqrt{\frac{2}{\pi}} \int_0^\infty F(\omega)\sin(\omega x)\, d\omega,$$

$$\mathcal{F}_C^{-1}[F](x) = f(x) = \sqrt{\frac{2}{\pi}} \int_0^\infty F(\omega)\cos(\omega x)\, d\omega.$$

The operators $\mathcal{F}_S$ and $\mathcal{F}_C$ are called the *Fourier sine transformation* and *Fourier cosine transformation*, respectively, while $\mathcal{F}_S^{-1}$ and $\mathcal{F}_C^{-1}$ are their inverse transformations. The comment made at the end of Remark 8.1(i) applies here as well.

8.15 Example. The Fourier sine and cosine transforms of the function

$$f(x) = \begin{cases} 1, & 0 \le x \le a, \\ 0, & x > a, \end{cases} \qquad a = \text{const},$$

are

$$\mathcal{F}_S[f](\omega) = \sqrt{\frac{2}{\pi}} \int_0^\infty f(x)\sin(\omega x)\, dx = \sqrt{\frac{2}{\pi}} \int_0^a \sin(\omega x)\, dx$$

$$= \sqrt{\frac{2}{\pi}}\, \frac{1}{\omega}\, [1 - \cos(a\omega)],$$

$$\mathcal{F}_C[f](\omega) = \sqrt{\frac{2}{\pi}} \int_0^\infty f(x)\cos(\omega x)\, dx = \sqrt{\frac{2}{\pi}} \int_0^a \cos(\omega x)\, dx$$

$$= \sqrt{\frac{2}{\pi}}\, \frac{1}{\omega}\, \sin(a\omega).$$

8.16 Remark. $\mathcal{F}_S$ ($\mathcal{F}_C$) can also be defined for functions on $\mathbb{R}$ if these functions are odd (even).

8.17 Theorem. (i) $\mathcal{F}_S$ and $\mathcal{F}_C$ are linear operators.

(ii) If $u = u(x,t)$ and $u(x,t) \to 0$ as $x \to \infty$, then
$$\mathcal{F}_S[u_x](\omega,t) = -\omega \mathcal{F}_C[u](\omega,t),$$
$$\mathcal{F}_C[u_x](\omega,t) = -\sqrt{\frac{2}{\pi}}\, u(0,t) + \omega \mathcal{F}_S[u](\omega,t).$$

(iii) If, in addition, $u_x(x,t) \to 0$ as $x \to \infty$, then
$$\mathcal{F}_S[u_{xx}](\omega,t) = \sqrt{\frac{2}{\pi}}\, \omega u(0,t) - \omega^2 \mathcal{F}_S[u](\omega,t),$$
$$\mathcal{F}_C[u_{xx}](\omega,t) = -\sqrt{\frac{2}{\pi}}\, u_x(0,t) - \omega^2 \mathcal{F}_C[u](\omega,t).$$

(iv) Time differentiation commutes with both the Fourier sine and cosine transformations:
$$\mathcal{F}_S[u_t](\omega,t) = \big(\mathcal{F}_S[u]\big)_t(\omega,t),$$
$$\mathcal{F}_C[u_t](\omega,t) = \big(\mathcal{F}_C[u]\big)_t(\omega,t).$$

As Theorem 8.17(iii) indicates, the choice between using the sine transformation or the cosine transformation in the solution of a given IBVP depends on the type of BC prescribed at $x = 0$.

Brief lists of Fourier sine and cosine transforms are given in Appendix A.4 and Appendix A.5, respectively.

8.2.1 Heat Conduction in a Semi-Infinite Rod

The process of heat conduction in a long rod for which the temperature at the near endpoint is prescribed while the effects of the conditions at the far endpoint are negligible is modeled by the IBVP
$$u_t(x,t) = ku_{xx}(x,t), \quad x > 0, \ t > 0,$$
$$u(0,t) = g(t), \quad u(x,t), \ u_x(x,t) \to 0 \ \text{as } x \to \infty, \ t > 0,$$
$$u(x,0) = f(x), \quad x > 0.$$

In view of the given BC, we use the Fourier sine transformation; thus, let
$$\mathcal{F}_S[u](\omega,t) = U(\omega,t), \quad \mathcal{F}_S[f](\omega) = F(\omega).$$

Applying $\mathcal{F}_S$ to the PDE and IC and using the properties in Theorem 8.17, we arrive at the ODE problem
$$U'(\omega,t) + k\omega^2 U(\omega,t) = \sqrt{\frac{2}{\pi}}\, k\omega g(t), \quad t > 0,$$
$$U(\omega,0) = F(\omega).$$

After finding U in the transform domain, we obtain the solution of the original IBVP as $u(x,t) = \mathcal{F}_S^{-1}[U](x,t)$, if U is invertible.

8.18 Example. The IBVP

$$u_t(x,t) = u_{xx}(x,t), \quad x > 0, \ t > 0,$$

$$u(0,t) = 1, \quad u(x,t), \ u_x(x,t) \to 0 \ \text{ as } x \to \infty, \ t > 0,$$

$$u(x,0) = 0, \quad x > 0$$

requires the use of the Fourier sine transformation because it is the temperature that is prescribed in the BC.

By Theorem 8.17, the transform $U(\omega,t) = \mathcal{F}_S[u](\omega,t)$ of u satisfies

$$U'(\omega,t) + \omega^2 U(\omega,t) = \sqrt{\frac{2}{\pi}}\,\omega, \quad t > 0,$$

$$U(\omega,0) = 0.$$

The solution of the transformed problem is

$$U(\omega,t) = \sqrt{\frac{2}{\pi}}\,\frac{1}{\omega}\left(1 - e^{-\omega^2 t}\right);$$

so, by formula 12 (with $a = 1/(2\sqrt{t})$) in Appendix A.4,

$$u(x,t) = \mathcal{F}_S^{-1}[U](x,t) = \operatorname{erfc}\frac{x}{2\sqrt{t}}.$$

VERIFICATION WITH MATHEMATICA®. The input

```
u = Erfc[x/(2*t^(1/2))];
{f, g} = {0, 1};
Simplify[{D[u,t] - D[u,x,x], Assuming[x>0, Limit[u, t->0,
    Direction -> -1]] - f, (u/. x->0) - g}]
```

generates the output $\{0,0,0\}$.

8.19 Example. The IBVP

$$u_t(x,t) = 2u_{xx}(x,t) + 2(x - 2t - 1)e^{-x}, \quad x > 0, \ t > 0,$$

$$u_x(0,t) = -2t - 1, \quad u(x,t), \ u_x(x,t) \to 0 \ \text{ as } x \to \infty, \ t > 0,$$

$$u(x,0) = -xe^{-x}, \quad x > 0$$

can be solved by means of the Fourier cosine transformation since the BC at $x = 0$ prescribes the value of u_x and not that of u. Thus, with the notation

$\mathcal{F}_C[u](\omega, t) = U(\omega, t)$, using formulas 2, 5, and 6 in Appendix A.5, and performing a simple algebraic manipulation of the coefficients, we convert the IBVP into the ODE problem

$$U'(\omega, t) + 2\omega^2 U(\omega, t) = 2\sqrt{\frac{2}{\pi}} \left\{ \frac{2\omega^2}{1 + \omega^2} t + \frac{1 + \omega^4}{(1 + \omega^2)^2} \right\}, \quad t > 0,$$

$$U(\omega, 0) = -\sqrt{\frac{2}{\pi}} \frac{1 - \omega^2}{(1 + \omega^2)^2}.$$

The general solution of the equation is of the form

$$U(\omega, t) = C(\omega)e^{-2\omega^2 t} + A(\omega)t + B(\omega).$$

Determining first the coefficients $A(\omega)$ and $B(\omega)$ and then applying the IC, we find that

$$C(\omega) = 0$$

and that

$$U(\omega, t) = \sqrt{\frac{2}{\pi}} \left\{ \frac{2}{1 + \omega^2} t - \frac{1 - \omega^2}{(1 + \omega^2)^2} \right\},$$

which, by formulas 5 and 6, yields

$$u(x, t) = \mathcal{F}_C^{-1}[U](x, t) = (2t - x)e^{-x}.$$

VERIFICATION WITH MATHEMATICA®. The input

```
u = (2 * t - x) * E^(-x);
{q, f, g} = {2 * (x - 2 * t - 1) * E^(-x), -x * E^(-x), -2 * t - 1};
Simplify[{D[u,t] - 2 * D[u,x,x] - q, (u /. t -> 0) - f,
    (D[u,x] /. x -> 0) - g}]
```

generates the output $\{0, 0, 0\}$.

Exercises

Use a suitable Fourier transformation to solve the IBVP consisting of the PDE

$$u_t(x, t) = ku_{xx}(x, t) + q(x, t), \quad x > 0, \ t > 0,$$

where

$$u(x, t), \ u_x(x, t) \to 0 \text{ as } x \to \infty,$$

and constant k, function q, BC, and IC as indicated.

1 $k = 1$, $q(x, t) = (1 - 3t + x)e^{-x}$, $u(0, t) = 3t$, $u(x, 0) = -xe^{-x}$.

2 $k = 1$, $q(x, t) = (4t - 4x - 5)e^{-2x}$, $u(0, t) = 2 - t$, $u(x, 0) = (2 + x)e^{-2x}$.

3 $k = 1$, $q(x,t) = (2x - t - 3)e^{-x}$, $u_x(0,t) = -t - 2$, $u(x,0) = -2xe^{-x}$.

4 $k = 1$, $q(x,t) = 2(2x + 4t - 5)e^{-2x}$, $u_x(0,t) = 4t - 3$,

$u(x,0) = (1 - x)e^{-2x}$.

5 $k = 2$, $q(x,t) = 2\big[4t - 1 + (2 - t)x\big]e^{-x/2}$, $u_x(0,t) = 2(2t - 1)$,

$u(x,0) = 4e^{-x/2}$.

6 $k = 2$, $q(x,t) = \big[(2t - 1)x - 4 - 4t\big]e^{-x}$, $u_x(0,t) = -t - 2$,

$u(x,0) = 2e^{-x}$.

7 $k = 2$, $q(x,t) = \big[4t - 2 + (2 - t)x\big]e^{-x/2}$ $u(0,t) = 4$, $u(x,0) = 4e^{-x/2}$.

8 $k = 2$, $q(x,t) = 2\big[4t + 1 + (1 - 2t)x\big]e^{-x}$, $u(0,t) = -1$, $u(x,0) = -e^{-x}$.

9 $k = 1$, $q(x,t) = -(8 + 5e^{-t})e^{-2x}$, $u(0,t) = 2 + e^{-t}$, $u(x,0) = 3e^{-2x}$.

10 $k = 1$, $q(x,t) = 2(5e^{-t} - 2)e^{-2x}$, $u_x(0,t) = 2(2e^{-t} - 1)$,

$u(x,0) = -e^{-2x}$.

11 $k = 2$, $q(x,t) = 2(4 - 3t)e^{2t - x/2}$, $u_x(0,t) = 2(t - 2)e^{2t}$,

$u(x,0) = 8e^{-x/2}$.

12 $k = 2$, $q(x,t) = -(2t + 3)e^{t - x/2}$, $u(0,t) = 2(1 - 2t)e^t$, $u(x,0) = 2e^{-x/2}$.

Answers to Odd-Numbered Exercises

1 $u(x,t) = (3t - x)e^{-x}$.

3 $u(x,t) = (t - 2x)e^{-x}$.

5 $u(x,t) = 4(1 + xt)e^{-x/2}$.

7 $u(x,t) = 2(2 + xt)e^{-x/2}$.

9 $u(x,t) = (2 + e^{-t})e^{-2x}$.

11 $u(x,t) = 4(2 - t)e^{2t - x/2}$.

8.2.2 Vibrations of a Semi-Infinite String

The method applied in this case is similar to that used above.

8.20 Example. The solution of the IBVP

$$u_{tt}(x,t) = 4u_{xx}(x,t), \quad x > 0, \ t > 0,$$
$$u_x(0,t) = 0, \quad u(x,t), \ u_x(x,t) \to 0 \ \text{ as } x \to \infty, \ t > 0,$$
$$u(x,0) = e^{-x}, \quad u_t(x,0) = 0, \quad x > 0$$

is obtained by means of the Fourier cosine transformation because the BC prescribes the x-derivative of u at $x = 0$. According to formula 5 (with

$a = 1$) in Appendix A.5, the Fourier cosine transform of the function e^{-x} is $\sqrt{2/\pi}\,(1+\omega^2)^{-1}$; hence, by Theorem 8.17, $U(\omega, t) = \mathcal{F}_C[u](\omega, t)$ satisfies

$$U''(\omega, t) + 4\omega^2 U(\omega, t) = 0, \quad t > 0,$$

$$U(\omega, 0) = \sqrt{\frac{2}{\pi}}\,\frac{1}{1+\omega^2}, \quad U'(\omega, 0) = 0,$$

with solution

$$U(\omega, t) = \sqrt{\frac{2}{\pi}}\,\frac{\cos(2\omega t)}{1+\omega^2}.$$

Then, by formula 14 (with $a = 2t$ and $b = 1$), we find that the solution of the IBVP is

$$u(x, t) = \mathcal{F}_C^{-1}[U](x, t) = \begin{cases} e^{-x}\cosh(2t), & t \le x/2, \\ e^{-2t}\cosh x, & t > x/2. \end{cases}$$

It is obvious that this solution is continuous across the line $x = 2t$.

VERIFICATION WITH MATHEMATICA®. Because u is defined piecewise, we use the input

```
{u1, u2} = {E^(-x) * Cosh[2*t], E^(-2*t) * Cosh[x]};
{q, f, g, h} = {0, E^(-x), 0, 0};
Simplify[{D[{u1,u2},t,t] - 4*D[{u1,u2},x,x] - q,
(D[u2,x] /. x->0) - h, (u1 /. t->0) - f, (D[u1,t] /. t->0) - g}]
```

to generate the output $\{\{0, 0\}, 0, 0, 0\}$.

8.21 Example. In view of the BC at $x = 0$, we solve the IBVP

$$u_{tt}(x, t) = u_{xx}(x, t) + (x + t - 3)e^{-x}, \quad x > 0, \ t > 0,$$

$$u(0, t) = 1 - t, \quad u(x, t), \ u_x(x, t) \to 0 \ \text{as} \ x \to \infty, \ t > 0,$$

$$u(x, 0) = (1 - x)e^{-x}, \quad u_t(x, 0) = -e^{-x}, \quad x > 0$$

by means of the Fourier sine transformation. Therefore, making the notation $\mathcal{F}_S[u](\omega, t) = U(\omega, t)$, using formulas 2, 6, and 7 in Appendix A.4, and reorganizing the coefficients, we reduce the IBVP to the ODE problem

$$U''(\omega, t) + \omega^2 U(\omega, t) = -\sqrt{\frac{2}{\pi}}\left\{\frac{\omega^3}{1+\omega^2}t + \frac{\omega^3(1-\omega^2)}{(1+\omega^2)^2}\right\}, \quad t > 0,$$

$$U(\omega, 0) = -\sqrt{\frac{2}{\pi}}\,\frac{\omega(1-\omega^2)}{(1+\omega^2)^2}, \quad U'(\omega, 0) = -\sqrt{\frac{2}{\pi}}\,\frac{\omega}{1+\omega^2}.$$

The general solution of the equation is

$$U(\omega, t) = C_1(\omega)\cos(\omega t) + C_2(\omega)\sin(\omega t) + A(\omega)t + B(\omega),$$

where, in fact, $A(\omega)t + B(\omega)$ is equal to the right-hand side divided by ω^2. Applying the ICs, we find that $C_1(\omega) = C_2(\omega) = 0$; hence, using partial fraction decomposition, we have

$$\frac{\omega(1 - \omega^2)}{(1 + \omega^2)^2} = -\frac{\omega}{1 + \omega^2} + \frac{2\omega}{(1 + \omega^2)^2},$$

so

$$U(\omega, t) = -\sqrt{\frac{2}{\pi}} \left\{ \frac{\omega}{1 + \omega^2} t + \frac{\omega(1 - \omega^2)}{(1 + \omega^2)^2} \right\}$$

$$= \sqrt{\frac{2}{\pi}} \left\{ \frac{\omega}{1 + \omega^2} (1 - t) - \frac{2\omega}{(1 + \omega^2)^2} \right\},$$

from which, by formulas 6 and 7,

$$u(x, t) = \mathcal{F}_S^{-1}[U](x, t) = (1 - t - x)e^{-x}.$$

VERIFICATION WITH MATHEMATICA®. The input

```
u = (1 - t - x) * E^( - x);
{q, f, g, h} = {(x + t - 3) * E^( - x), (1 - x) * E^( - x), - E^( - x), 1 - t};
Simplify[{D[u,t,t] - D[u,x,x] - q, (u /. x -> 0) - h, {(u /. t -> 0) - f,
    (D[u,t] /. t -> 0) - g}}]
```

generates the output $\{0, 0, \{0, 0\}\}$.

Exercises

Use a suitable Fourier transformation to solve the IBVP consisting of the PDE

$$u_{tt}(x, t) = c^2 u_{xx}(x, t) + q(x, t), \quad x > 0, \ t > 0,$$

where

$$u(x, t), \ u_x(x, t) \to 0 \text{ as } x \to \infty,$$

and constant c, function q, BC, and ICs as indicated.

1 $c = 2$, $q(x, t) = (8 + t - 2x)e^{-x/2}$, $u(0, t) = -t$,
$u(x, 0) = 2xe^{-x/2}$, $u_t(x, 0) = -e^{-x/2}$.

2 $c = 2$, $q(x, t) = 4(3t - x)e^{-x}$, $u(0, t) = 2 - 3t$,
$u(x, 0) = (x + 2)e^{-x}$, $u_t(x, 0) = -3e^{-x}$.

3 $c = 2$, $q(x, t) = 2(8 + 3t - 2x)e^{-x/2}$, $u_x(0, t) = 4 + 3t$,
$u(x, 0) = 4xe^{-x/2}$, $u_t(x, 0) = -6e^{-x/2}$.

4 $c = 2$, $q(x,t) = 4(1 + t - 2x)e^{-x}$, $u_x(0,t) = t - 1$,
  $u(x,0) = (2x + 3)e^{-x}$, $u_t(x,0) = -e^{-x}$.

5 $c = 1$, $q(x,t) = (2 + t^2 - 2x)e^{-x}$, $u_x(0,t) = t^2 + 2$,
  $u(x,0) = 2xe^{-x}$, $u_t(x,0) = 0$.

6 $c = 1$, $q(x,t) = 2(4x - 2t^2 - 2t - 3)e^{-2x}$, $u_x(0,t) = -2(t^2 + t + 1)$,
  $u(x,0) = -2xe^{-2x}$, $u_t(x,0) = e^{-2x}$.

7 $c = 1$, $q(x,t) = (4 - t^2 - x)e^{-x}$, $u(0,t) = t^2$,
  $u(x,0) = xe^{-x}$, $u_t(x,0) = 0$.

8 $c = 1$, $q(x,t) = (4t^2 - 8t - 6 + 4x)e^{-2x}$, $u(0,t) = 2t - t^2$,
  $u(x,0) = -xe^{-2x}$, $u_t(x,0) = 2e^{-2x}$.

9 $c = 1$, $q(x,t) = 4(xt - t - 3)e^{-2x}$, $u(0,t) = 3$,
  $u(x,0) = 3e^{-2x}$, $u_t(x,0) = -xe^{-2x}$.

10 $c = 1$, $q(x,t) = 4(3xt - 3t - 1)e^{-2x}$, $u_x(0,t) = -2 - 3t$,
  $u(x,0) = e^{-2x}$, $u_t(x,0) = -3xe^{-2x}$.

11 $c = 2$, $q(x,t) = 2(2 - 3t)e^{2t-x/2}$, $u_x(0,t) = (t - 2)e^{2t}$,
  $u(x,0) = 4e^{-x/2}$, $u_t(x,0) = 6e^{-x/2}$.

12 $c = 2$, $q(x,t) = -4e^{-t-x/2}$, $u(0,t) = (2t + 1)e^{-t}$,
  $u(x,0) = e^{-x/2}$, $u_t(x,0) = e^{-x/2}$.

Answers to Odd-Numbered Exercises

1 $u(x,t) = (2x - t)e^{-x/2}$.

3 $u(x,t) = 2(2x - 3t)e^{-x/2}$.

5 $u(x,t) = (2x - t^2)e^{-x}$.

7 $u(x,t) = (x + t^2)e^{-x}$.

9 $u(x,t) = (3 - xt)e^{-2x}$.

11 $u(x,t) = 2(2 - t)e^{2t-x/2}$.

8.2.3 Equilibrium Temperature in a Semi-Infinite Strip

The steady-state temperature distribution in a semi-infinite strip $0 \le x \le L$, $y \ge 0$ is modeled by the BVP

$$u_{xx}(x,y) + u_{yy}(x,y) = 0, \quad 0 < x < L, \ y > 0,$$
$$u(0,y) = g_1(y), \quad u(L,y) = g_2(y), \quad y > 0,$$
$$g_1(y), g_2(y) \to 0 \quad \text{as } y \to \infty,$$
$$u(x,0) = f(x), \quad u(x,y), u_y(x,y) \to 0 \quad \text{as } y \to \infty, \ 0 < x < L.$$

Using the principle of superposition, we write the solution in the form

$$u(x, y) = u_1(x, y) + u_2(x, y),$$

where u_1 satisfies the given BVP with $g_1 = 0$ and $g_2 = 0$, and u_2 satisfies it with $f = 0$.

Thus, the first BVP is

$$(u_1)_{xx}(x, y) + (u_1)_{yy}(x, y) = 0, \quad 0 < x < L, \ y > 0,$$

$$u_1(0, y) = 0, \quad u_1(L, y) = 0, \quad y > 0,$$

$$u_1(x, 0) = f(x), \quad u_1(x, y), \ (u_1)_y(x, y) \to 0 \ \text{ as } y \to \infty, \ 0 < x < L,$$

which is solved by the method of separation of variables. Proceeding as in Section 5.3, from the PDE and BCs at $x = 0$ and $x = L$ we obtain

$$u_1(x, y) = \sum_{n=1}^{\infty} \sin \frac{n\pi x}{L} \left(A_n \cosh \frac{n\pi y}{L} + B_n \sinh \frac{n\pi y}{L} \right)$$

$$= \sum_{n=1}^{\infty} \sin \frac{n\pi x}{L} \left(a_n \, e^{n\pi y/L} + b_n \, e^{-n\pi y/L} \right).$$

Since $u_1(x, y) \to 0$ as $y \to \infty$, we must have $a_n = 0$, $n = 1, 2, \ldots$; hence,

$$u_1(x, y) = \sum_{n=1}^{\infty} b_n \sin \frac{n\pi x}{L} e^{-n\pi y/L}. \tag{8.11}$$

The BC at $y = 0$ now yields

$$u_1(x, 0) = f(x) = \sum_{n=1}^{\infty} b_n \sin \frac{n\pi x}{L},$$

where, by (2.12), the Fourier sine series coefficients b_n are

$$b_n = \frac{2}{L} \int_0^L f(x) \sin \frac{n\pi x}{L} \, dx, \quad n = 1, 2, \ldots.$$

Consequently, the solution of the first BVP is given by (8.11) with the b_n computed by means of the above formula.

The second BVP is

$$(u_2)_{xx}(x, y) + (u_2)_{yy}(x, y) = 0, \quad 0 < x < L, \ y > 0,$$

$$u_2(0, y) = g_1(y), \quad u_2(L, y) = g_2(y), \ y > 0,$$

$$g_1(y), \, g_2(y) \to 0 \quad \text{as } y \to \infty,$$

$$u_2(x, 0) = 0, \quad u_2(x, y) \to 0, \ (u_2)_y(x, y) \to 0 \ \text{ as } y \to \infty, \ 0 < x < L.$$

In view of the BC at $y = 0$, we use $\mathcal{F}_S$ with respect to y and, assuming that g_1 and g_2 are transformable, make the notation

$$\mathcal{F}_S[u_2](x,\omega) = U(x,\omega), \quad \mathcal{F}_S[g_1](\omega) = G_1(\omega), \quad \mathcal{F}_S[g_2](\omega) = G_2(\omega).$$

Then the above BVP reduces to

$$U''(x,\omega) - \omega^2 U(x,\omega) = 0, \quad 0 < x < L,$$
$$U(0,\omega) = G_1(\omega), \quad U(L,\omega) = G_2(\omega).$$

The general solution of the equation can be written in terms of hyperbolic functions as (see Remark 1.4(i))

$$U(x,\omega) = C_1(\omega)\sinh(\omega x) + C_2(\omega)\sinh\big(\omega(x - L)\big),$$

where C_1 and C_2 are arbitrary functions of ω. Using the BCs at $x = 0$ and $x = L$, we find that

$$U(x,\omega) = \operatorname{csch}(\omega L)\big[G_2(\omega)\sinh(\omega x) - G_1(\omega)\sinh\big(\omega(x - L)\big)\big]. \qquad (8.12)$$

The solution u_2 is now obtained by applying $\mathcal{F}_S^{-1}$ to the above equality, if U is invertible.

Remarks 8.11 remain valid in this case as well.

8.22 Example. For the BVP

$$u_{xx}(x,y) + u_{yy}(x,y) = \big[1 - 2x^2 + (2 + x^2)y\big]e^{-y},$$
$$0 < x < 1, \quad y > 0,$$
$$u(0,y) = e^{-y}, \quad u(1,y) = (1 + y)e^{-y}, \quad y > 0,$$
$$u(x,0) = 1, \quad u(x,y), u_y(x,y) \to 0 \text{ as } y \to \infty, \ 0 < x < 1,$$

formula (8.12) proves rather awkward, so we prefer to solve the problem from first principles. Thus, the BC at $y = 0$ indicates that we need to use the Fourier sine transformation with respect to y; therefore, using formulas 6 and 7 in Appendix A.4 and sorting out the terms, we see that $U(x,\omega) = \mathcal{F}_S[u](x,\omega)$ is the solution of the ODE problem

$$U''(x,\omega) - \omega^2 U(x,\omega) = \sqrt{\frac{2}{\pi}}\left\{\frac{\omega(4 - \omega^2 - \omega^4)}{(1 + \omega^2)^2} - \frac{2\omega^3}{(1 + \omega^2)^2}x^2\right\},$$
$$0 < x < 1,$$
$$U(0,\omega) = \sqrt{\frac{2}{\pi}}\frac{\omega}{1 + \omega^2}, \quad U(1,\omega) = \sqrt{\frac{2}{\pi}}\frac{\omega(3 + \omega^2)}{(1 + \omega^2)^2}.$$

The general solution of the equation above is

$$U(x, \omega) = C_1(\omega) \sinh(\omega x) + C_2(\omega) \sinh\big(\omega(1-x)\big) + A(\omega)x^2 + B(\omega)x + C(\omega)$$

$$= C_1(\omega) \sinh(\omega x) + C_2(\omega) \sinh\big(\omega(1-x)\big)$$

$$+ \sqrt{\frac{2}{\pi}}\left\{ \frac{\omega}{1+\omega^2} + \frac{2\omega}{(1+\omega^2)^2}x^2 \right\},$$

and the BCs at $x = 0$ and $x = 1$ yield $C_1(\omega) = C_2(\omega) = 0$, which means that

$$U(x, \omega) = \sqrt{\frac{2}{\pi}}\left\{ \frac{\omega}{1+\omega^2} + \frac{2\omega}{(1+\omega^2)^2}x^2 \right\};$$

hence,

$$u(x, y) = \mathcal{F}_S^{-1}[U](x, y) = (1 + x^2 y)e^{-y}.$$

VERIFICATION WITH MATHEMATICA®. The input

```
u = (1+x^2*y)*E^(-y);
{f,g,h}={(1-2*x^2+(2+x^2)*y)*E^(-y), E^(-y),
   (1+y)*E^(-y), 1};
Simplify[{D[u,x,x]+D[u,y,y]-q, {(u/.x->0)-f, (u/.x->1)-g},
   (u/.y->0)-h}]
```

generates the output $\{0, \{0, 0\}, 0\}$.

8.23 Example. Formula (8.12) is also unhelpful in the case of the BVP

$$u_{xx}(x, y) + u_{yy}(x, y) = 4(3 + 2y - 2xy)e^{-2x}, \quad x > 0, \ 0 < y < 1,$$

$$u_x(0, y) = -6 - 2y, \quad u(x, y), \ u_x(x, y) \to 0 \ \text{as } x \to \infty, \ 0 < y < 1,$$

$$u(x, 0) = 3e^{-2x}, \quad u_y(x, 1) = -2xe^{-2x}, \quad x > 0;$$

consequently, we resort again to solution from first principles. First, the BC at $x = 0$ tells us that we have to use the Fourier cosine transformation with respect to x, so, by formulas 6 and 7 in Appendix A.5, $U(\omega, y) = \mathcal{F}_C[u](\omega, y)$ satisfies the ODE problem

$$U''(\omega, y) - \omega^2 U(\omega, y) = \sqrt{\frac{2}{\pi}}\left\{ -\frac{6\omega^2}{4+\omega^2} + \frac{2\omega^2(4-\omega^2)}{(4+\omega^2)^2}y \right\}, \quad 0 < y < 1,$$

$$U(\omega, 0) = \sqrt{\frac{2}{\pi}}\frac{6}{4+\omega^2}, \quad U'(\omega, 1) = -\sqrt{\frac{2}{\pi}}\frac{2(4-\omega^2)}{(4+\omega^2)^2}.$$

Here, the GS of the equation is best written as

$$U(\omega, y) = C_1(\omega) \sinh(\omega y) + C_2(\omega) \cosh\big(\omega(y-1)\big) + A(\omega)y + B(\omega)$$

$$= C_1(\omega) \sinh(\omega y) + C_2(\omega) \cosh\big(\omega(y-1)\big)$$

$$+ \sqrt{\frac{2}{\pi}}\left\{ \frac{6}{4+\omega^2} - \frac{2(4-\omega^2)}{(4+\omega^2)^2}y \right\},$$

which allows us to find that $C_1(\omega) = C_2(\omega) = 0$; therefore,

$$U(\omega, y) = \sqrt{\frac{2}{\pi}} \left\{ 3 \cdot \frac{2}{4 + \omega^2} - \frac{4 - \omega^2}{(4 + \omega^2)^2} \cdot 2y \right\},$$

and, by inversion,

$$u(x, y) = \mathcal{F}_C^{-1}[U](x, y) = 3e^{-2x} - 2yxe^{-2x} = (3 - 2xy)e^{-2x}.$$

VERIFICATION WITH MATHEMATICA®. The input

```
u = (3 - 2*x*y) * E^(-2*x);
{q,f,g,h}={4*(3+2*y-2*x*y)*E^(-2*x), -6-2*y,
  3*E^(-2*x), -2*x*E^(-2*x),
Simplify[{D[u,x,x]+D[u,y,y]-q, (D[u,x]/.x->0)-f,
  {(u/.y->0)-g, (D[u,y]/.y->1)-h}}]
```

generates the output $\{0, 0, \{0, 0\}\}$.

Exercises

In **1–8**, use a suitable Fourier transformation to solve the BVP consisting of the PDE

$$u_{xx}(x, y) + u_{yy}(x, y) = q(x, y), \quad 0 < x < 1, \ y > 0,$$

where

$$u(x, y), \ u_y(x, y) \to 0 \text{ as } y \to \infty,$$

and function q and BCs as indicated.

1 $q(x, y) = (2 + 2x - y)e^{-y}, \quad u(0, y) = -ye^{-y}, \quad u(1, y) = (2 - y)e^{-y},$
$u(x, 0) = 2x.$

2 $q(x, y) = (2y - x^2 - 5)e^{-y}, \quad u(0, y) = (1 + 2y)e^{-y}, \quad u(1, y) = 2ye^{-y},$
$u(x, 0) = 1 - x^2.$

3 $q(x, y) = 4(3x + y - 1)e^{-2y}, \quad u(0, y) = ye^{-2y}, \quad u(1, y) = (3 + y)e^{-2y},$
$u_y(x, 0) = 1 - 6x.$

4 $q(x, y) = (14 + x^2 - y)e^{-y/2},$
$u(0, y) = 4(2 - y)e^{-y/2}, \quad u(1, y) = 4(3 - y)e^{-y/2},$
$u_y(x, 0) = -2(4 + x^2).$

5 $q(x, y) = (1 - 6x + 3xy)e^{-y}, \quad u(0, y) = e^{-y}, \quad u_x(1, y) = 3ye^{-y},$

   $u_y(x, 0) = 3x - 1.$

6 $q(x, y) = (xy - 4x - 2)e^{-y/2}, \quad u_x(0, y) = 4ye^{-y/2}, \quad u(1, y) = 4(y - 2)e^{-y/2},$

   $u_y(x, 0) = 4(1 + x).$

7 $q(x, y) = 4(3xy - 3x - 2)e^{-2y}, \quad u_x(0, y) = 3ye^{-2y}, \quad u(1, y) = (3y - 2)e^{-2y},$

   $u(x, 0) = -2.$

8 $q(x, y) = (2 - 4x + xy)e^{-y/2}, \quad u(0, y) = 8e^{-y/2}, \quad u_x(1, y) = 4ye^{-y/2},$

   $u(x, 0) = 8.$

In **9–16**, use a suitable Fourier transformation to solve the BVP consisting of
the PDE

$$u_{xx}(x, y) + u_{yy}(x, y) = q(x, y), \quad x > 0, \ 0 < y < 1,$$

where

$$u(x, y), \ u_x(x, y) \to 0 \text{ as } x \to \infty,$$

and function q and BCs as indicated.

9 $q(x, y) = 2(2x - 3 + 4y - 2y^2)e^{-2x}, \quad u(x, 0) = xe^{-2x}, \quad u(x, 1) = (1 + x)e^{-2x},$

   $u(0, y) = 2y - y^2.$

10 $q(x, y) = (3x - 2 + 2y^2)e^{-x}, \quad u(x, 0) = 3xe^{-x}, \quad u(x, 1) = (3x + 2)e^{-x},$

   $u(0, y) = 2y^2.$

11 $q(x, y) = (9 - x + 2y^2)e^{-x}, \quad u(x, 0) = (3 - x)e^{-x}, \quad u(x, 1) = (5 - x)e^{-x},$

   $u_x(0, y) = -4 - 2y^2.$

12 $q(x, y) = 4(3 - x + 4y^2)e^{-2x}, \quad u(x, 0) = -xe^{-2x}, \quad u(x, 1) = (4 - x)e^{-2x},$

   $u_x(0, y) = -1 - 8y^2.$

13 $q(x, y) = (3 - 10e^y)e^{-x/2}, \quad u_y(x, 0) = -8e^{-x/2}, \quad u(x, 1) = 4(3 - 2e)e^{-x/2},$

   $u_x(0, y) = 4e^y - 6.$

14 $q(x, y) = (6 + 5y)e^{-x - 2y}, \quad u(x, 0) = 2e^{-x}, \quad u_y(x, 1) = -5e^{-2 - x},$

   $u_x(0, y) = -(2 + y)e^{-2y}.$

15 $q(x, y) = -(31 + 51y)e^{2y - x/2}, \quad u(x, 0) = 4e^{-x/2}, \quad u_y(x, 1) = -28e^{2 - x/2},$

   $u(0, y) = 4(1 - 3y)e^{2y}.$

16 $q(x, y) = 2(1 - e^{-y})e^{-x}, \quad u_y(x, 0) = e^{-x}, \quad u(x, 1) = (2 - e^{-1})e^{-x},$

   $u(0, y) = 2 - e^{-y}.$

Answers to Odd-Numbered Exercises

1 $u(x, y) = (2x - y)e^{-y}$.

3 $u(x, y) = (3x + y)e^{-2y}$.

5 $u(x, y) = (1 + 3xy)e^{-y}$.

7 $u(x, y) = (3xy - 2)e^{-2y}$.

9 $u(x, y) = (x + 2y - y^2)e^{-2x}$.

11 $u(x, y) = (3 - x + 2y^2)e^{-x}$.

13 $u(x, y) = 4(3 - 2e^y)e^{-x/2}$.

15 $u(x, y) = 4(1 - 3y)e^{2y-x/2}$.

8.3 Other Applications

The Fourier transformation method may also be applied to other suitable problems, such as those mentioned in Section 4.4, including some with non-homogeneous PDEs.

8.24 Example. The IVP (Cauchy problem)

$$u_t(x, t) = u_{xx}(x, t) + 2u(x, t) + (1 - 4x^2t)e^{-x^2},$$

$$-\infty < x < \infty, \ t > 0,$$

$$u(x, t), \ u_x(x, t) \to 0 \quad \text{as } x \to \pm\infty, \ t > 0,$$

$$u(x, 0) = 0, \quad -\infty < x < \infty$$

models a one-dimensional diffusion process with a chain reaction and a source.
By formulas 7 and 9 (with $a = 1$) in Appendix A.3,

$$\mathcal{F}\left[(1 - 4x^2t)e^{-x^2}\right] = \frac{1}{\sqrt{2}}\left[(\omega^2 - 2)t + 1\right]e^{-\omega^2/4};$$

hence, setting $\mathcal{F}[u](\omega, t) = U(\omega, t)$ and using formula 2, we see that U is the solution of the transformed problem

$$U'(\omega, t) + (\omega^2 - 2)U(\omega, t) = \frac{1}{\sqrt{2}}\left[(\omega^2 - 2)t + 1\right]e^{-\omega^2/4}, \quad t > 0,$$

$$U(\omega, 0) = 0.$$

Direct calculation shows that

$$U(\omega, t) = \frac{1}{\sqrt{2}} te^{-\omega^2/4},$$

so, by formula 8, the solution of the given IVP is

$$u(x, t) = \mathcal{F}^{-1}[U](x, t) = te^{-x^2}.$$

VERIFICATION WITH MATHEMATICA®. The input

```
u = t * E^( -x^2);
{q, f} = {(1 - 4 * x^2 * t) * E^( -x^2), 0};
Simplify[{D[u,t] - D[u,x,x] - 2 * u - q, (u /. t -> 0) - f}]
```

generates the output $\{0, 0\}$.

8.25 Example. The IBVP

$$u_{tt}(x,t) + 2u_t(x,t) + u(x,t) = u_{xx}(x,t) + (2 + 4t - 3t^2)e^{-2x},$$
$$x > 0, \ t > 0,$$

$$u(0,t) = t^2, \quad u(x,t), \ u_x(x,t) \to 0 \ \text{as} \ x \to \infty, \quad t > 0,$$
$$u(x,0) = 0, \quad u_t(x,0) = 0, \quad x > 0$$

describes the propagation of a dissipative wave along a semi-infinite string initially at rest, under the action of a prescribed displacement of the near endpoint and an external force. In view of the BC, we use the Fourier sine transformation to find the solution.

First, by formula 6 (with $a = 2$) in Appendix A.4,

$$\mathcal{F}_S\left[(2 + 4t - 3t^2)e^{-2x}\right] = \sqrt{\frac{2}{\pi}}\frac{\omega}{4 + \omega^2}(2 + 4t - 3t^2);$$

hence, by Theorem 8.17 and the ICs, the transform $U(\omega, t) = \mathcal{F}_S[u](\omega, t)$ of u satisfies

$$U''(\omega, t) + 2U'(\omega, t) + (1 + \omega^2)U(\omega, t)$$
$$= \sqrt{\frac{2}{\pi}}\frac{\omega}{4 + \omega^2}\left[2 + 4t + (1 + \omega^2)t^2\right], \quad t > 0,$$

$$U(\omega, 0) = 0, \quad U'(\omega, 0) = 0.$$

Solving this ODE problem in the usual way, we find that

$$U(\omega, t) = \sqrt{\frac{2}{\pi}}\frac{\omega}{4 + \omega^2}t^2,$$

so, again by formula 6,

$$u(x,t) = \mathcal{F}_S^{-1}[U](x,t) = t^2 e^{-2x}.$$

VERIFICATION WITH MATHEMATICA®. The input

```
u = t^2 * E^( -2 * x);
{q, f, g, h} = {(2 + 4 * t - 3 * t^2) * E^( -2 * x), 0, 0, t^2};
Simplify[{D[u,t,t] + 2 * D[u,t] + u - D[u,x,x] - q, {(u /. t -> 0) - f,
   (D[u,t] /. t -> 0) - g}, (u /. x -> 0) - h}]
```

generates the output $\{0, \{0, 0\}, 0\}$.

8.26 Example. The BVP

$$u_{xx}(x, y) + u_{yy}(x, y) - u(x, y) = \tfrac{1}{4}\left[4x^2 - 6 + (3x^2 - 8)y\right]e^{-y/2},$$

$$0 < x < 1, \ y > 0,$$

$$u(0, y) = 2e^{-y/2}, \quad u(1, y) = (2 - y)e^{-y/2}, \quad y > 0,$$

$$u_y(x, 0) = -1 - x^2, \quad u(x, y), \ u_y(x, y) \to 0 \quad \text{as } y \to \infty, \ 0 < x < 1$$

models the steady-state distribution of heat in a semi-infinite strip with a time-dependent source and flux specified on the base. Since the BC at $y = 0$ prescribes the derivative u_y, we use the Fourier cosine transformation with respect to y to find the solution.

By formulas 2, 5, and 6 (with $a = 1/2$) in Appendix A.5 and Theorem 8.17(iii), the transform $U(x, \omega) = \mathcal{F}_C[u](x, \omega)$ of u satisfies

$$U''(x, \omega) - (1 + \omega^2)U(x, \omega)$$

$$= \sqrt{\frac{\pi}{2}}\left\{-\frac{4(3 - 3\omega^2 + 4\omega^4)}{(1 + 4\omega^2)^2} + \frac{4(1 - 3\omega^2 - 4\omega^4)}{(1 + 4\omega^2)^2}x^2\right\}, \quad 0 < x < 1,$$

$$U(0, \omega) = \sqrt{\frac{2}{\pi}}\frac{4}{1 + 4\omega^2}, \quad U(1, \omega) = \sqrt{\frac{\pi}{2}}\frac{32\omega^2}{(1 + 4\omega^2)^2},$$

with solution

$$U(x, \omega) = \sqrt{\frac{\pi}{2}}\left\{\frac{4}{1 + 4\omega^2} - \frac{4(1 - 4\omega^2)}{(1 + 4\omega^2)^2}x^2\right\};$$

so, by inversion,

$$u(x, y) = \mathcal{F}_C^{-1}[U](x, y) = (2 - x^2 y)e^{-y/2}.$$

VERIFICATION WITH MATHEMATICA®. The input

```
u = (2 - x^2 * y) * E^(- y /2);
{q, f1, f2, g} = {(1 /4) * (4 * x^2 - 6 + (3 * x^2 - 8) * y) * E^(- y /2),
 2 * E^(- y /2), (2 - y) * E^(- y /2)};
Simplify[{D[u,x,x] + D[u,y,y] - u - q, {(u /.x -> 0) - f1,
 (u /.x -> 1) - f2}, (D[u,y] /.y -> 0) - g}]
```

generates the output $\{0, \{0, 0\}, 0\}$.

8.27 Example. The same solution method can be applied to the BVP

$$u_{xx}(x, y) + u_{yy}(x, y) - 3u(x, y) = 2e^{2y-x}, \quad x > 0, \ 0 < y < 1,$$

$$u(0, y) = e^{2y}, \quad u(x, y), \ u_x(x, y) \to 0 \quad \text{as } x \to \infty, \ 0 < y < 1,$$

$$u_y(x, 0) = 2e^{-x}, \quad u(x, 1) = e^{2-x}, \quad x > 0,$$

where, since u is prescribed at $x = 0$, we need to use the Fourier sine transformation with respect to x.

Setting $U(\omega, y) = \mathcal{F}_S[u](\omega, y)$ and making use of formulas 2 and 6 (with $a = 1$) in Appendix A.4, we convert the given BVP to its transformed version

$$U''(\omega, y) - (3 + \omega^2)U(\omega, y) = \sqrt{\frac{2}{\pi}} \frac{\omega(1 - \omega^2)}{1 + \omega^2} e^{2y}, \quad 0 < y < 1,$$

$$U'(\omega, 0) = \sqrt{\frac{2}{\pi}} \frac{2\omega}{1 + \omega^2}, \quad U(\omega, 1) = \sqrt{\frac{2}{\pi}} \frac{e^2\omega}{1 + \omega^2},$$

which yields

$$U(\omega, y) = \sqrt{\frac{2}{\pi}} \frac{\omega}{1 + \omega^2} e^{2y};$$

hence, by inversion,

$$u(x, y) = \mathcal{F}_S^{-1}[U](x, y) = e^{-x}e^{2y} = e^{2y-x}.$$

VERIFICATION WITH MATHEMATICA®. The input

```
u = E^(2*y-x);
{q, f, g1, g2} = {2*E^(2*y-x), E^(2*y), 2*E^(-x), E^(2-x)};
Simplify[{D[u,x,x] + D[u,y,y] - 3*u - q, (u/.x->0) - f,
   {(D[u,y]/.y->0) - g1, (u/.y->1) - g2}}]
```

generates the output $\{0, 0, \{0, 0\}\}$.

Exercises

In **1–6**, solve the IVP (IBVP) consisting of the PDE

$$u_t(x, t) = k u_{xx}(x, t) + b u(x, t) + q(x, t),$$

domain in the (x, t)-plane, numbers k, b, function q, IC (IC and BC), and decay of the solution as specified.

1 $-\infty < x < \infty, \ t > 0, \ k = 1, \ b = -2,$
$q(x, t) = \left[9 + 4t - 4(2 + t)x^2\right]e^{-x^2},$
$u(x, t), u_x(x, t) \to 0$ as $x \to \pm\infty, \ u(x, 0) = 2e^{-x^2}.$

2 $-\infty < x < \infty, \ t > 0, \ k = 1, \ b = -1,$
$q(x, t) = \left[8 - 9t + 64(t - 1)x^2\right]e^{-4x^2},$
$u(x, t), u_x(x, t) \to 0$ as $x \to \pm\infty, \ u(x, 0) = e^{-4x^2}.$

3 $x > 0, \ t > 0, \ k = 2, \ b = 1, \ q(x, t) = \left[4t + (1 - 3t)x\right]e^{-x},$
$u(0, t) = 0, \ u(x, t), u_x(x, t) \to 0$ as $x \to \infty, \ u(x, 0) = 0.$

4 $x > 0$, $t > 0$, $k = 1$, $b = -1$, $q(x,t) = \left[4t - 3 + (4 + 3t)x\right]e^{-x/2}$,

$u(0,t) = -4$, $u(x,t)$, $u_x(x,t) \to 0$ as $x \to \infty$, $u(x,0) = -4e^{-x/2}$.

5 $x > 0$, $t > 0$, $k = 2$, $b = -2$, $q(x,t) = -3(x + 4t)e^{-x}$,

$u_x(0,t) = -1 - 3t$, $u(x,t)$, $u_x(x,t) \to 0$ as $x \to \infty$, $u(x,0) = e^{-x}$.

6 $x > 0$, $t > 0$, $k = 1$, $b = 2$, $q(x,t) = 2\left[9 + 4t + (1 - 6t)x\right]e^{-2x}$,

$u_x(0,t) = 2(3 + t)$, $u(x,t)$, $u_x(x,t) \to 0$ as $x \to \infty$, $u(x,0) = -3e^{-2x}$.

In **7–12**, solve the IVP (IBVP) consisting of the PDE

$$u_{tt}(x,t) + au_t(x,t) + bu(x,t) = c^2 u_{xx}(x,t) + q(x,t),$$

domain in the (x,t)-plane, numbers a, b, c, function q, ICs (ICs and BC), and decay of the solution as specified.

7 $-\infty < x < \infty$, $t > 0$, $a = 0$, $b = 1$, $c = 1$,

$q(x,t) = \left[5 + 3t^2 - 4(1 + t^2)x^2\right]e^{-x^2}$,

$u(x,t)$, $u_x(x,t) \to 0$ as $x \to \pm\infty$, $u(x,0) = e^{-x^2}$, $u_t(x,0) = 0$.

8 $-\infty < x < \infty$, $t > 0$, $a = 0$, $b = 1$, $c = 1$, $q(x,t) = (10 - x^2)e^{-t-x^2/4}$,

$u(x,t)$, $u_x(x,t) \to 0$ as $x \to \pm\infty$,

$u(x,0) = 4e^{-x^2/4}$, $u_t(x,0) = -4e^{-x^2/4}$.

9 $x > 0$, $t > 0$, $a = 4$, $b = 4$, $c = 1$, $q(x,t) = (5 + 6t)e^{-x}$,

$u(0,t) = 2t - 1$, $u(x,t)$, $u_x(x,t) \to 0$ as $x \to \infty$,

$u(x,0) = -e^{-x}$, $u_t(x,0) = 2e^{-x}$.

10 $x > 0$, $t > 0$, $a = 2$, $b = 1$, $c = 1$, $q(x,t) = -(7 + 3t)e^{-2x}$,

$u(0,t) = 3 + t$, $u(x,t)$, $u_x(x,t) \to 0$ as $x \to \infty$,

$u(x,0) = 3e^{-2x}$, $u_t(x,0) = e^{-2x}$.

11 $x > 0$, $t > 0$, $a = 2$, $b = 2$, $c = 2$, $q(x,t) = 2(9 + t + 2x)e^{-x/2}$,

$u_x(0,t) = 4 - t$, $u(x,t)$, $u_x(x,t) \to 0$ as $x \to \infty$,

$u(x,0) = 4xe^{-x/2}$, $u_t(x,0) = 2e^{-x/2}$.

12 $x > 0$, $t > 0$, $a = 2$, $b = 1$, $c = 2$, $q(x,t) = (5 + 12t - 6x)e^{-x}$,

$u_x(0,t) = 1 + 4t$, $u(x,t)$, $u_x(x,t) \to 0$ as $x \to \infty$,

$u(x,0) = (1 + 2x)e^{-x}$, $u_t(x,0) = -4e^{-x}$.

In **13–20**, solve the BVP consisting of the PDE

$$u_{xx}(x,y) + u_{yy}(x,y) - au_x(x,y) - bu_y(x,y) + cu(x,y) = q(x,y),$$

domain in the (x,y)-plane, numbers a, b, c, function q, BCs, and decay of the solution as specified.

13 $0 < x < 1$, $-\infty < y < \infty$, $a = 1$, $b = 0$, $c = 0$,

$q(x,y) = \left[3 + 2x - 4(1+x)y^2\right]e^{-y^2}$,

$u(0,y) = -e^{-y^2}$, $u(1,y) = -2e^{-y^2}$, $u(x,y)$, $u_y(x,y) \to 0$ as $y \to \pm\infty$.

14 $0 < x < 1$, $-\infty < y < \infty$, $a = 2$, $b = 0$, $c = -1$,

$q(x,y) = \left[18x - 5 + 64(1 - 2x)y^2\right]e^{-4y^2}$,

$u(0,y) = e^{-4y^2}$, $u_x(1,y) = -2e^{-4y^2}$, $u(x,y)$, $u_y(x,y) \to 0$ as $y \to \pm\infty$.

15 $-\infty < x < \infty$, $0 < y < 1$, $a = 0$, $b = -1$, $c = -2$,

$q(x,y) = \left[-2 + 8y - 10y^2 + (1+y^2)x^2\right]e^{-x^2/4}$,

$u_y(x,0) = 0$, $u(x,1) = 8e^{-x^2/4}$, $u(x,y)$, $u_x(x,y) \to 0$ as $x \to \pm\infty$.

16 $-\infty < x < \infty$, $0 < y < 1$, $a = 0$, $b = 3$, $c = 1$,

$q(x,y) = \left[2y^2 + 12y - 5 + 4(1 - 2y^2)x^2\right]e^{-x^2}$,

$u_y(x,0) = 0$, $u(x,1) = -e^{-x^2}$, $u(x,y)$, $u_x(x,y) \to 0$ as $x \to \pm\infty$.

17 $0 < x < 1$, $y > 0$, $a = 2$, $b = 0$, $c = -1$,

$q(x,y) = 2\left[x - x^2 + (2 - 2x)y\right]e^{-y}$, $u(0,y) = 0$, $u(1,y) = 0$,

$u(x,0) = 0$, $u(x,y)$, $u_y(x,y) \to 0$ as $y \to \infty$.

18 $0 < x < 1$, $y > 0$, $a = 3$, $b = 0$, $c = 2$,

$q(x,y) = \left[4x^2 - 4 + (1 + 24x - 9x^2)y\right]e^{-y/2}$, $u(0,y) = 4ye^{-y/2}$, $u(1,y) = 0$,

$u_y(x,0) = 4(1 - x^2)$, $u(x,y)$, $u_y(x,y) \to 0$ as $y \to \infty$.

19 $x > 0$, $0 < y < 1$, $a = 0$, $b = 0$, $c = -1$,

$q(x,y) = (3x + 3y^2 - 2)e^{-2x}$, $u(x,0) = xe^{-2x}$, $u_y(x,1) = 2e^{-2x}$,

$u_x(0,y) = 1 - 2y^2$, $u(x,y)$, $u_x(x,y) \to 0$ as $x \to \infty$.

20 $x > 0$, $0 < y < 1$, $a = 0$, $b = -4$, $c = 1$,

$q(x,y) = 2(3 + 4y + y^2 - x)e^{-x}$, $u_y(x,0) = 0$, $u(x,1) = (2 - x)e^{-x}$,

$u(0,y) = 1 + y^2$, $u(x,y)$, $u_x(x,y) \to 0$ as $x \to \infty$.

Answers to Odd-Numbered Exercises

1 $u(x,t) = (t + 2)e^{-x^2}$.

3 $u(x,t) = xte^{-x}$.

5 $u(x,t) = (1 - 3xt)e^{-x}$.

7 $u(x,t) = (1 + t^2)e^{-x^2}$.

9 $u(x,t) = (2t - 1)e^{-x}$.

11 $u(x,t) = 2(t + 2x)e^{-x/2}$.

13 $u(x,y) = -(x + 1)e^{-y^2}$.

15 $u(x,y) = 4(y^2 + 1)e^{-x^2/4}$.

17 $u(x,y) = (x^2 - x)ye^{-y}$.

19 $u(x,y) = (x + y^2)e^{-2x}$.

CHAPTER 9

The Laplace Transformation

The Fourier transformations are used mainly with respect to the space variables. In certain circumstances, however, for reasons of expedience or necessity, it is desirable to eliminate time as an active variable. This is achieved by means of the Laplace transformation. Problems where the spatial part of the domain is unbounded but the solution is not expected to decay fast enough away from the origin are particularly suited to this method.

9.1 Definition and Properties

First, we introduce a couple of useful mathematical entities.

9.1 Definition. The function H defined by

$$H(t) = \begin{cases} 0, & t < 0, \\ 1, & t \geq 0 \end{cases}$$

is called the *Heaviside (unit step) function*. It is clear that, more generally, for any real number t_0,

$$H(t - t_0) = \begin{cases} 0, & t < t_0, \\ 1, & t \geq t_0. \end{cases}$$

9.2 Remark. H is piecewise continuous. As mentioned in Remark 2.4(ii), the type of analysis we are performing is not affected by the value of a piecewise continuous function at its points of discontinuity. For our purposes, therefore, the value $H(0) = 1$ is chosen purely for convenience, to have H correctly defined as a function on the entire real line, but it is otherwise unimportant.

249

In mathematical modeling, it is often necessary to cater for a special type of physical data, such as unit impulses and point sources. Let us assume, for example, that a unit impulse is produced by a constant force of magnitude $1/\varepsilon$ acting over a very short time interval $(t_0 - \varepsilon/2, t_0 + \varepsilon/2)$, $\varepsilon > 0$. We may express this mathematically by taking the force to be

$$g_{t_0,\varepsilon}(t) = \begin{cases} 1/\varepsilon, & t_0 - \varepsilon/2 < t < t_0 + \varepsilon/2, \\ 0 & \text{otherwise} \end{cases}$$

and computing the total impulse as

$$\int\limits_{-\infty}^{\infty} g_{t_0,\varepsilon}(t)\, dt = \int\limits_{t_0-\varepsilon/2}^{t_0+\varepsilon/2} \frac{1}{\varepsilon}\, dt = \frac{1}{\varepsilon} \cdot \varepsilon = 1.$$

We note that the above integral is equal to 1 irrespective of the value of ε. If we now want to regard the impulse as being produced at the single moment $t = t_0$, we need to consider a limiting process and introduce some sort of limit of $g_{t_0,\varepsilon}$ as $\varepsilon \to 0$, which we denote by $\delta(t - t_0)$.

9.3 Definition. The mathematical object δ defined by
   (i) $\delta(t - t_0) = 0$ for all $t \neq t_0$,

   (ii) $\int\limits_{-\infty}^{\infty} \delta(t - t_0)\, dt = 1$

is called the *Dirac delta*.

9.4 Remarks.   (i) From Definition 9.3 it is obvious that δ cannot be ascribed a finite value at $t = t_0$, because then its integral over $\mathbb{R}$ would be 0, not 1. Consequently, δ is not a function. Strictly speaking, δ is a so-called *distribution* (generalized function) and its proper handling requires a special formalism that goes beyond the scope of this book.

   (ii) If $t < t_0$, we have

$$\int\limits_{-\infty}^{t} \delta(\tau - t_0)\, d\tau = \int\limits_{-\infty}^{t} 0\, d\tau = 0\,;$$

if $t > t_0$, we can find $\varepsilon > 0$ sufficiently small so that $t_0 + \varepsilon/2 < t$; hence, using the function $g_{t_0,\varepsilon}$ introduced above, we see that

$$\int\limits_{-\infty}^{t} \delta(\tau - t_0)\, d\tau = \lim_{\varepsilon \to 0} \int\limits_{-\infty}^{t} g_{t_0,\varepsilon}(\tau)\, d\tau = \lim_{\varepsilon \to 0} \int\limits_{t_0-\varepsilon/2}^{t_0+\varepsilon/2} \frac{1}{\varepsilon}\, d\tau = 1.$$

Therefore, combining these results, we may write

$$\int\limits_{-\infty}^{t} \delta(\tau - t_0)\, d\tau = H(t - t_0),$$

which means that, in a certain generalized (distributional) sense,

$$H'(t - t_0) = \delta(t - t_0).$$

(iii) If f is a continuous function, then, according to the mean value theorem, there is t', $t - \varepsilon/2 < t' < t + \varepsilon/2$, such that

$$\int_{-\infty}^{\infty} f(\tau)\delta(t - \tau)\,d\tau = \lim_{\varepsilon \to 0} \int_{-\infty}^{\infty} f(\tau)g_{t,\varepsilon}(\tau)\,d\tau = \lim_{\varepsilon \to 0} \int_{t-\varepsilon/2}^{t+\varepsilon/2} f(\tau)\frac{1}{\varepsilon}\,d\tau$$

$$= \lim_{\varepsilon \to 0} \frac{1}{\varepsilon}\left[\left(t + \frac{\varepsilon}{2}\right) - \left(t - \frac{\varepsilon}{2}\right)\right] f(t') = f(t). \qquad (9.1)$$

In distribution theory, δ is, in fact, defined rigorously by a formula of this type and not as in Definition 9.3. Like differentiation, integration on the left-hand side above is understood in a generalized, distributional sense.

(iv) The Dirac delta may also be used in problems formulated on semi-infinite or finite intervals. In such cases, its symbol stands for the 'restriction' of this distribution to the corresponding interval.

9.5 Definition. The *Laplace transform* of a function $f(t)$, $0 < t < \infty$, is defined by

$$\mathcal{L}[f](s) = F(s) = \int_0^{\infty} f(t)e^{-st}\,dt.$$

Here, s is the *transformation parameter*. The corresponding *inverse Laplace transform*, computed by means of complex variable techniques (see Section 14.1), is

$$\mathcal{L}^{-1}[F](t) = f(t) = \frac{1}{2\pi i} \int_{c-i\infty}^{c+i\infty} F(s)e^{st}\,ds.$$

The operators $\mathcal{L}$ and $\mathcal{L}^{-1}$ are called the *Laplace transformation* and *inverse Laplace transformation*, respectively.

9.6 Remark. $\mathcal{L}$ is applicable to a wider class of functions than $\mathcal{F}$.

The following assertion gives sufficient conditions for the existence of the Laplace transform of a function.

9.7 Theorem. *If*
   (i) *f is piecewise continuous on $[0, \infty)$;*
   (ii) *there are constants C and α such that*

$$|f(t)| \le Ce^{\alpha t}, \quad 0 < t < \infty,$$

then $\mathcal{L}[f](s) = F(s)$ exists for all $s > \alpha$, and

$$F(s) \to 0 \quad as \quad s \to \infty.$$

9.8 Examples.   (i) The function $f(t) = 1$, $t > 0$, satisfies the conditions in Theorem 9.7 with $C = 1$ and $\alpha = 0$, and we have

$$F(s) = \int_0^\infty e^{-st}\, dt = -\frac{1}{s}\left[e^{-st}\right]_0^\infty = \frac{1}{s}, \quad s > 0.$$

(ii) For $f(t) = e^{2t}$, $t > 0$, Theorem 9.7 holds with $C = 1$ and $\alpha = 2$. In this case, we have

$$F(s) = \int_0^\infty e^{2t}\, e^{-st}\, dt = \int_0^\infty e^{(2-s)t}\, dt = \frac{1}{s-2}, \quad s > 2.$$

(iii) The rate of growth of the function $f(t) = e^{t^2}$, $t > 0$, as $t \to \infty$ exceeds the exponential growth prescribed in Theorem 9.7. It turns out that this function does not have a Laplace transform.

(iv) According to Remarks 9.4(iii),(iv),

$$\int_0^\infty \delta(t - a)e^{-st}\, dt = e^{-as}, \quad a > 0;$$

therefore, using a slightly different notation, we may write

$$\mathcal{L}[\delta(t - a)] = e^{-as}, \quad a > 0, \tag{9.2}$$

and, by extension, as $a \to 0$,

$$\mathcal{L}[\delta(t)] = 1. \tag{9.3}$$

However, the last equality shows that the transform of $\delta(t)$ does not tend to 0 when $s \to \infty$, as specified in Theorem 9.7. This is because δ is not a function, so the statement of that theorem does not apply to it. Rigorously speaking, the constructs in (9.2) and (9.3) are distributional Laplace transforms, which have been included in Appendix A.6 because of their practical usefulness.

9.9 Theorem.   (i) $\mathcal{L}$ *is linear; that is,*

$$\mathcal{L}[c_1 f_1 + c_2 f_2] = c_1 \mathcal{L}[f_1] + c_2 \mathcal{L}[f_2]$$

for any functions f_1, f_2 *(to which* $\mathcal{L}$ *can be applied) and any numbers* c_1, c_2.

(ii) *If* $u = u(x,t)$ *and* $\mathcal{L}[u](x,s) = U(x,s)$, *then*

$$\mathcal{L}[u_t](x,s) = sU(x,s) - u(x,0),$$
$$\mathcal{L}[u_{tt}](x,s) = s^2 U(x,s) - su(x,0) - u_t(x,0).$$

(iii) *For the same type of function u, differentiation with respect to x and the Laplace transformation commute:*

$$\mathcal{L}[u_x](x,s) = (\mathcal{L}[u])_x(x,s) = U'(x,s).$$

(iv) *If we adopt a definition of the convolution $f * g$ of two functions f and g which is slightly different from Definition 8.3, namely,*

$$(f * g)(t) = \int_0^t f(\tau)g(t-\tau)\,d\tau$$

$$= \int_0^t f(t-\tau)g(\tau)\,d\tau = (g * f)(t), \tag{9.4}$$

then

$$\mathcal{L}[f * g] = \mathcal{L}[f]\mathcal{L}[g].$$

9.10 Remarks. (i) As in the case of the Fourier transformations, in general $\mathcal{L}[fg] \neq \mathcal{L}[f]\mathcal{L}[g]$.

(ii) Clearly, $\mathcal{L}^{-1}$ is also linear.

The Laplace transforms of some frequently used functions are listed in Appendix A.6.

9.11 Example. To find the inverse Laplace transform of the function $1/(s(s^2+1))$, we first see from Appendix A.6 that $1/s$ is the transform of the constant function 1 and $1/(s^2+1)$ is the transform of the function $\sin t$. Therefore, by Theorem 9.9(iv), we can write symbolically

$$\frac{1}{s(s^2+1)} = \frac{1}{s}\frac{1}{s^2+1} = \mathcal{L}[1]\mathcal{L}[\sin t] = \mathcal{L}\big[1 * (\sin t)\big],$$

so

$$\mathcal{L}^{-1}\left[\frac{1}{s(s^2+1)}\right] = 1 * (\sin t) = \int_0^t \sin\tau\,d\tau = 1 - \cos t.$$

Alternatively, we can split the given function into partial fractions as

$$\frac{1}{s(s^2+1)} = \frac{1}{s} - \frac{s}{s^2+1}$$

and then use the linearity of $\mathcal{L}^{-1}$ to obtain the above result.

VERIFICATION WITH MATHEMATICA®. The input

```
F = 1/s - s/(s^2 + 1);

InverseLaplaceTransform[F - 1/(s * (s^2 + 1)), s, t]
```

generates the output 0.

9.12 Example. The simplest way to find the inverse Laplace transform of $(3s^2 + 2s + 12)/(s(s^2 + 4))$ by direct calculation is to establish the partial fraction decomposition

$$\frac{3s^2 + 2s + 12}{s(s^2 + 4)} = \frac{3}{s} + \frac{2}{s^2 + 2^2}$$

and then apply formulas 5 and 8 in Appendix A.6 to arrive at

$$\mathcal{L}^{-1}\left[\frac{3s^2 + 2s + 12}{s(s^2 + 4)}\right] = 3 + \sin(2t).$$

VERIFICATION WITH MATHEMATICA®. The input

```
f = 3 + Sin[2 * t];
Simplify[LaplaceTransform[f,t,s] -
   (3 * s^2 + 2 * s + 12)/(s * (s^2 + 4))]
```

generates the output 0.

The next assertion lists two other helpful properties of the Laplace transformation.

9.13 Theorem. *If* $\mathcal{L}[f](s) = F(s)$, *then*

(i) $\mathcal{L}[e^{at}f](s) = F(s - a)$, $s > a = \text{const}$;

(ii) $\mathcal{L}[f(t - b)H(t - b)](s) = e^{-bs}F(s)$, $b = \text{const} > 0$.

9.14 Example. Since

$$\mathcal{L}[\sin(2t)] = \frac{2}{s^2 + 4}, \quad \mathcal{L}[\cos(3t)] = \frac{s}{s^2 + 9}, \quad \mathcal{L}[t^2] = \frac{2}{s^3},$$

from Theorem 9.13 it follows that

$$\mathcal{L}\left[e^{-t}\sin(2t) + e^{5t}\cos(3t) - 2(t - 2)^2 H(t - 2)\right]$$
$$= \frac{2}{(s + 1)^2 + 4} + \frac{s - 5}{(s - 5)^2 + 9} - \frac{4}{s^3}e^{-2s}.$$

VERIFICATION WITH MATHEMATICA®. The input

```
F = 2/((s + 1)^2 + 4) + (s - 5)/((s - 5)^2 + 9) - (4/s^3) * E^(-2 * s);
Simplify[InverseLaplaceTransform[F,s,t] - E^(-t) * Sin[2 * t] -
   E^(5 * t) * Cos[3 * t] + 2 * (t - 2)^2 * HeavisideTheta[t - 2]]
```

generates the output 0.

9.15 Example. Similarly, we have

$$\mathcal{L}^{-1}\left[\frac{s-1}{s^2-2s+10} - \frac{1}{s^2+4}e^{-s}\right]$$

$$= \mathcal{L}^{-1}\left[\frac{s-1}{(s-1)^2+3^2} - \frac{1}{s^2+2^2}e^{-s}\right]$$

$$= e^t \cos(3t) - \tfrac{1}{2}\sin\big(2(t-1)\big)H(t-1).$$

VERIFICATION WITH MATHEMATICA®. The input

```
f = E^t * Cos[3*t] - (1/2)*HeavisideTheta[t-1]*Sin[2*(t-1)];
Simplify[LaplaceTransform[f,t,s] - (s-1)/(s^2-2*s+10) +
    (1/(s^2+4))*E^(-s)]
```

generates the output 0.

Exercises

In **1–4**, use Appendix A.6 to compute the Laplace transform of the given function f.

1 $f(t) = e^{-t}\sin(3t) - 3t^4$. **2** $f(t) = e^{4t}\cos(2t) + 4(t-3)^3 H(t-3)$.

3 $f(t) = t^3 e^{t-2} - 3\sin^2(t/2)$. **4** $f(t) = e^{-2t}(\cos t - 3\sin t) - 2t^2 H(t-1)$.

In **5–8**, use Appendix A.6 to compute the inverse Laplace transform of the given function F.

5 $F(s) = \dfrac{2s+1}{s^2-2s+26}$. **6** $F(s) = \dfrac{3s+2}{s^2+6s+25} - \dfrac{2s}{(s-1)^2}e^{-s}$.

7 $F(s) = \dfrac{3s+1}{s}e^{-\sqrt{s/2}}$. **8** $F(s) = \dfrac{2-s}{s^2}e^{-3\sqrt{s}}$.

Answers to Odd-Numbered Exercises

1 $3/(s^2+2s+10) - 72/s^5$.

3 $6/(e^2(s-1)^4) - 3/(2(s^3+s))$.

5 $e^t[2\cos(5t) + (3/5)\sin(5t)]$.

7 $(3/(2\sqrt{2\pi}))t^{-3/2}e^{-1/(8t)} + \mathrm{erfc}(1/(2\sqrt{2t}))$.

9.2 Applications

In this section, we examine the use of the Laplace transformation in the study of some typical linear mathematical models involving partial differential equations.

9.2.1 The Signal Problem for the Wave Equation

Consider a very long elastic string of negligible weight, initially at rest, where the vertical displacement (signal) is given at the near endpoint. This is described by an IBVP of the form

$$u_{tt}(x,t) = c^2 u_{xx}(x,t), \quad x > 0, \ t > 0,$$
$$u(0,t) = f(t), \quad t > 0,$$
$$u(x,0) = 0, \quad u_t(x,0) = 0, \quad x > 0.$$

Making the notation

$$\mathcal{L}[u](x,s) = U(x,s), \quad \mathcal{L}[f](s) = F(s),$$

applying $\mathcal{L}$ to the PDE and BC, and using the properties of $\mathcal{L}$ listed in Theorem 9.9, we arrive at the transformed problem

$$s^2 U(x,s) = c^2 U''(x,s), \quad x > 0,$$
$$U(0,s) = F(s).$$

The ODE above can be rewritten in the form

$$U''(x,s) - (s/c)^2 U(x,s) = 0,$$

and its general solution is

$$U(x,s) = C_1(s)e^{(s/c)x} + C_2(s)e^{-(s/c)x},$$

where $C_1(s)$ and $C_2(s)$ are arbitrary functions of the transformation parameter. Since x, $c > 0$ and, according to Theorem 9.7, $U(x,s) \to 0$ as $s \to \infty$, we must have $C_1(s) = 0$. Then the BC yields $C_2(s) = F(s)$, so

$$U(x,s) = F(s)e^{-(s/c)x} = F(s)e^{-(x/c)s}.$$

Consequently, by Theorem 9.13(ii), the solution of the original IBVP is

$$u(x,t) = \mathcal{L}^{-1}\big[F(s)e^{-(x/c)s}\big] = f(t - x/c)H(t - x/c)$$

$$= \begin{cases} 0, & t < x/c, \\ f(t - x/c), & t \geq x/c. \end{cases}$$

This solution can also be expressed as

$$u(x,t) = \begin{cases} f(t - x/c), & x \leq ct, \\ 0, & x > ct. \end{cases} \tag{9.5}$$

We see that $u(x, t)$ is constant when $x - ct = $ const. Physically, this means that the solution is a wave of fixed shape (determined by the BC function f) with velocity $dx/dt = c$. Formula (9.5) indicates that at time t the signal originating from $x = 0$ has not reached the points $x > ct$, which are still in the initial state of rest.

9.16 Remarks. (i) The above procedure works only if the function f is transformable—that is, if it has a Laplace transform F.

(ii) The same inversion result can also be obtained by means of convolution. Since, by formula 12 in Appendix A.6, $e^{-(x/c)s} = \mathcal{L}\big[\delta(t - x/c)\big]$, we can write

$$U(x, s) = F(s)e^{-(x/c)s} = \mathcal{L}[f]\mathcal{L}\big[\delta(t - x/c)\big] = \mathcal{L}\big[f * \delta(t - x/c)\big],$$

from which we conclude that

$$u(x, t) = \mathcal{L}^{-1}[U](x, t) = f * \delta(t - x/c) = \int_0^t f(\tau)\delta(t - x/c - \tau)\, d\tau.$$

If $t < x/c$, then $t - x/c - \tau < 0$ for $0 \le \tau \le t$, so $\delta(t - x/c - \tau) = 0$; hence, $u(x, t) = 0$ for $t < x/c$, or, equivalently, for $x > ct$.

If $t > x/c$, then $t - x/c - \tau = 0$ at $0 < \tau = t - x/c < t$; therefore, by (9.1), for $x < ct$,

$$u(x, t) = \int_0^t f(\tau)\delta(t - x/c - \tau)\, d\tau = \int_0^\infty f(\tau)\delta(t - x/c - \tau)\, d\tau = f(t - x/c).$$

This confirms the result expressed by (9.5).

(iii) The above method operates equally well when the PDE and ICs are nonhomogeneous. In this case, the solution may or may not be continuous across the line $x = ct$.

9.17 Example. Consider the IBVP

$$u_{tt}(x, t) = \tfrac{1}{4}\, u_{xx}(x, t) + x(t - 2)e^{-t}, \quad x > 0,\ t > 0,$$
$$u(0, t) = -2, \quad t > 0,$$
$$u(x, 0) = -2, \quad u_t(x, 0) = x, \quad x > 0.$$

The GS of the equation in the transformed problem

$$\tfrac{1}{4}\, U''(x, s) - s^2 U(x, s) = 2s + \left[\frac{2}{s + 1} - \frac{1}{(2 + 1)^2} - 1\right]x, \quad x > 0,$$
$$U(0, s) = -\frac{2}{s},$$

where $U(x, s) = \mathcal{L}[u](x, s)$, is

$$U(x, s) = C_1(s)e^{2sx} + C_2(s)e^{-2sx} - \frac{2}{s} + \frac{1}{(s + 1)^2}\, x.$$

Applying the property that $U(x, s) \to 0$ as $s \to \infty$, and then the BC, we find that $C_1(\omega) = C_2(\omega) = 0$, so

$$U(x, s) = -\frac{2}{s} + \frac{1}{(s+1)^2}x;$$

therefore,

$$u(x, t) = \mathcal{L}^{-1}[U](x, t) = -2 + xte^{-t}.$$

VERIFICATION WITH MATHEMATICA®. The input

```
u = - 2 + x * t * E^(- t);
{q,f,g,h} = {x * (t - 2) * E^(- t), - 2, x, - 2};
{D[u,t,t] - (1/4) * D[u,x,x] - q, {(u /. t -> 0) - f, (D[u,t] /. t -> 0) - g},
 (u /. x -> 0) - h}
```

generates the output $\{0, \{0, 0\}, 0\}$.

9.18 Example. Using formulas 1 and 5 in Appendix A.6, we reduce the IBVP

$$u_{tt}(x, t) = u_{xx}(x, t) - 2, \quad x > 0, \ t > 0,$$
$$u(0, t) = -1, \quad t > 0,$$
$$u(x, 0) = 1, \quad u_t(x, 0) = x, \quad x > 0$$

to the transformed problem

$$U''(x, s) - s^2 U(x, s) = \frac{2}{s} - s - x, \quad x > 0,$$
$$U(0, s) = -\frac{1}{s},$$

where $U(x, s) = \mathcal{L}[u](x, s)$. The general solution of the ODE is

$$U(x, s) = C_1(s)e^{sx} + C_2(s)e^{-sx} + \frac{s^2 - 2}{s^3} + \frac{1}{s^2}x,$$

which the property that $U(x, s) \to 0$ as $s \to \infty$, the BC, and partial fraction decomposition bring to the simpler form

$$U(x, s) = \frac{1}{s} - \frac{2}{s^3} + \frac{1}{s^2}x + \left(\frac{2}{s^3} - \frac{2}{s}\right)e^{-sx};$$

hence,

$$u(x, t) = \mathcal{L}^{-1}[U](x, t) = 1 - t^2 + xt + \left[(t - x)^2 - 2\right]H(t - x)$$
$$- \begin{cases} 1 - t^2 + xt, & t < x, \\ x^2 - xt - 1, & t \geq x. \end{cases}$$

It is easily seen that the solution is discontinuous across the line $x = t$.

VERIFICATION WITH MATHEMATICA®. The input

```
{u1,u2}={1-t^2+x*t,x^2-x*t-1};
{q,f,g,h}={-2,-1,1,x};
{D[{u1,u2},t,t]-D[{u1,u2},x,x]-q, (u2/.x->0)-f,(u1/.t->0)-g,
  (D[u1,t]/.t->0)-h}
```

generates the output $\{\{0,0\},0,0,0\}$.

9.19 Example. Writing $\mathcal{L}[u](x, s) = U(x, s)$, we convert the IBVP

$$u_{tt}(x, t) = u_{xx}(x, t) + 6e^{2t-x}, \quad x > 0, \ t > 0,$$

$$u(0, t) = t - 3t^2 + 2e^{2t}, \quad t > 0,$$

$$u(x, 0) = 2e^{-x}, \quad u_t(x, 0) = 4e^{-x}, \quad x > 0$$

to the ODE problem

$$U''(x, s) - s^2 U(x, s) = -\frac{2s^2 - 2}{s - 2} e^{-x}, \quad x > 0,$$

$$U(0, s) = \frac{1}{s^2} - \frac{6}{s^3} + \frac{2}{s - 2}.$$

Standard procedure now leads to the general solution

$$U(x, s) = C_1(s)e^{sx} + C_2(s)e^{-sx} + \frac{2}{s - 2} e^{-x},$$

which, subjected to the requirement that $U(x, s) \to 0$ as $s \to \infty$ followed by the application of the BC, reduces to

$$U(x, s) = \left(\frac{1}{s^2} - \frac{6}{s^3}\right)e^{-sx} + \frac{2}{s - 2} e^{-x};$$

hence,

$$u(x, t) = \mathcal{L}^{-1}[U](x, t) = 2e^{2t-x} + \left[t - x - 3(t - x)^2\right]H(t - x)$$

$$= \begin{cases} 2e^{2t-x}, & t < x, \\ 2e^{2t-x} + t - x - 3(t - x)^2, & t \geq x. \end{cases}$$

Here, the solution is continuous across the line $x = t$.

VERIFICATION WITH MATHEMATICA®. The input

```
{u1, u2}={2*E^(2*t-x),t-x-3*(t-x)^2};
u=u1+u2*HeavisideTheta[t-x];
{q,f,g1,g2}={6*E^(2*t-x),t-3*t^2+2*E^(2*t),
```

```
    2*E^(-x),4*E^(-x)};
  Simplify[{Simplify[D[u,t,t]-D[u,x,x]-q,t<x||t>x],
    ((u1+u2)/.x->0)-f,{(u1/.t->0)-g1,
    (D[u1,t]/.t->0)-g2}}]
```

generates the output $\{0, 0, \{0, 0\}\}$.

9.20 Example. The same treatment applied to the IBVP

$$u_{tt}(x,t) = 4u_{xx}(x,t) - 8 - e^t, \quad x > 0, \ t > 0,$$

$$u_x(0,t) = 0, \quad t > 0,$$

$$u(x,0) = x^2 - 1, \quad u_t(x,0) = 1, \quad x > 0$$

leads to the transformed problem

$$4U''(x,s) - s^2 U(x,s) = -1 + \frac{1}{s-1} + \frac{8}{s} - s(x^2 - 1), \quad x > 0,$$

$$U'(0,s) = 0,$$

with general solution

$$U(x,s) = C_1(s)e^{sx/2} + C_2(s)e^{-sx/2} - \frac{s^2 + 2s - 2}{s^2(s-1)} + \frac{1}{s}x^2,$$

which, on the basis of the arguments used in the preceding examples, yields

$$U(x,s) = \frac{2}{s^2} - \frac{1}{s-1} + \frac{1}{s}x^2.$$

Consequently, by inversion,

$$u(x,t) = \mathcal{L}^{-1}[U](x,t) = 2t - e^t + x^2.$$

VERIFICATION WITH MATHEMATICA®. The input

```
  u=2*t-E^t+x^2;
  {q,f,g,h}={-8-E^t,0,x^2-1,1};
  Simplify[{D[u,t,t]-4*D[u,x,x]-q,(D[u,x]/.x->0)-f,
    {(u/.t->0)-g,(D[u,t]/.t->0)-h}}]
```

generates the output $\{0, 0, \{0, 0\}\}$.

9.21 Example. Using the Laplace transformation in the IBVP

$$u_{tt}(x,t) = u_{xx}(x,t) - 4, \quad x > 0, \ t > 0,$$

$$u_x(0,t) = -1 - 3t - \sin t, \quad t > 0,$$

$$u(x,0) = 2x^2 + 1, \quad u_t(x,0) = -3x, \quad x > 0,$$

we arrive at the ODE problem

$$U''(x,s) - s^2 U(x,s) = \frac{4}{s} - s + 3x - 2sx^2 \quad x > 0,$$

$$U'(0,s) = -\frac{1}{s} - \frac{3}{s^2} - \frac{1}{s^2+1}.$$

The GS of the equation is

$$U(x,s) = C_1(s)e^{sx} + C_2(s)e^{-sx} + \frac{1}{s} - \frac{3}{s^2}x + \frac{2}{s}x^2,$$

from which, in the way described earlier, we eventually obtain the particular solution

$$U(x,s) = \left(\frac{1}{s} + \frac{1}{s^2} - \frac{s}{s^2+1} \right)e^{-sx} + \frac{1}{s} - \frac{3}{s^2}x + \frac{2}{s}x^2.$$

By inversion, this leads to

$$u(x,t) = \mathcal{L}^{-1}[U](x,t) = 1 + 2x^2 - 3xt + \left[1 + t - x - \cos(t-x)\right]H(t-x)$$

$$= \begin{cases} 1 + 2x^2 - 3xt, & t < x, \\ 2 + t - x + 2x^2 - 3xt - \cos(t-x), & t \geq x. \end{cases}$$

This solution is continuous across the line $x = t$.

VERIFICATION WITH MATHEMATICA®. The input

```
{u1,u2} = {1+2*x^2-3*x*t, 2+t-x+2*x^2-3*x*t-Cos[t-x]};
{q,f,g,h} = {-4, 2*x^2+1, -3*x, -1-3*t-Sin[t]};
{D[{u1,u2},t,t] - D[{u1,u2},x,x] - {q,q}, {(u1/.t->0)-f,
  (D[u1,t]/.t->0)-g}, (D[u2,x]/.x->0)-h}
```

generates the output $\{\{0,0\}, \{0,0\}, 0\}$.

9.22 Remark. If the solution of an IBVP with u_x prescribed at the boundary point $x = 0$ has the form $u_1(x,t) + u_2(t-x/c)H(t-x/c)$, it is not difficult to verify that this solution is unique up to an arbitrary constant term in u_2. The Laplace transformation method generates the unique solution that is continuous across the line $x = ct$. This property also holds for problems involving linear wave-type PDEs that include terms with the first-order derivatives of u but not u itself.

Exercises

Use the Laplace transformation to solve the IBVP consisting of the PDE

$$u_{tt}(x,t) = c^2 u_{xx}(x,t) + q(x,t), \quad x > 0, \ t > 0$$

and coefficient c, function q, BC, and ICs as indicated.

1 $c = 1$, $q(x,t) = -(2 + 3e^{-2t})e^{-x}$, $u(0,t) = 2 - e^{-2t}$, $u(x,0) = e^{-x}$,
   $u_t(x,0) = 2e^{-x}$.

2 $c = 2$, $q(x,t) = -8e^{-x}\sin(2t)$, $u(0,t) = \sin(2t)$, $u(x,0) = 0$,
   $u_t(x,0) = 2e^{-x}$.

3 $c = 2$, $q(x,t) = 2x^2 - 8t^2$, $u(0,t) = 3$, $u(x,0) = 3$, $u_t(x,0) = -2x$.

4 $c = 1$, $q = 2x$, $u(0,t) = t$, $u(x,0) = 0$, $u_t(x,0) = 0$.

5 $c = \frac{1}{2}$, $q(x,t) = -1 - 2x$, $u(0,t) = 4t$, $u(x,0) = 2x^2$, $u_t(x,0) = 3$.

6 $c = 1$, $q(x,t) = 4\sin(2x)$, $u(0,t) = e^t - 1$, $u(x,0) = \sin(2x)$,
   $u_t(x,0) = 0$.

7 $c = 1$, $q(x,t) = 1$, $u(0,t) = t$, $u(x,0) = 0$, $u_t(x,0) = -1$.

8 $c = 1$, $q(x,t) = -1$, $u(0,t) = t + 1$, $u(x,0) = 2$, $u_t(x,0) = 0$.

9 $c = 1$, $q = xt$, $u(0,t) = t + 1$, $u(x,0) = 2x - 1$, $u_t(x,0) = -2x$.

10 $c = 1$, $q(x,t) = 4xe^{-2t} - 2$, $u_x(0,t) = e^{-2t}$, $u(x,0) = x + 1$,
    $u_t(x,0) = -2x$.

11 $c = 1$, $q(x,t) = 3x(3t - 2)e^{-3t}$, $u_x(0,t) = te^{-3t}$, $u(x,0) = 4$,
    $u_t(x,0) = x - 2$.

12 $c = 2$, $q(x,t) = 2(2t + 5e^{-t})\sin x$, $u_x(0,t) = 2e^{-t} + t$,
    $u(x,0) = 2\sin x$, $u_t(x,0) = -\sin x$.

13 $c = 2$, $q(x,t) = e^{-3t}$, $u_x(0,t) = 2t$, $u(x,0) = 0$, $u_t(x,0) = -1$.

14 $c = 2$, $q(x,t) = t$, $u_x(0,t) = 2e^{-t}$, $u(x,0) = 1$, $u_t(x,0) = 0$.

15 $c = 2$, $q(x,t) = 2$, $u_x(0,t) = \cos t$, $u(x,0) = 0$, $u_t(x,0) = 1$.

16 $c = 1$, $q(x,t) = -2 - 4H(t - x)$, $u(0,t) = 4 - t^2$, $u(x,0) = 4 + x^2$,
    $u_t(x,0) = -2x$.

17 $c = 1$, $q(x,t) = -2 - e^x H(t - x)$, $u(0,t) = 2 - t^2$, $u(x,0) = 1 + 2x$,
    $u_t(x,0) = 0$.

18 $c = 1$, $q(x,t) = -6t + 4H(t - x)$, $u_x(0,t) = 1$, $u(x,0) = x - 1$,
    $u_t(x,0) = 3x^2$.

Answers to Odd-Numbered Exercises

1 $u(x,t) = (2 - e^{-2t})e^{-x}$.

3 $u(x,t) = x^2t^2 - 2xt + 3$.

5 $u(x,t) = (t - 2x)H(t - 2x) + 2x^2 - xt^2 + 3t$.

7 $u(x,t) = (1/2)\big[t^2 - 2t - (t - x)(t - x - 4)H(t - x)\big]$.

9 $u(x,t) = -1 + (2 - 2t + t^3/6)x + (2 + t - x)H(t - x)$.

11 $u(x,t) = 4 - 2t + xte^{-3t}$.

13 $u(x,t) = (1/9)(e^{-3t} - 6t - 1) - (1/2)(x - 2t)^2 H(t - x/2)$.

15 $u(x,t) = t^2 + t - 2\sin(t - x/2)H(t - x/2)$.

17 $u(x,t) = \left[e^x - e^{(t+x)/2} + e^{(t-x)/2} \right] H(t - x) + 1 + 2x - t^2$.

9.2.2 Heat Conduction in a Semi-Infinite Rod

As usual, we look at a few specific cases.

9.23 Example. Consider the IBVP

$$u_t(x,t) = u_{xx}(x,t) + \sin x, \quad x > 0, \ t > 0,$$
$$u(0,t) = -3, \quad t > 0,$$
$$u(x,0) = 1, \quad x > 0.$$

Setting $\mathcal{L}[u](x,s) = U(x,s)$ and applying the Laplace transformation to the PDE and BC, we arrive at the ODE problem

$$U''(x,s) - sU(x,s) = -1 - \frac{1}{s}\sin x, \quad x > 0,$$
$$U(0,s) = -\frac{3}{s}.$$

The general solution of the equation, written as the sum of the complementary function and a particular integral, is

$$U(x,s) = C_1(s)e^{\sqrt{s}\,x} + C_2(s)e^{-\sqrt{s}\,x} + \frac{1}{s} + \frac{1}{s(s+1)}\sin x.$$

Since $U(x,s) \to 0$ as $s \to \infty$, it follows that $C_1(s) = 0$. Then, applying the BC, we find that $C_2(s) = -4/s$; hence, given the partial fraction decomposition

$$\frac{1}{s(s+1)} = \frac{1}{s} - \frac{1}{s+1},$$

we arrive at

$$U(x,s) = -\frac{4}{s}e^{-\sqrt{s}\,x} + \frac{1}{s} + \left(\frac{1}{s} - \frac{1}{s+1}\right)\sin x.$$

By formulas 5–7 and 15 in Appendix A.6, the solution of the original IBVP is

$$u(x,t) = \mathcal{L}^{-1}[U](x,t) = -4\,\mathrm{erfc}\left(\frac{x}{2\sqrt{t}}\right) + 1 + (1 - e^{-t})\sin x.$$

VERIFICATION WITH MATHEMATICA®. The input

```
u = -4 * Erfc[x/(2 * t^(1/2))] + 1 + (1 - E^(-t)) * Sin[x];
{q, f, g} = {Sin[x], -3, 1};
Simplify[{D[u,t] - D[u,x,x] - q, (u/. x -> 0) - f, Assuming[x>0,
  Limit[u, t -> 0, Direction -> -1]] - g}]
```

generates the output $\{0, 0, 0\}$.

9.24 Example. The IBVP

$$u_t(x, t) = 9u_{xx}(x, t) - x - 2e^{2t}, \quad x > 0, \ t > 0,$$

$$u_x(0, t) = -t, \quad t > 0,$$

$$u(x, 0) = 1, \quad x > 0$$

is treated similarly. With $U(x, s) = \mathcal{L}[u](x, s)$, we reduce this IBVP to the ODE problem

$$9U''(x, s) - sU(x, s) = -1 + \frac{2}{s - 2} + \frac{1}{s}x, \quad x > 0,$$

$$U'(0, s) = -\frac{1}{s^2}.$$

The general solution of the equation is

$$U(x, s) = C_1(s)e^{\sqrt{s}\,x/3} + C_2(s)e^{-\sqrt{s}\,x/3} + \frac{1}{s}\left(1 - \frac{2}{s - 2}\right) - \frac{1}{s^2}x,$$

where we must have $C_1(s) = 0$ to eliminate the first exponential term, which does not vanish as $s \to \infty$. Then the IC yields $C_2(s) = 0$, so

$$U(x, s) = \frac{1}{s}\left(1 - \frac{2}{s - 2}\right) - \frac{1}{s^2}x = \frac{2}{s} - \frac{1}{s - 2} - \frac{1}{s^2}x;$$

hence,

$$u(x, t) = \mathcal{L}^{-1}[U](x, t) = 2 - e^{2t} - xt.$$

VERIFICATION WITH MATHEMATICA®. The input

```
u = 2 - E^(2 * t) - x * t;
{q, f, g} = {-x - 2 * E^(2 * t), -t, 1};
Simplify[{D[u,t] - 9 * D[u,x,x] - q, (D[u,x] /.x -> 0) - f,
  (u/.t -> 0) - g}]
```

generates the output $\{0, 0, 0\}$.

9.25 Example. A very long rod without sources, with the near endpoint kept in open air of zero temperature and with a uniform initial temperature distribution, is modeled by the IBVP

$$u_t(x, t) = u_{xx}(x, t), \quad x > 0, \ t > 0,$$
$$u_x(0, t) - u(0, t) = 0, \quad t > 0,$$
$$u(x, 0) = u_0 = \text{const}, \quad x > 0.$$

Let $\mathcal{L}[u](x, s) = U(x, s)$. Applying $\mathcal{L}$ to the PDE and BCs, we arrive at the transformed problem

$$U''(x, s) - sU(x, s) + u_0 = 0, \quad x > 0,$$
$$U'(0, s) - U(0, s) = 0.$$

The general solution of the equation is

$$U(x, s) = C_1(s)e^{\sqrt{s}\,x} + C_2(s)e^{-\sqrt{s}\,x} + \frac{1}{s}u_0,$$

where $C_1(s)$ and $C_2(s)$ are arbitrary functions of the transformation parameter. Since $U(x, s) \to 0$ as $s \to \infty$, we must have $C_1(s) = 0$. Then, differentiating U and using the BC at $x = 0$, we see that

$$-C_2(s)\sqrt{s} - C_2(s) - \frac{1}{s}u_0 = 0,$$

from which $C_2(s) = -u_0/(s(\sqrt{s}+1))$; hence,

$$U(x, s) = u_0 \left[-\frac{1}{s(\sqrt{s}+1)}\,e^{-\sqrt{s}\,x} + \frac{1}{s} \right].$$

After some manipulation (coupled with the use of a more comprehensive Laplace transform table than A.6), we find that the solution of the original IBVP is

$$u(x, t) = \mathcal{L}^{-1}[U](x, t) = u_0 \left[1 - \text{erfc}\left(\frac{x}{2\sqrt{t}}\right) + e^{x+t}\,\text{erfc}\left(\sqrt{t} + \frac{x}{2\sqrt{t}}\right) \right].$$

VERIFICATION WITH MATHEMATICA®. The input

```
u = u0 * (1 - Erfc[x/(2*t^(1/2))] + Erfc[t^(1/2) +
    x/(2*t^(1/2))] * E^(x+t));
Simplify[{D[u,t] - D[u,x,x], (D[u,x]-u) /.x->0,
    Assuming[x>0, Limit[u,t->0, Direction->"FromAbove"]] - u0}]
```

generates the output $\{0, 0, 0\}$.

Exercises

Use the Laplace transformation to solve the IBVP consisting of the PDE

$$u_t(x,t) = ku_{xx}(x,t) + q(x,t), \quad x > 0, \ t > 0$$

and coefficient k, function q, BC, and IC as indicated.

1 $k = 1$, $q(x,t) = 1$, $u(0,t) = t + 1$, $u(x,0) = \sin(2x)$.

2 $k = 4$, $q(x,t) = 1$, $u(0,t) = 2 - t$, $u(x,0) = -2$.

3 $k = 4$, $q(x,t) = 3e^{-3t}$, $u(0,t) = 3 - e^{-3t}$, $u(x,0) = 1$.

4 $k = 2$, $q(x,t) = -(2t + 3)e^{-x}$, $u(0,t) = t + 2$, $u(x,0) = 2e^{-x}$.

5 $k = 1$, $q(x,t) = 5e^{-x}\left[\cos(2t) + \sin(2t)\right]$,

   $u(0,t) = -3\cos(2t) + \sin(2t)$, $u(x,0) = -3e^{-x}$.

6 $k = 1$, $q(x,t) = -2x + 2t - 6$, $u(0,t) = t^2$, $u(x,0) = 3x^2$.

7 $k = 1$, $q(x,t) = (2t + 1)\cos x$, $u_x(0,t) = 0$,

   $u(x,0) = -\cos x$.

8 $k = 1$, $q(x,t) = e^{-2t}(4\cos t + 3\sin t)\sin(2x)$,

   $u_x(0,t) = 2e^{-2t}(\cos t + 2\sin t)$, $u(x,0) = \sin(2x)$.

9 $k = 1$, $q(x,t) = x^2 - 6t$, $u_x(0,t) = 1$, $u(x,0) = x + 1$.

10 $k = 4$, $q(x,t) = -2e^{2t-x}$, $u_x(0,t) = -e^{2t} + 1/\sqrt{\pi t}$,

   $u(x,0) = e^{-x}$.

11 $k = 1$, $q(x,t) = (1 - t)e^{-t} - 4$, $u_x(0,t) = 3/\sqrt{\pi t}$,

   $u(x,0) = 2x^2$.

12 $k = 2$, $q(x,t) = 8xe^{2t} + 6$, $u_x(0,t) = 4e^{2t} - \sqrt{2/(\pi t)}$,

   $u(x,0) = 4x - x^2$.

Answers to Odd-Numbered Exercises

1 $u(x,t) = t + e^{-4t}\sin(2x) + \mathrm{erfc}\left(x/(2\sqrt{t})\right)$.

3 $u(x,t) = 2 - e^{-3t} + \mathrm{erfc}\left(x/(4\sqrt{t})\right)$.

5 $u(x,t) = e^{-x}\left[\sin(2t) - 3\cos(2t)\right]$.

7 $u(x,t) = (2t - 1)\cos x$.

9 $u(x,t) = x^2 t - 2t^2 + x + 1$.

11 $u(x,t) = 2x^2 + te^{-t} - 3\,\mathrm{erfc}\left(x/(2\sqrt{t})\right)$.

9.2.3 Finite Rod with Temperature Prescribed on the Boundary

The IBVP

$$w_t(x, t) = w_{xx}(x, t), \quad 0 < x < 1, \ t > 0,$$

$$w(0, t) = 0, \quad w(1, t) = 1, \quad t > 0,$$

$$w(x, 0) = 0, \quad 0 < x < 1$$

is of a type that we have already encountered. The equilibrium solution in this case, computed as in Section 6.1, is $w_\infty(x) = x$. Using this solution, we reduce the problem to a similar one where both BCs are homogeneous and which can thus be solved by the method of separation of variables. Putting all the results together, we obtain

$$w(x, t) = x + \sum_{n=1}^{\infty} (-1)^n \frac{2}{n\pi} \sin(n\pi x) e^{-n^2 \pi^2 t}. \tag{9.6}$$

The same IBVP can also be solved by using the Laplace transformation with respect to t. If we write $\mathcal{L}[w](x, s) = W(x, s)$, then from the PDE and BCs we find that W satisfies

$$W''(x, s) - sW(x, s) = 0, \quad 0 < x < 1,$$

$$W(0, s) = 0, \quad W(1, s) = \frac{1}{s}.$$

The general solution of the transformed equation can be written in the form (see Remark 1.4(i))

$$W(x, s) = C_1(s) \sinh(\sqrt{s}\, x) + C_2(s) \sinh\left(\sqrt{s}\,(x - 1)\right),$$

with $C_1(s)$ and $C_2(s)$ determined from the BCs. Applying these conditions leads to

$$C_2(s) = 0, \quad C_1(s) \sinh\sqrt{s} = \frac{1}{s},$$

from which

$$C_1(s) = \frac{1}{s \sinh\sqrt{s}};$$

hence,

$$W(x, s) = \frac{\sinh\left(\sqrt{s}\, x\right)}{s \sinh\sqrt{s}}.$$

Then, using the inverse transformation and comparing with (9.6), we see that

$$w(x, t) = \mathcal{L}^{-1}\left[\frac{1}{s} \frac{\sinh\left(\sqrt{s}\, x\right)}{\sinh\sqrt{s}}\right] = x + \sum_{n=1}^{\infty} (-1)^n \frac{2}{n\pi} \sin(n\pi x) e^{-n^2 \pi^2 t}.$$

Now consider the more general IBVP

$$u_t(x, t) = u_{xx}(x, t), \quad 0 < x < 1, \ t > 0,$$
$$u(0, t) = 0, \quad u(1, t) = f(t), \quad t > 0,$$
$$u(x, 0) = 0, \quad 0 < x < 1,$$

and let $\mathcal{L}[u](x, s) = U(x, s)$ and $\mathcal{L}[f](s) = F(s)$. Applying $\mathcal{L}$ to the PDE and BCs, we arrive at the BVP

$$U''(x, s) - sU(x, s) = 0, \quad 0 < x < 1,$$
$$U(0, s) = 0, \quad U(1, s) = F(s).$$

Proceeding exactly as above, we find that

$$U(x, s) = F(s) \frac{\sinh\left(\sqrt{s}\, x\right)}{\sinh \sqrt{s}} = F(s)\left\{ s\left[\frac{1}{s} \frac{\sinh\left(\sqrt{s}\, x\right)}{\sinh \sqrt{s}}\right]\right\}$$

$$= F(s)\big[sW(x, s)\big]. \tag{9.7}$$

Since $w(x, 0) = 0$ in the IBVP for w, it follows that

$$\mathcal{L}[w_t](x, s) = sW(x, s) - w(x, 0) = sW(x, s);$$

therefore, by (9.7) and Theorem 9.9(iv),

$$\mathcal{L}[u] = U = F(sW) = \mathcal{L}[f]\mathcal{L}[w_t] = \mathcal{L}[f * w_t].$$

Using (9.4) and integration by parts, we now obtain

$$u(x, t) = (f * w_t)(x, t)$$

$$= \int_0^t f(t - \tau)w_\tau(x, \tau)\, d\tau$$

$$= f(t - \tau)w(x, \tau)\big|_{\tau=0}^{\tau=t} + \int_0^t w(x, \tau)f'(t - \tau)\, d\tau$$

$$= f(0)w(x, t) - f(t)w(x, 0) + \int_0^t w(x, t - \tau)f'(\tau)\, d\tau$$

$$= \int_0^t w(x, t - \tau)f'(\tau)\, d\tau + f(0)w(x, t), \tag{9.8}$$

where we have used the condition $w(x, 0) = 0$ and the commutativity of the convolution operation.

This result shows how the solution of a problem with more general BCs can sometimes be obtained from that of a problem with simpler ones.

9.26 Example. If $f(t) = 3 - 2t$, then, by (9.8),

$$u(x,t) = -2 \int_0^t w(x, t - \tau)\, d\tau + 3w(x,t),$$

where w is given by (9.6). Since

$$\int_0^t e^{-n^2\pi^2(t-\tau)}\, d\tau = \frac{1}{n^2\pi^2} e^{-n^2\pi^2(t-\tau)}\Big|_{\tau=0}^{\tau=t} = \frac{1}{n^2\pi^2}\left(1 - e^{-n^2\pi^2 t}\right),$$

we have

$$u(x,t) = -2\left\{ xt + \sum_{n=1}^\infty (-1)^n \frac{2}{n^3\pi^3}\left(1 - e^{-n^2\pi^2 t}\right)\sin(n\pi x) \right\}$$

$$+ 3\left\{ x + \sum_{n=1}^\infty (-1)^n \frac{2}{n\pi} e^{-n^2\pi^2 t} \sin(n\pi x) \right\}$$

$$= x(3 - 2t) + \sum_{n=1}^\infty (-1)^n \sin(n\pi x)\left\{ -\frac{4}{n^3\pi^3} + \left(\frac{6}{n\pi} + \frac{4}{n^3\pi^3}\right)e^{-n^2\pi^2 t} \right\}.$$

9.2.4 Diffusion–Convection Problems

Suppose that a chemical substance is being poured at a constant rate into a straight, narrow, clean river that flows with a constant velocity. The concentration $u(x,t)$ of the substance at a distance x downstream at time t is the solution of the IBVP

$$u_t(x,t) = \sigma u_{xx}(x,t) - v u_x(x,t), \quad x > 0, \ t > 0,$$
$$u(0,t) = \alpha = \text{const}, \quad t > 0,$$
$$u(x,0) = 0, \quad x > 0,$$

where σ is the diffusion coefficient, $v = \text{const} > 0$ is the velocity of the river, $\alpha = \text{const} > 0$ is related to the substance discharge rate, and the second term on the right-hand side in the PDE accounts for the convection effect of the water flow on the substance.

If the river is slow, then the convection term is much smaller than the diffusion term and the PDE assumes the approximate form

$$u_t(x,t) = \sigma u_{xx}(x,t), \quad x > 0, \ t > 0,$$

which is the diffusion equation. If the river is fast, then the approximation is given by the convection equation

$$u_t(x,t) = -v u_x(x,t), \quad x > 0, \ t > 0.$$

Since we have already studied the diffusion (heat) equation, we now turn our attention to the convection and combined cases.

(i) The IBVP for pure convection is

$$u_t(x,t) = -vu_x(x,t), \quad x > 0, \ t > 0,$$
$$u(0,t) = \alpha, \quad t > 0,$$
$$u(x,0) = 0, \quad x > 0.$$

Let $\mathcal{L}[u](x,s) = U(x,s)$. Applying the Laplace transformation to the PDE and BC, we arrive at the problem

$$vU'(x,s) + sU(x,s) = 0, \quad x > 0,$$
$$U(0,s) = \frac{1}{s}\alpha,$$

with solution

$$U(x,s) = \frac{1}{s}\alpha e^{-(s/v)x} = \frac{1}{s}\alpha e^{-(x/v)s}.$$

Since

$$\mathcal{L}^{-1}\left[e^{-as}/s\right] = H(t-a),$$

we set $a = x/v$ to find that the solution of the original IBVP is

$$u(x,t) = \mathcal{L}^{-1}[U](x,t) = \mathcal{L}^{-1}\left[\frac{1}{s}\alpha e^{-(x/v)s}\right]$$

$$= \alpha H(t - x/v) = \begin{cases} 0, & t < x/v, \\ \alpha, & t \geq x/v \end{cases}$$

Thus, the substance reaches a fixed position x at time $t = x/v$; after that, the concentration of the substance at x remains constant (equal to the concentration of the substance at the point where it is poured into the river). The line $t = x/v$ in the (x,t)-plane is the advancing wave front of the substance (see Fig. 9.1).

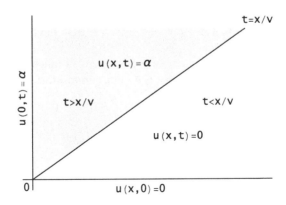

Figure 9.1: The front wave and the regions behind it and ahead of it.

(ii) We now consider a very long river with the substance already uniformly distributed in it from the source of the river up to the observation point $x = 0$, and assume that both diffusion and convection effects are significant. This mixed diffusion–convection problem in an infinite one-dimensional medium is modeled by the IVP

$$u_t(x,t) = \sigma u_{xx}(x,t) - v u_x(x,t), \quad -\infty < x < \infty, \ t > 0,$$

$$u(x,t), \ u_x(x,t) \to 0 \ \text{ as } x \to \pm\infty, \ t > 0,$$

$$u(x,0) = 1 - H(x), \quad -\infty < x < \infty.$$

We already know two possible methods for solving this problem: we can apply the Laplace transformation with respect to t or the (full) Fourier transformation with respect to x. In light of the discussion in (i) above, however, we indicate a third one, which consists in changing the x-coordinate by connecting it to the wave front through the combination

$$\xi = x - vt. \tag{9.9}$$

Clearly, $\xi = 0$ means that the point (x,t) is on the wave front, $\xi > 0$ means that (x,t) is ahead of the wave front, and $\xi < 0$ means that (x,t) is behind the wave front. We also write

$$u(x,t) = u(\xi + vt, t) = w(\xi,t).$$

Hence, by the chain rule,

$$u_t = w_\xi \xi_t + w_t = -v w_\xi + w_t,$$

$$u_x = w_\xi \xi_x = w_\xi,$$

$$u_{xx} = (w_\xi)_\xi \xi_x = w_{\xi\xi}.$$

Since $t = 0$ yields $x = \xi$, the above IVP becomes

$$w_t(\xi,t) = \sigma w_{\xi\xi}(\xi,t), \quad -\infty < \xi < \infty, \ t > 0,$$

$$w(\xi,t), \ w_\xi(\xi,t) \to 0 \ \text{ as } \xi \to \pm\infty, \ t > 0,$$

$$w(\xi,0) = 1 - H(\xi), \quad -\infty < \xi < \infty.$$

This problem was solved earlier by means of the Fourier transformation (see Section 8.1), and its solution is

$$w(\xi,t) = \frac{1}{2\sqrt{\pi\sigma t}} \int_{-\infty}^{\infty} [1 - H(y)] e^{-(\xi-y)^2/(4\sigma t)} \, dy$$

$$= \frac{1}{2\sqrt{\pi\sigma t}} \int_{-\infty}^{0} e^{-(\xi-y)^2/(4\sigma t)} \, dy.$$

Then, by (9.9), we find that the solution of our IVP in terms of the original variables x and t is

$$u(x,t) = \frac{1}{2\sqrt{\pi\sigma t}} \int_{-\infty}^{0} e^{-(x-vt-y)^2/(4\sigma t)} \, dy.$$

9.27 Remark. The above method is inadequate when $x > 0$ and the PDE is nonhomogeneous and contains a term of the form $cu(x,t)$, $c = $ const. In that case, the solution procedure relies on the direct use of the Laplace transformation.

9.28 Example. Setting $\mathcal{L}[u](x,s) = U(x,s)$, we reduce the IBVP

$$u_t(x,t) = u_{xx}(x,t) - 2u_x(x,t) + u(x,t) + (3-t)e^{2x}, \quad x > 0, \ t > 0,$$
$$u(0,t) = t - 2, \ t > 0,$$
$$u(x,0) = -2e^{2x}, \ x > 0$$

to the transformed problem

$$U''(x,s) - 2U'(x,s) + (1-s)U(x,s) = \frac{2s^2 - 3s + 1}{s^2} e^{2x},$$
$$U(0,s) = \frac{1}{s^2} - \frac{2}{s},$$

with general solution

$$U(x,s) = C_1(s)e^{(1+\sqrt{s})x} + C_2(s)e^{(1-\sqrt{s})x} + \frac{2s^2 - 3s + 1}{s^2(1-s)} e^{2x}.$$

The standard argument consisting of the vanishing of $U(x,s)$ as $s \to \infty$, the BC, and partial fraction decomposition now yield $C_2(s) = 0$ and then

$$U(x,s) = \left(\frac{1}{s^2} - \frac{2}{s} \right) e^{2x},$$

from which
$$u(x,t) = \mathcal{L}^{-1}[U](x,t) = (t-2)e^{2x}.$$

VERIFICATION WITH MATHEMATICA®. The input

```
u = (t - 2) * E^(2 * x);
{q,f,g}={(3 - t) * E^(2 * x), t-2, -2 * E^(2 * x)};
Simplify[{D[u,t] - D[u,x,x] + 2 * D[u,x] - u - q, (u /. x -> 0) - f,
   (u /. t -> 0) - g}]
```

generates the output $\{0, 0, 0\}$.

9.29 Example. The IBVP

$$u_t(x,t) = u_{xx}(x,t) - 4u_x(x,t) + 4u(x,t) + (3t-4)e^x,$$
$$x > 0, \ t > 0,$$
$$u_x(0,t) = 1 - 3t, \ \ t > 0,$$
$$u(x,0) = e^x, \ \ x > 0$$

is handled in similar fashion. With the notation $\mathcal{L}[u](x,s) = U(x,s)$ and the application of the appropriate formulas from Appendix A.6, we change this problem to its transformed version

$$U''(x,s) - 4U'(x,s) + (4-s)U(x,s) = \frac{-s^2 + 4s - 3}{s^2} e^x, \ \ x > 0,$$
$$U'(0,s) = \frac{1}{s} - \frac{3}{s^2},$$

whose GS is

$$U(x,s) = C_1(s)e^{(2+\sqrt{s})x} + C_2(s)e^{(2-\sqrt{s})x} + \left(\frac{1}{s} - \frac{3}{s^2}\right)e^x.$$

The same technique applied in earlier examples, but now with a BC for the derivative of U, produces the solution

$$U(x,s) = \left(\frac{1}{s} - \frac{3}{s^2}\right)e^x,$$

so

$$u(x,t) = \mathcal{L}^{-1}[U](x,t) = (1 - 3t)e^x.$$

VERIFICATION WITH MATHEMATICA®. The input

```
u = (1 - 3 * t) * E^x;
{q,f,g} = {(3 * t - 4) * E^x, 1 - 3 * t, E^x};
Simplify[{D[u,t] - D[u,x,x] + 4 * D[u,x] - 4 * u - q,
    (D[u,x] /.x -> 0) - f, (u /.t -> 0) - g}]
```

generates the output $\{0, 0, 0\}$.

9.30 Example. An identical solution scheme can be applied to the IBVP

$$u_t(x,t) = \tfrac{1}{4}u_{xx}(x,t) - u_x(x,t) + u(x,t) + 2 + x + t - xt,$$
$$x > 0, \ t > 0,$$
$$u(0,t) = -1, \ \ t > 0,$$
$$u(x,0) = -2, \ \ x > 0.$$

Writing $\mathcal{L}[u](x,s) = U(x,s)$ and using the necessary formulas from Appendix A.6, we convert this problem to the simpler one

$$U''(x,s) - 4U'(x,s) + 4(1-s)U(x,s)$$

$$= 8 - \frac{8}{s} - \frac{4}{s^2} - \left(\frac{4}{s} - \frac{4}{s^2}\right)x, \quad x > 0,$$

$$U(0,s) = -\frac{1}{s},$$

with GS

$$U(x,s) = C_1(s)e^{(2+2\sqrt{s})x} + C_2(s)e^{(2-2\sqrt{s})x} - \frac{2}{s} + \frac{1}{s^2}x.$$

In this case, the arguments set out earlier lead to

$$U(x,s) = \frac{1}{s}e^{2x}e^{-2x\sqrt{s}} - \frac{2}{s} + \frac{1}{s^2}x,$$

from which

$$u(x,t) = \mathcal{L}^{-1}[U](x,t) = e^{2x}\operatorname{erfc}\frac{x}{\sqrt{t}} + xt - 2.$$

VERIFICATION WITH MATHEMATICA®. The input

```
u = E^(-2*x)*Erfc[x/t^(1/2)] + x*t - 2;
{q,f,g} = {2+x+t-x*t, -1, -2};
Simplify[{D[u,t] - (1/4)*D[u,x,x] + D[u,x] - u - q,
   (u/.x->0) - f, (u/.t->0) - g}]
```

generates the output $\{0, 0, 0\}$.

Exercises

Use the Laplace transformation to solve the IBVP consisting of the PDE

$$u_t(x,t) = ku_{xx}(x,t) - au_x(x,t) + bu(x,t) + q(x,t), \quad x > 0, \ t > 0$$

and coefficients k, a, b, function q, BC, and IC as indicated.

1 $k = 1$, $a = 2$, $b = 1$, $q(x,t) = (6t+1)\cos x + (2t+3)\sin x$,
$u(0,t) = -t$, $u(x,0) = \sin x$.

2 $k = 1$, $a = 4$, $b = 4$, $q(x,t) = (3-2t)e^{2t} + (1-9t)e^{-x}$,
$u_x(0,t) = -t$, $u(x,0) = -1$.

3 $k = 2$, $a = 1$, $b = 1/8$, $q(x,t) = 0$, $u(0,t) = 1$, $u(x,0) = 0$.

4 $k = 2$, $a = 2$, $b = 1/2$, $q(x,t) = 0$, $u_x(0,t) = 1 - \sqrt{2/(\pi t)}$, $u(x,0) = 1$.

Answers to Odd-Numbered Exercises

1 $u(x,t) = -t\cos x + (3t+1)\sin x.$

3 $u(x,t) = e^{x/4}\operatorname{erfc}\left(x/\sqrt{8t}\right).$

9.2.5 Dissipative Waves

Once again, we illustrate the treatment of this type of model by solving a few specific IBVPs.

9.31 Example. Making the notation $\mathcal{L}[u](x,s) = U(x,s)$ and applying the Laplace transformation to the PDE and BC, we reduce the IBVP

$$u_{tt}(x,t) + 6u_t(x,t) = 4u_{xx}(x,t) - 12u_x(x,t) + 24 + (16t - 6)e^{-x},$$
$$x > 0, \quad t > 0,$$
$$u(0,t) = 1 - t - e^t, \quad t > 0,$$
$$u(x,0) = 2x, \quad u_t(x,0) = -e^{-x}, \quad x > 0$$

to the ODE problem

$$4U''(x,s) - 12U'(x,s) - (s^2 + 6s)U(x,s)$$
$$= -(2s+12)x - \frac{24}{s} + \left(1 - \frac{16}{s^2} + \frac{6}{s}\right)e^{-x}, \quad x > 0,$$
$$U(0,s) = \frac{1}{s} - \frac{1}{s^2} - \frac{1}{s-1},$$

whose general solution, constructed as the sum of the complementary function and a particular integral, is

$$U(x,s) = C_1(s)e^{(3+s/2)x} + C_2(s)e^{-sx/2} + \frac{2}{s}x - \frac{1}{s^2}e^{-x}.$$

First, the limit $U(x,s) \to 0$ as $s \to \infty$ requires that $C_1(s) = 0$, after which the BC leads to

$$U(x,s) = \left(\frac{1}{s} - \frac{1}{s-1}\right)e^{-sx/2} + \frac{2}{s}x - \frac{1}{s^2}e^{-x}.$$

To find the solution of the given IBVP, we now use formulas 5, 7, and 6 (with $n = 1$) in Appendix A.6 and conclude that

$$u(x,t) = \mathcal{L}^{-1}[U](x,t) = (1 - e^{t-x/2})H(t - x/2) + 2x - te^{-x}$$
$$= \begin{cases} 2x - te^{-x}, & t < x, \\ 2x - te^{-x} + 1 - e^{t-x/2}, & t \geq x. \end{cases}$$

We see that the solution is continuous across the line $x = 2t$.

VERIFICATION WITH MATHEMATICA®. The input

```
{u1, u2} = {2*x-t*E^(-x), 1-E^(t-x/2)};
u = u1 + u2 * HeavisideTheta[t-x];
{q, f, g1, g2} = {24 + (16*t-6) * E^(-x), 1-t-E^t,
   2*x, -E^(-x)};
Simplify[{Simplify[D[u,t,t] + 6*D[u,t] - 4*D[u,x,x] +
   12*D[u,x] - q, t>x||t<x], ((u1+u2)/.x->0) - f,
   (u1/.t->0) - g1, (D[u1,t]/.t->0) - g2}]
```

generates the output $\{0, 0, 0, 0\}$.

9.32 Example. Consider the IBVP

$$u_{tt}(x,t) + 3u_t(x,t) + 2u(x,t)$$
$$= u_{xx}(x,t) - u_x(x,t) + 2x^2 - 2x - 2t - 7, \quad x > 0, \ t > 0,$$
$$u_x(0,t) = -2, \quad t > 0,$$
$$u(x,0) = x^2 - 2x, \quad u_t(x,0) = -1, \quad x > 0.$$

If we write, as usual, $\mathcal{L}[u](x,s) = U(x,s)$ and apply the Laplace transformation to the PDE and BC, we arrive at the ODE problem

$$U''(x,s) - U'(x,s) - (s^2 + 3s + 2)U(x,s)$$
$$= -\frac{s^2 + 3s + 2}{s}x^2 + \frac{2s^2 + 6s + 2}{s}x + \frac{s^2 + 7s + 2}{s^2}, \quad x > 0,$$
$$U'(0,s) = -\frac{2}{s}.$$

After some standard algebraic manipulation, the above ODE yields the general solution

$$U(x,s) = C_1(s)e^{(2+s)x} + C_2(s)e^{-(1+s)x} + \frac{1}{s}x^2 - \frac{2}{s}x - \frac{1}{s^2}.$$

Next, the property that $U(x,s) \to 0$ as $s \to \infty$ implies that $C_1(s) = 0$, and then the BC yields $C_2(s) = 0$, so

$$U(x,s) = \frac{1}{s}x^2 - \frac{2}{s}x - \frac{1}{s^2}.$$

Applying formulas 5 and 6 in Appendix A.6, we now find that

$$u(x,t) = \mathcal{L}^{-1}[U](x,t) = x^2 - 2x - t.$$

VERIFICATION WITH MATHEMATICA®. The input

```
u = x^2 - 2*x - t;
{q, f, g1, g2} = {2*x^2 - 2*x - 2*t - 7, -2, x^2 - 2*x, -1};
Simplify[{[D[u,t,t] + 3*D[u,t] + 2*u - D[u,x,x] + D[u,x] - q,
    (D[u,x] /.x -> 0) - f, (u /.t -> 0) - g1, (D[u,t] /.t -> 0) - g2}]
```

generates the output $\{0, 0, 0, 0\}$.

9.33 Example. The IBVP

$$u_{tt}(x, t) + 5u_t(x, t) + 6u(x, t)$$
$$= u_{xx}(x, t) - u_x(x, t) + 7 - 5t + (5 + 6t)x, \quad x > 0, \ t > 0,$$
$$u(0, t) = 2 - t + e^{-3t/2}, \quad t > 0,$$
$$u(x, 0) = 2, \quad u_t(x, 0) = x - 1, \quad x > 0$$

is reduced by the Laplace transformation to the ODE problem

$$U''(x, s) - U'(x, s) - (s^2 + 5s + 6)U(x, s)$$
$$= -2s - 9 - \frac{7}{s} + \frac{5}{s^2} - \left(1 + \frac{5}{s} + \frac{6}{s^2}\right)x, \quad x > 0,$$
$$U(0, s) = \frac{2}{s} - \frac{1}{s^2} + \frac{2}{2s + 3},$$

where $U(x, s) = \mathcal{L}[u](x, s)$. The general solution of the linear, second-order differential equation in this case turns out to be

$$U(x, s) = C_1(s)e^{(3+s)x} + C_2(s)e^{-(2+s)x} + \frac{1}{s^2}x - \frac{1}{s^2} + \frac{2}{s},$$

from which, repeating the steps taken in the preceding examples, we ultimately arrive at the particular solution

$$U(x, s) = \frac{1}{s + \frac{3}{2}}e^{-sx}e^{-2x} + \frac{1}{s^2}x - \frac{1}{s^2} + \frac{2}{s}.$$

By inversion, we now find that

$$u(x, t) = \mathcal{L}^{-1}[U](x, t) = e^{-2x}e^{-3(t-x)/2}H(t - x) + xt - t + 2$$
$$= \begin{cases} xt - t + 2, & t < x, \\ xt - t + 2 + e^{-(3t+x)/2}, & t \geq x. \end{cases}$$

As can be seen, this function is discontinuous across the line $x = t$.

VERIFICATION WITH MATHEMATICA®. The input

```
{u1, u2} = x * t - t + 2, E^( - (3 * t + x) /2);
u = u1 + u2 * HeavisideTheta[t - x];
{q,f,g1,g2} = {7 - 5 * t + (5 + 6 * t) * x, 2 - t + E^( - 3 * t /2),
  2, x - 1};
Simplify[{D[u,t,t] + 5 * D[u,t] + 6 * u - D[u,x,x] + D[u,x] - q,
  (u /.x -> 0) - f, (u /.t -> 0) - g1, (D[u,t] /.t -> 0) - g2}]
```

generates the output $\{0, 0, 0, 0\}$.

Exercises

Use the Laplace transformation to find the solution of the IBVP consisting of the PDE

$$u_{tt}(x,t) + au_t(x,t) + bu(x,t) = c^2 u_{xx}(x,t) - du_x(x,t) + q(x,t),$$
$$x > 0, \ t > 0,$$

coefficients a, b, c, d, function q, BC, and ICs as indicated.

1 $a = 1$, $b = 0$, $c = 2$, $d = 1$, $q(x,t) = 2x^2 + 4xt - 16t$,
  $u(0,t) = e^{-t}$, $u(x,0) = 1$, $u_t(x,0) = 2x^2 - 1$.

2 $a = 4$, $b = 4$, $c = 1$, $d = 0$, $q(x,t) = 2e^{-2t}$,
  $u(0,t) = 4t$, $u(x,0) = 0$, $u_t(x,0) = 1$.

3 $a = 2$, $b = 1$, $c = 1$, $d = 0$, $q(x,t) = 1$,
  $u(0,t) = t + 2$, $u(x,0) = -1$, $u_t(x,0) = 0$.

4 $a = 1$, $b = 2$, $c = 1$, $d = 1$, $q(x,t) = (8t + 5)e^{x+2t}$,
  $u_x(0,t) = te^{2t}$, $u(x,0) = 0$, $u_t(x,0) = e^x$.

5 $a = 3$, $b = 2$, $c = 1$, $d = 1$, $q(x,t) = 4t$,
  $u_x(0,t) = -1 - e^{-2t}$, $u(x,0) = 2$, $u_t(x,0) = -4$.

6 $a = 2$, $b = 3/4$, $c = 1/2$, $d = 1/2$, $q(x,t) = 12$,
  $u_x(0,t) = -t - e^{-3t/2}$, $u(x,0) = 1$, $u_t(x,0) = 1$.

Answers to Odd-Numbered Exercises

1 $u(x,t) = e^{-t} + 2x^2 t$.

3 $u(x,t) = 1 - 2(t + 1)e^{-t} + (1 + t - x)(e^{-x} + 2e^{-t})H(t - x)$.

5 $u(x,t) = 2t - 3 + e^{-2t} + 4e^{-t} + (e^{-x} - e^{-2t+x})H(t - x)$.

CHAPTER 10

The Method of Green's Functions

The types of problems we have considered for the heat, wave, and Laplace equations have solutions that are determined uniquely by the prescribed data (boundary conditions, initial conditions, and any nonhomogeneous term in the equation). It is natural, therefore, to seek a formula that gives the solution directly in terms of the data. Such closed-form solutions are constructed by means of the so-called Green's function of the given problem, and are of great importance in practical applications.

10.1   The Heat Equation

We begin by examining the simplest case.

10.1.1   The Time-Independent Problem

The equilibrium temperature distribution in a finite rod with internal sources and zero temperature at the endpoints is modeled by a BVP of the form (see Section 6.1)

$$u''(x) = -\frac{1}{k}q(x), \quad 0 < x < L,$$
$$u(0) = 0, \quad u(L) = 0.$$

$$(10.1)$$

For convenience, we have omitted the subscript ∞ from the symbol of the steady-state solution, but have kept the factor $-1/k$ since we will later make a comparison between the solutions of the equilibrium and time-dependent problems.

If we have just one unit source concentrated at a point ξ, $0 < \xi < L$, then
$$q(x) = \delta(x - \xi),$$
and the corresponding two-point solution $G(x, \xi)$ of the above BVP satisfies
$$G_{xx}(x, \xi) = -\frac{1}{k} \delta(x - \xi), \quad 0 < x < L, \tag{10.2}$$
$$G(0, \xi) = 0, \quad G(L, \xi) = 0.$$

The function $G(x, \xi)$ can be computed explicitly. Since (see Remark 9.4(ii))
$$H_x(x - \xi) = \delta(x - \xi),$$
from (10.2) it follows that
$$G_x(x, \xi) = -\frac{1}{k} H(x - \xi) + C_1(\xi) = \begin{cases} C_1(\xi), & x < \xi, \\ -\dfrac{1}{k} + C_1(\xi), & x > \xi, \end{cases}$$
from which
$$G(x, \xi) = \begin{cases} xC_1(\xi) + C_2(\xi), & x < \xi, \\ x\left[C_1(\xi) - \dfrac{1}{k}\right] + C_3(\xi), & x > \xi, \end{cases} \tag{10.3}$$
where C_1, C_2, and C_3 are arbitrary functions of ξ. Using the BCs in the BVP, we find that
$$C_2(\xi) = 0, \quad L\left[C_1(\xi) - \frac{1}{k}\right] + C_3(\xi) = 0.$$
Hence, $C_3(\xi) = -L[C_1(\xi) - 1/k]$, and (10.3) becomes
$$G(x, \xi) = \begin{cases} xC_1(\xi), & x < \xi, \\ (x - L)\left[C_1(\xi) - \dfrac{1}{k}\right], & x > \xi. \end{cases} \tag{10.4}$$

If $G(x, \xi)$ had a jump (H-type) discontinuity at $x = \xi$, then G_x would have a δ-type singularity at $x = \xi$. Since this is not the case, we must conclude that $G(x, \xi)$ is continuous at $x = \xi$; in other words, $G(\xi-, \xi) = G(\xi+, \xi)$, which, in view of (10.4), leads to
$$\xi C_1(\xi) = (\xi - L)\left[C_1(\xi) - \frac{1}{k}\right].$$

Thus, $C_1(\xi) = (L - \xi)/(kL)$, and (10.4) yields
$$G(x, \xi) = \begin{cases} \dfrac{x}{kL}(L - \xi), & x \leq \xi, \\ \dfrac{\xi}{kL}(L - x), & x > \xi. \end{cases} \tag{10.5}$$

Clearly, $G(x, \xi) = G(\xi, x)$.

Using integration by parts, we find that for two smooth functions u and v on $[0, L]$,

$$\int_0^L (u''v - v''u)\, dx = u'v\Big|_0^L - \int_0^L u'v'\, dx - v'u\Big|_0^L + \int_0^L v'u'\, dx$$

$$= \left[u'(L)v(L) - v'(L)u(L)\right] - \left[u'(0)v(0) - v'(0)u(0)\right]. \tag{10.6}$$

This is known as *Green's formula*. If u is now the solution of (10.1) and $v = G$ is the solution of (10.2), then the right-hand side in (10.6) vanishes and we arrive at

$$\int_0^L \left[u(x)\delta(x - \xi) - G(x, \xi)q(x)\right] dx = 0.$$

By (9.1), interchanging x and ξ and recalling that $G(x, \xi) = G(\xi, x)$, we obtain

$$u(x) = \int_0^L G(x, \xi)q(\xi)\, d\xi. \tag{10.7}$$

$G(x, \xi)$, called the *Green's function* of the BVP (10.1), is the temperature at x due to a concentrated unit heat source at ξ. Formula (10.7) shows the aggregate influence of all the sources $q(\xi)$ in the rod on the temperature at x.

A representation formula similar to (10.7) can also be derived for nonhomogeneous BCs. Suppose that the BCs in (10.1) are replaced by $u(0) = a$ and $u(L) = b$. Then (10.6) with the same choice of u and v as above becomes

$$\int_0^L \left[u(x)\delta(x - \xi) - G(x, \xi)q(x)\right] dx = -k\left[u(x)G_x(x, \xi)\right]_{x=0}^{x=L},$$

so

$$u(x) = \int_0^L G(x, \xi)q(\xi)\, d\xi - k\left[bG_\xi(x, L) - aG_\xi(x, 0)\right].$$

By (10.5),

$$G_\xi(x, \xi) = \begin{cases} -\dfrac{x}{kL}, & x < \xi, \\ -\dfrac{x - L}{kL}, & x > \xi; \end{cases}$$

consequently, the desired representation formula is

$$u(x) = \int_0^L G(x, \xi)q(\xi)\, d\xi + b\frac{x}{L} + a\left(1 - \frac{x}{L}\right). \tag{10.8}$$

10.1 Example. To compute the steady-state solution for the IBVP

$$u_t(x,t) = u_{xx}(x,t) + x - 1, \quad 0 < x < 1, \ t > 0,$$
$$u(0,t) = 2, \quad u(1,t) = -1, \quad t > 0,$$
$$u(x,0) = f(x), \quad 0 < x < 1,$$

we use (10.8) with $k = 1$, $L = 1$, $a = 2$, $b = -1$, and $q(x) = x - 1$. First, by (10.5),

$$G(x,\xi) = \begin{cases} x(1-\xi), & x \le \xi, \\ \xi(1-x), & x > \xi; \end{cases}$$

hence,

$$\int_0^1 G(x,\xi)q(\xi)\,d\xi = \int_0^x \xi(1-x)(\xi-1)\,d\xi + \int_x^1 x(1-\xi)(\xi-1)\,d\xi$$

$$= (1-x)\int_0^x (\xi^2 - \xi)\,d\xi - x\int_x^1 (\xi-1)^2\,d\xi$$

$$= -\frac{1}{6}x^3 + \frac{1}{2}x^2 - \frac{1}{3}x.$$

Since

$$bx/L + a(1 - x/L) = -x + 2(1-x) = 2 - 3x,$$

the equilibrium solution (10.8) of the given IBVP is

$$u(x) = -\frac{1}{6}x^3 + \frac{1}{2}x^2 - \frac{10}{3}x + 2.$$

VERIFICATION WITH MATHEMATICA®. The input

```
u = - (1/6) * x^3 + (1/2) * x^2 - (10/3) * x + 2;
{q, f1, f2} = {x - 1, 2, - 1};
Simplify[{D[u,x,x] + q, {(u /. x -> 0) - f1, (u /. x -> 1) - f2}}]
```

generates the output $\{0, \{0, 0\}\}$.

10.2 Example. In the case of the IBVP

$$u_t(x,t) = u_{xx}(x,t) + q(x), \quad 0 < x < 1, \ t > 0,$$
$$u(0,t) = 1, \quad u(1,t) = 3, \quad t > 0,$$
$$u(x,0) = f(x), \quad 0 < x < 1$$

with

$$q(x) = \begin{cases} 2, & 0 < x \le 1/2, \\ -1, & 1/2 < x < 1, \end{cases}$$

we notice that the function G is the same as in the preceding example, whereas $a = 1$ and $b = 3$. Given that the expressions of q and G change at $x = 1/2$ and $x = \xi$, respectively, we split the computation of the equilibrium solution (10.8) into two parts.

(i) If $0 < x < 1/2$, the first term on the right-hand side in (10.8) is written as a sum of three integrals, one for each of the intervals $0 < \xi \le x$, $x < \xi \le 1/2$, and $1/2 < \xi < 1$. Thus, (10.8) yields

$$u(x) = \int_0^x 2\xi(1-x)\,d\xi + \int_x^{1/2} 2x(1-\xi)\,d\xi + \int_{1/2}^1 -x(1-\xi)\,d\xi + 2x + 1$$

$$= -x^2 + \tfrac{21}{8}x + 1.$$

(ii) If $1/2 < x < 1$, the three integrals are over the intervals $0 < \xi \le 1/2$, $1/2 < \xi \le x$, and $x < \xi < 1$, so

$$u(x) = \int_0^{1/2} 2\xi(1-x)\,d\xi + \int_{1/2}^x -\xi(1-x)\,d\xi + \int_x^1 -x(1-\xi)\,d\xi + 2x + 1$$

$$= \tfrac{1}{2}x^2 + \tfrac{9}{8}x + \tfrac{11}{8}.$$

It is easy to verify that

$$u\left(\tfrac{1}{2}-\right) = u\left(\tfrac{1}{2}+\right) = \tfrac{33}{16}, \quad u'\left(\tfrac{1}{2}-\right) = u'\left(\tfrac{1}{2}+\right) = \tfrac{13}{8},$$

but

$$u''\left(\tfrac{1}{2}-\right) = -2, \quad u''\left(\tfrac{1}{2}+\right) = 1,$$

which confirms that, as expected, the discontinuity of q at $x = 1/2$ has reduced the smoothness of the solution.

VERIFICATION WITH MATHEMATICA®. In view of the split form of the source term and solution, we use the input

```
{u1, u2  = {-x^2+ (21/8) *x+1, (1/2) *x^2+ (9/8) *x+ (11/8)};
{{q1, q2}, {f1, f2}}={{2,-1}, {1,3}};
Simplify[{D[{u1, u2},x,x] +{q1, q2}, (u1 /. x->0) -f1,
   (u2 /. x->1) -f2}]
```

which generates the output $\{\{0,0\},0,0\}$.

10.3 Remark. The Green's function can be expanded in a double Fourier series. In view of the eigenfunctions of the Sturm–Liouville problem associated with (10.1) (see Section 5.1), the continuity of G, and the symmetry relation $G(x,\xi) = G(\xi,x)$, it seems reasonable to seek a series representation of the form

$$G(x,\xi) = \sum_{m=1}^{\infty}\left(\sum_{n=1}^{\infty} b_{mn} \sin\frac{n\pi x}{L}\right)\sin\frac{m\pi\xi}{L}. \tag{10.9}$$

Differentiating (10.9) term by term twice with respect to x and substituting in the ODE in (10.2), we obtain

$$\sum_{m=1}^{\infty} \left[\sum_{n=1}^{\infty} \left(\frac{n\pi}{L} \right)^2 b_{mn} \sin \frac{n\pi x}{L} \right] \sin \frac{m\pi\xi}{L} = \frac{1}{k} \delta(x - \xi).$$

Next, we multiply this equality by $\sin(p\pi x/L)$, $p = 1, 2, \ldots$, integrate over $[0, L]$, and use (2.5) and (9.1) to find that

$$\left(\frac{p\pi}{L} \right)^2 \frac{kL}{2} \sum_{m=1}^{\infty} b_{mp} \sin \frac{m\pi\xi}{L} = \sin \frac{p\pi\xi}{L}, \quad p = 1, 2, \ldots.$$

Theorem 3.26(ii) now implies that the only nonzero coefficients above are $b_{pp} = 2L/(kp^2\pi^2)$, so series (10.9) is

$$G(x, \xi) = \sum_{n=1}^{\infty} \frac{2L}{kn^2\pi^2} \sin \frac{n\pi x}{L} \sin \frac{n\pi\xi}{L}.$$

Exercises

In **1–10**, construct the appropriate Green's function and solve the BVP for the one-dimensional steady-state heat equation

$$ku''(x) + q(x) = 0, \quad 0 < x < L$$

with constants k, L, function q, and BCs as indicated.

1 $k = 2$, $L = 1$, $q(x) = 2x - 1$, $u(0) = 1$, $u(1) = -2$.

2 $k = 1$, $L = 2$, $q(x) = \begin{cases} 1, & 0 < x \le 1, \\ x, & 1 < x < 2, \end{cases}$ $u(0) = 3$, $u(2) = 1$.

3 $k = 1$, $L = 2$, $q(x) = 4$, $u(0) = 2$, $u'(2) = -3$.

4 $k = 2$, $L = 1$, $q(x) = 1 - x$, $u(0) = -1$, $u'(1) = 1$.

5 $k = 2$, $L = 2$, $q(x) = \begin{cases} -1, & 0 < x \le 1, \\ 2, & 1 < x < 2, \end{cases}$ $u(0) = -2$, $u'(2) = 3$.

6 $k = 1$, $L = 1$, $q(x) = \begin{cases} 1, & 0 < x \le 1/2, \\ x + 1, & 1/2 < x < 1, \end{cases}$ $u(0) = 4$, $u'(1) = 2$.

7 $k = 3$, $L = 1$, $q(x) = -2$, $u'(0) = -3$, $u(1) = 4$.

8 $k = 2$, $L = 2$, $q(x) = x - 2$, $u'(0) = 1$, $u(2) = -3$.

9 $k = 1$, $L = 1$, $q(x) = \begin{cases} 2, & 0 < x \le 1/2, \\ -3, & 1/2 < x < 1, \end{cases}$ $u'(0) = -2$, $u(1) = 5$.

10 $k = 1$, $L = 2$, $q(x) = \begin{cases} x - 1, & 0 < x \leq 1, \\ 1, & 1 < x < 2, \end{cases}$ $u'(0) = 3$, $u(2) = -3$.

Answers to Odd-Numbered Exercises

1 $G(x, \xi) = \begin{cases} x(1 - \xi)/2, & x \leq \xi, \\ \xi(1 - x)/2, & x > \xi, \end{cases}$ $u(x) = (1/12)(12 - 37x + 3x^2 - 2x^3)$.

3 $G(x, \xi) = \begin{cases} x, & x \leq \xi, \\ \xi, & x > \xi, \end{cases}$ $u(x) = 2 + 5x - 2x^2$.

5 $G(x, \xi) = \begin{cases} x/2, & x \leq \xi, \\ \xi/2, & x > \xi, \end{cases}$ $u(x) = \begin{cases} x^2/4 + 7x/2 - 2, & 0 \leq x \leq 1, \\ -x^2/2 + 5x - 11/4, & 1 < x \leq 2. \end{cases}$

7 $G(x, \xi) = \begin{cases} (1 - \xi)/3, & x \leq \xi, \\ (1 - x)/3, & x > \xi, \end{cases}$ $u(x) = (1/3)(20 - 9x + x^2)$.

9 $G(x, \xi) = \begin{cases} 1 - \xi, & x \leq \xi, \\ 1 - x, & x > \xi, \end{cases}$ $u(x) = \begin{cases} 59/8 - 2x - x^2, & 0 \leq x \leq 1/2, \\ 8 - 9x/2 + 3x^2/2, & 1/2 < x \leq 1. \end{cases}$

10.1.2 The Time-Dependent Problem

A finite rod with internal sources and zero temperature at the endpoints is modeled by the IBVP

$$u_t(x, t) = ku_{xx}(x, t) + q(x, t), \quad 0 < x < L, \ t > 0,$$
$$u(0, t) = 0, \quad u(L, t) = 0, \quad t > 0,$$
$$u(x, 0) = f(x), \quad 0 < x < L.$$

(According to the arguments presented in Chapter 6, we may consider homogeneous BCs without loss of generality.) This problem can be solved by the method of eigenfunction expansion (see Section 7.1), so let

$$u(x, t) = \sum_{n=1}^{\infty} u_n(t) \sin \frac{n\pi x}{L},$$

$$q(x, t) = \sum_{n=1}^{\infty} q_n(t) \sin \frac{n\pi x}{L},$$

$$f(x) = \sum_{n=1}^{\infty} f_n \sin \frac{n\pi x}{L},$$

where

$$q_n(t) = \frac{2}{L} \int_0^L q(x, t) \sin \frac{n\pi x}{L} \, dx, \quad f_n = \frac{2}{L} \int_0^L f(x) \sin \frac{n\pi x}{L} \, dx. \tag{10.10}$$

Replacing these series in the PDE, we find in the usual way that the u_n, $n = 1, 2, \ldots$, satisfy

$$u_n'(t) + k\left(\frac{n\pi}{L}\right)^2 u_n(t) = q_n(t), \quad t > 0,$$

$$u_n(0) = f_n.$$

This problem is solved, for example, by means of the integrating factor

$$\exp\left\{\int k\left(\frac{n\pi}{L}\right)^2 dt\right\} = e^{k(n\pi/L)^2 t}.$$

Thus, taking the IC for u_n into account, we obtain

$$u_n(t) = e^{-k(n\pi/L)^2 t}\left[\int_0^t q_n(\tau)e^{k(n\pi/L)^2 \tau} d\tau + C\right]$$

$$= f_n e^{-k(n\pi/L)^2 t} + e^{-k(n\pi/L)^2 t}\int_0^t q_n(\tau)e^{k(n\pi/L)^2 \tau} d\tau;$$

so, by (10.10),

$$u(x,t) = \sum_{n=1}^\infty \left[f_n e^{-k(n\pi/L)^2 t} + e^{-k(n\pi/L)^2 t}\int_0^t q_n(\tau)e^{k(n\pi/L)^2 \tau} d\tau\right]\sin\frac{n\pi x}{L}$$

$$= \sum_{n=1}^\infty \left\{\left[\frac{2}{L}\int_0^L f(\xi)\sin\frac{n\pi\xi}{L} d\xi\right]e^{-k(n\pi/L)^2 t}\right.$$

$$\left. + e^{-k(n\pi/L)^2 t}\int_0^t \left[\frac{2}{L}\int_0^L q(\xi,\tau)\sin\frac{n\pi\xi}{L} d\xi\right]e^{k(n\pi/L)^2 \tau} d\tau\right\}$$

$$\times \sin\frac{n\pi x}{L}$$

$$= \int_0^L f(\xi)\left[\sum_{n=1}^\infty \frac{2}{L}\sin\frac{n\pi x}{L}\sin\frac{n\pi\xi}{L} e^{-k(n\pi/L)^2 t}\right] d\xi$$

$$+ \int_0^L \int_0^t q(\xi,\tau)\left[\sum_{n=1}^\infty \frac{2}{L}\sin\frac{n\pi x}{L}\sin\frac{n\pi\xi}{L} e^{-k(n\pi/L)^2 (t-\tau)}\right] d\tau\, d\xi.$$

If we now define the Green's function of this problem by

$$G(x,t;\xi,\tau) = \sum_{n=1}^\infty \frac{2}{L}\sin\frac{n\pi x}{L}\sin\frac{n\pi\xi}{L} e^{-k(n\pi/L)^2 (t-\tau)}, \quad \tau < t, \qquad (10.11)$$

then the solution of the IBVP can be written in the form

$$u(x,t) = \int_0^L G(x,t;\xi,0)f(\xi)\,d\xi + \int_0^t\int_0^L G(x,t;\xi,\tau)q(\xi,\tau)\,d\tau\,d\xi. \qquad (10.12)$$

The first term in this formula represents the influence of the initial temperature in the rod on the subsequent temperature at any point x and any time t. The second term represents the influence of all the sources in the rod at all times $0 < \tau < t$ on the temperature at x and t. This expresses what is known as the *causality principle*.

10.4 Example. In the IBVP

$$u_t(x,t) = u_{xx}(x,t) + e^{-\pi^2 t}\sin(\pi x), \quad 0 < x < 1,\ t > 0,$$
$$u(0,t) = 0, \quad u(1,t) = 0, \quad t > 0,$$
$$u(x,0) = -\sin(3\pi x), \quad 0 < x < 1,$$

we have $k = 1$, $L = 1$, $q(x,t) = e^{-\pi^2 t}\sin(\pi x)$, and $f(x) = -\sin(3\pi x)$; therefore, by (10.11), (10.12), and the orthogonality formulas in Appendix A.1,

$$G(x,t;\xi,\tau) = \sum_{n=1}^{\infty} 2\sin(n\pi x)\sin(n\pi\xi)e^{-n^2\pi^2(t-\tau)}$$

and

$$u(x,t) = -\int_0^1 G(x,t;\xi,0)\sin(3\pi\xi)\,d\xi + \int_0^t\int_0^1 G(x,t;\xi,\tau)e^{-\pi^2\tau}\sin(\pi\xi)\,d\tau\,d\xi$$

$$= -\sum_{n=1}^{\infty} 2\sin(n\pi x)\left[\int_0^1 \sin(n\pi\xi)\sin(3\pi\xi)\,d\xi\right]e^{-n^2\pi^2 t}$$

$$+ \sum_{n=1}^{\infty} 2\sin(n\pi x)\left[\int_0^1 \sin(n\pi\xi)\sin(\pi\xi)\,d\xi\right]\left[\int_0^t e^{(n^2-1)\pi^2\tau}\,d\tau\right]e^{-n^2\pi^2 t}$$

$$= -2\cdot\tfrac{1}{2}\sin(3\pi x)e^{-9\pi^2 t} + 2\cdot\tfrac{1}{2}\left[\int_0^t d\tau\right]e^{-\pi^2 t}$$

$$= -e^{-9\pi^2 t}\sin(3\pi x) + te^{-\pi^2 t}\sin(\pi x).$$

VERIFICATION WITH MATHEMATICA®. The input

```
u = -E^(-9*Pi^2*t)*Sin[3*Pi*x] +t*E^(-Pi^2*t)*Sin[Pi*x];
{q,f}={E^(-Pi^2*t)*Sin[Pi*x], -Sin[3*Pi*x]};
Simplify[{D[u,t]-D[u,x,x]-q, {u/.x->0, u/.x->1},
    (u/.t->0)-f}]
```

generates the output $\{0, \{0,0\}, 0\}$.

10.5 Example. We use (10.11) and (10.12) once again to compute the solution of the IBVP

$$u_t(x,t) = u_{xx}(x,t) + t(x-1), \quad 0 < x < 1, \ t > 0,$$

$$u(0,t) = 0, \quad u(1,t) = 0, \quad t > 0,$$

$$u(x,0) = x, \quad 0 < x < 1.$$

Here, $k = 1$, $L = 1$, $q(x,t) = t(x-1)$, and $f(x) = x$, so

$$G(x,t;\xi,\tau) = \sum_{n=1}^{\infty} 2\sin(n\pi x)\sin(n\pi\xi)e^{-n^2\pi^2(t-\tau)},$$

which leads to

$$u(x,t) = \int_0^1 G(x,t;\xi,0)\xi \, d\xi + \int_0^1 \int_0^t G(x,t;\xi,\tau)\tau(\xi-1) \, d\tau \, d\xi$$

$$= \sum_{n=1}^{\infty} 2\sin(n\pi x)\left[\int_0^1 \xi\sin(n\pi\xi) \, d\xi\right]e^{-n^2\pi^2 t}$$

$$+ \sum_{n=1}^{\infty} 2\sin(n\pi x)\left[\int_0^1 (\xi-1)\sin(n\pi\xi) \, d\xi\right]\left[\int_0^t \tau e^{-n^2\pi^2(t-\tau)} \, d\tau\right].$$

Integrating by parts, we have

$$\int_0^1 \xi\sin(n\pi\xi) \, d\xi = (-1)^{n+1}\frac{1}{n\pi},$$

$$\int_0^1 (\xi-1)\sin(n\pi\xi) \, d\xi = -\frac{1}{n\pi},$$

$$\int_0^t \tau e^{-n^2\pi^2(t-\tau)} \, d\tau = \frac{1}{n^2\pi^2}t - \frac{1}{n^4\pi^4}(1 - e^{-n^2\pi^2 t}).$$

The combination of these results now yields the solution of the given IBVP in the form

$$u(x,t) = \sum_{n=1}^{\infty} \frac{2}{n\pi}\left[(-1)^{n+1}e^{-n^2\pi^2 t}\right.$$

$$\left. - \frac{1}{n^2\pi^2}t + \frac{1}{n^4\pi^4}(1 - e^{-n^2\pi^2 t})\right]\sin(n\pi x). \qquad (10.13)$$

VERIFICATION WITH MATHEMATICA®. Since (10.13) is the Fourier sine series representation of u, we check that the nth term satisfies the PDE and the IC with $t(x-1)$ and x replaced, respectively, by the corresponding terms in their own expansions. The BCs are also tested. Specifically, the input

```
un = ((2/(n*Pi))*((-1)^(n+1)*E^(-n^2*Pi^2*t) - (1/(n^2
    *Pi^2))*t + (1/(n^4*Pi^4))*(1 - E^(-n^2*Pi^2*t))))
    *Sin[n*Pi*x];
{q, f} = {t*(x-1), x};
qn = Simplify[2*Integrate[q*Sin[n*Pi*x], {x,0,1}], n ∈ Integers]
    *Sin[n*Pi*x];
fn = Simplify[2*Integrate[f*Sin[n*Pi*x], {x,0,1}], n ∈ Integers]
    *Sin[n*Pi*x];
Simplify[{D[un,t] - D[un,x,x] - qn, {un/.x->0, un/.x->1},
    (un/.t->0) - fn}, n ∈ Integers]
```

generates the output $\{0, \{0, 0\}, 0\}$.

10.6 Remark. Green's functions and representation formulas in terms of such functions can also be constructed for IBVPs with other types of BCs, and for IBVPs where the space variable takes values in a semi-infinite or infinite interval (see, for example, (8.8)).

Exercises

Construct a series representation of the appropriate Green's function and solve the IBVP consisting of the PDE

$$u_t(x,t) = ku_{xx}(x,t) + q(x,t), \quad 0 < x < L, \ t > 0,$$

constants k, L, function q, BCs, and IC as indicated.

1 $k = 2$, $L = 1$, $q(x,t) = t\sin(\pi x)$,

$u(0,t) = 0$, $u(1,t) = 0$, $u(x,0) = -\sin(2\pi x)$.

2 $k = 1$, $L = 2$, $q(x,t) = x(t-2)$,

$u(0,t) = 0$, $u(2,t) = 0$, $u(x,0) = 1$.

3 $k = 1$, $L = 2$, $q(x,t) = \begin{cases} 1, & 0 < x \leq 1, \\ -1, & 1 < x < 2, \end{cases}$

$u(0,t) = 0$, $u(2,t) = 0$, $u(x,0) = 1 - x$.

4 $k = 3$, $L = 1$, $q(x,t) = \begin{cases} x, & 0 < x \le 1/2, \\ 0, & 1/2 < x < 1, \end{cases}$

$u(0,t) = 0$, $u(1,t) = 0$, $u(x,0) = \begin{cases} 0, & 0 < x \le 1/2, \\ 2, & 1/2 < x < 1. \end{cases}$

5 $k = 1$, $L = 1$, $q(x,t) = 2xt$,

$u_x(0,t) = 0$, $u_x(1,t) = 0$, $u(x,0) = x$.

6 $k = 2$, $L = 1$, $q(x,t) = \begin{cases} -1, & 0 < x \le 1/2, \\ 2, & 1/2 < x < 1, \end{cases}$

$u_x(0,t) = 0$, $u_x(1,t) = 0$, $u(x,0) = x - 1$.

7 $k = 2$, $L = 2$, $q(x,t) = t$,

$u(0,t) = 0$, $u_x(2,t) = 0$, $u(x,0) = -1$.

8 $k = 1$, $L = 2$, $q(x,t) = \begin{cases} 0, & 0 < x \le 1, \\ 1, & 1 < x < 2, \end{cases}$

$u(0,t) = 0$, $u_x(2,t) = 0$, $u(x,0) = x$.

9 $k = 1$, $L = 1$, $q(x,t) = \begin{cases} -1, & 0 < x \le 1/2, \\ 0, & 1/2 < x < 1, \end{cases}$

$u_x(0,t) = 0$, $u(1,t) = 0$, $u(x,0) = \begin{cases} 0, & 0 < x \le 1/2, \\ 2, & 1/2 < x < 1. \end{cases}$

10 $k = 1$, $L = 2$, $q(x,t) = 1$,

$u_x(0,t) = 0$, $u(2,t) = 0$, $u(x,0) = x - 2$.

Answers to Odd-Numbered Exercises

1 $G(x,t;\xi,\tau) = \sum\limits_{n=1}^{\infty} 2e^{-2n^2\pi^2(t-\tau)} \sin(n\pi x) \sin(n\pi\xi)$,

$u(x,t) = (1/(4\pi^4))(e^{-2\pi^2 t} + 2\pi^2 t - 1) \sin(\pi x) - e^{-8\pi^2 t} \sin(2\pi x)$.

3 $G(x,t;\xi,\tau) = \sum\limits_{n=1}^{\infty} e^{-n^2\pi^2(t-\tau)/4} \sin(n\pi x/2) \sin(n\pi\xi/2)$,

$u(x,t) = \sum\limits_{n=1}^{\infty} (2/(n^3\pi^3))\{[1 + (-1)^n]n^2\pi^2 + 4(e^{n^2\pi^2 t/4} - 1)$

$\times [1 + (-1)^n - 2\cos(n\pi/2)]\}e^{-n^2\pi^2 t/4} \sin(n\pi x/2)$.

5 $G(x, t; \xi, \tau) = 1 + \sum_{n=1}^{\infty} 2e^{-n^2\pi^2(t-\tau)}\cos(n\pi x)\cos(n\pi\xi)$,

$u(x, t) = 1/2 + t^2/2 + \sum_{n=1}^{\infty}[(-1)^n - 1](2/(n^6\pi^6))$

$\qquad \times [2 + n^4\pi^4 + 2(n^2\pi^2t - 1)e^{n^2\pi^2t}]e^{-n^2\pi^2t}\cos(n\pi x)$.

7 $G(x, t; \xi, \tau) = \sum_{n=1}^{\infty} e^{-(2n-1)^2\pi^2(t-\tau)/8}\sin((2n - 1)\pi x/4)\sin((2n - 1)\pi\xi/4)$,

$u(x, t) = \sum_{n=1}^{\infty}(4/((2n - 1)^5\pi^5))$

$\qquad \times \{(2n - 1)^4\pi^4 - 8[8 + ((2n - 1)^2\pi^2t - 8)e^{(2n-1)^2\pi^2t/8}]\}$

$\qquad \times e^{-(2n-1)^2\pi^2t/8}\sin((2n - 1)\pi x/4)$.

9 $G(x, t; \xi, \tau) = \sum_{n=1}^{\infty} 2e^{-(2n-1)^2\pi^2(t-\tau)/4}\cos((2n - 1)\pi x/2)$

$\qquad\qquad\qquad \times \cos((2n - 1)\pi\xi/2)$,

$u(x, t) = \sum_{n=1}^{\infty} -[8/((2n - 1)^3\pi^3)]\{(-1)^n(2n - 1)^2\pi^2$

$\qquad\qquad - [2 - (2n - 1)^2\pi^2 - 2e^{(2n-1)^2\pi^2t/4}]\sin((2n - 1)\pi/4)\}$

$\qquad\qquad \times e^{-(2n-1)^2\pi^2t/4}\cos((2n - 1)\pi x/2)$.

10.2 The Laplace Equation

The equilibrium temperature in a thin, uniform rectangular plate with time-independent sources and zero temperature on the boundary is the solution of the BVP

$$(\Delta u)(x, y) = q(x, y), \quad 0 < x < L, \ 0 < y < K,$$
$$u(0, y) = 0, \quad u(L, y) = 0, \quad 0 < y < K,$$
$$u(x, 0) = 0, \quad u(x, K) = 0, \quad 0 < x < L.$$

As in Section 10.1, let $G(x, y; \xi, \eta)$ be the effect at (x, y) produced by just one unit source located at a point (ξ, η), $0 < \xi < L$, $0 < \eta < K$. Then G is the solution of the BVP

$$\Delta(x, y)G(x, y; \xi, \eta) = \delta(x - \xi, y - \eta), \quad 0 < x < L, \ 0 < y < K,$$
$$G(0, y; \xi, \eta) = 0, \quad G(L, y; \xi, \eta) = 0, \quad 0 < y < K,$$
$$G(x, 0; \xi, \eta) = 0, \quad G(x, K; \xi, \eta) = 0, \quad 0 < x < L,$$

where $\delta(x-\xi, y-\eta) = \delta(x-\xi)\delta(y-\eta)$ and $\Delta(x, y)$ indicates that the Laplacian is applied with respect to the variables x, y. Also, let D be the rectangle where the problem is formulated—that is,

$$D = \{(x, y) : 0 < x < L,\ 0 < y < K\},$$

and let ∂D be the four-sided boundary of D.

Since ∂D is piecewise smooth, we use the divergence theorem to find that for a pair of smooth functions u and v,

$$\int_D (u\Delta v - v\Delta u)\, da = \int_D (u\, \mathrm{div}\, \mathrm{grad}\, v - v\, \mathrm{div}\, \mathrm{grad}\, u)\, da$$

$$= \int_D \big[\, \mathrm{div}(u\, \mathrm{grad}\, v) - (\mathrm{grad}\, u)\cdot(\mathrm{grad}\, v)$$

$$- \mathrm{div}(v\, \mathrm{grad}\, u) + (\mathrm{grad}\, v)\cdot(\mathrm{grad}\, u)\big]\, da$$

$$= \int_{\partial D} \big[(u\, \mathrm{grad}\, v)\cdot n - (v\, \mathrm{grad}\, u)\cdot n\big]\, ds$$

$$= \int_{\partial D} (uv_n - vu_n)\, ds, \tag{10.14}$$

where da and ds are the elements of area and arc, respectively, and the subscript n denotes the derivative in the direction of the unit outward normal to the boundary. (The normal is not defined at the four corner points, but this does not influence the outcome.)

Equality (10.14) is Green's formula for functions of two space variables. If u is the solution of the given BVP and v is replaced by G, then the homogeneous BCs satisfied by both u and G make the right-hand side in (10.14) vanish, and the formula reduces to

$$\int_D \big[u(x, y)\delta(x - \xi)\delta(y - \eta) - q(x, y)G(x, y; \xi, \eta)\big]\, da(x, y) = 0,$$

where $da(x, y)$ indicates that integration is performed with respect to x, y. By (9.1), this yields

$$u(\xi, \eta) = \int_D G(x, y; \xi, \eta)q(x, y)\, da(x, y). \tag{10.15}$$

At the same time, applying (10.14) with $u(x, y)$ replaced by $G(x, y; \xi, \eta)$ and $v(x, y)$ replaced by $G(x, y; \rho, \sigma)$ and making use once more of (9.1), we obtain the symmetry

$$G(\xi, \eta; \rho, \sigma) = G(\rho, \sigma; \xi, \eta).$$

Then, as a simple interchange of variables shows, (10.15) becomes the representation formula

$$u(x, y) = \int_D G(x, y; \xi, \eta) q(\xi, \eta) \, da(\xi, \eta). \tag{10.16}$$

$G(x, y; \xi, \eta)$ is called the *Green's function* of the given BVP. Formula (10.16) shows the effect of all the sources in D on the temperature at the point (x, y).

10.7 Remark. To find a Fourier series representation for G, we recall that the two-dimensional eigenvalue problem (5.55) associated with our BVP has the eigenvalue-eigenfunction pairs

$$\lambda_{nm} = \left(\frac{n\pi}{L}\right)^2 + \left(\frac{m\pi}{K}\right)^2,$$

$$S_{nm} = \sin\frac{n\pi x}{L} \sin\frac{m\pi y}{K}, \quad n, m = 1, 2, \ldots.$$

Consequently, it seems reasonable to seek an expansion of the form

$$G(x, y; \xi, \eta) = \sum_{n=1}^{\infty} \sum_{m=1}^{\infty} c_{nm}(\xi, \eta) S_{nm}(x, y)$$

$$= \sum_{n=1}^{\infty} \sum_{m=1}^{\infty} c_{nm}(\xi, \eta) \sin\frac{n\pi x}{L} \sin\frac{m\pi y}{K}.$$

If we replace this series in the equation satisfied by G, then

$$\Delta(x, y) G(x, y; \xi, \eta) = \sum_{n=1}^{\infty} \sum_{m=1}^{\infty} c_{nm}(\xi, \eta)(\Delta S_{nm})(x, y)$$

$$= -\sum_{n=1}^{\infty} \sum_{m=1}^{\infty} \lambda_{nm} c_{nm}(\xi, \eta) S_{nm}(x, y)$$

$$= \delta(x - \xi) \delta(y - \eta).$$

Multiplying both sides above by $S_{pq}(x, y)$, integrating over D, and taking (9.1) into account, we arrive at

$$c_{pq}(\xi, \eta) = -\frac{4}{LK\lambda_{pq}} S_{pq}(\xi, \eta).$$

Hence, the double Fourier sine series for G is

$$G(x, y; \xi, \eta)$$

$$= -\frac{4}{LK} \sum_{n=1}^{\infty} \sum_{m=1}^{\infty} \frac{\sin(n\pi\xi/L)\sin(m\pi\eta/K)}{(n\pi/L)^2 + (m\pi/K)^2} \sin\frac{n\pi x}{L} \sin\frac{m\pi y}{K}. \tag{10.17}$$

10.8 Example. To compute the solution of the BVP

$$u_{xx}(x, y) + u_{yy}(x, y) = -5\pi^2 \sin(\pi x) \sin(2\pi y),$$
$$0 < x < 1,\ 0 < y < 2,$$
$$u(0, y) = 0,\quad u(1, y) = 0,\quad 0 < y < 2,$$
$$u(x, 0) = 0,\quad u(x, 2) = 0,\quad 0 < x < 1,$$

we notice that here

$$L = 1,\quad K = 2,\quad q(x, y) = -5\pi^2 \sin(\pi x) \sin(2\pi y).$$

Consequently, by (10.17),

$$G(x, y; \xi, \eta) = -2 \sum_{n=1}^{\infty} \sum_{m=1}^{\infty} \frac{\sin(n\pi\xi) \sin(m\pi\eta/2)}{n^2\pi^2 + m^2\pi^2/4} \sin(n\pi x) \sin\frac{m\pi y}{2},$$

so from (10.16) and (2.5) it follows that

$$u(x, y) = -2 \int_0^2 \int_0^1 \sum_{n=1}^{\infty} \sum_{m=1}^{\infty} \frac{4\sin(n\pi\xi) \sin(m\pi\eta/2)}{\pi^2(4n^2 + m^2)} \sin(n\pi x) \sin\frac{m\pi y}{2}$$
$$\times (-5\pi^2) \sin(\pi\xi) \sin(2\pi\eta)\, d\xi\, d\eta$$

$$= 40 \sum_{n=1}^{\infty} \sum_{m=1}^{\infty} \frac{1}{4n^2 + m^2} \left(\int_0^1 \sin(\pi\xi) \sin(n\pi\xi)\, d\xi \right)$$

$$\times \left(\int_0^2 \sin(2\pi\eta) \sin\frac{m\pi\eta}{2}\, d\eta \right) \sin(n\pi x) \sin\frac{m\pi y}{2}$$

$$= \left(\frac{40}{4 \cdot 1^2 + 4^2} \cdot \frac{1}{2} \cdot 1 \right) \sin(\pi x) \sin(2\pi y) = \sin(\pi x) \sin(2\pi y).$$

VERIFICATION WITH MATHEMATICA®. The input

```
u = Sin[Pi * x] * Sin[2 * Pi * y];
{q, f1, f2, g1, g2} = {-5 * Pi^2 * Sin[Pi * x] * Sin[2 * Pi * y],
  0, 0, 0, 0};
Simplify[{D[u,x,x] + D[u,y,y] - q, {(u /. x -> 0) - f1, (u /. x -> 1) - f2},
  {(u /. y -> 0) - g1, (u /. y -> 2) - g2}}]
```

generates the output $\{0, \{0, 0\}, \{0, 0\}\}$.

10.9 Example. Consider the BVP

$$u_{xx}(x,y) + u_{yy}(x,y) = xy, \quad 0 < x < 2, \ 0 < y < 1,$$
$$u(0,y) = 0, \quad u(2,y) = 0, \quad 0 < y < 1,$$
$$u(x,0) = 0, \quad u(x,1) = 0, \quad 0 < x < 2,$$

where $L = 2$, $K = 1$, and $q(x,y) = xy$. By (10.17), (10.16), and integration by parts, we have

$$G(x,y;\xi,\eta) = -2 \sum_{n=1}^{\infty} \sum_{m=1}^{\infty} \frac{\sin(n\pi\xi/2)\sin(m\pi\eta)}{n^2\pi^2/4 + m^2\pi^2} \sin\frac{n\pi x}{2} \sin(m\pi y)$$

and

$$u(x,y) = \int_0^1 \int_0^2 (-2) \sum_{n=1}^{\infty} \sum_{m=1}^{\infty} \frac{4\sin(n\pi\xi/2)\sin(m\pi\eta)}{\pi^2(n^2 + 4m^2)} \sin\frac{n\pi x}{2} \sin(m\pi y)$$
$$\times \xi\eta\, d\xi\, d\eta$$

$$= -\frac{8}{\pi^2} \sum_{n=1}^{\infty} \sum_{m=1}^{\infty} \frac{1}{n^2 + 4m^2} \left(\int_0^2 \xi \sin\frac{n\pi\xi}{2}\, d\xi \right) \left(\int_0^1 \eta \sin(m\pi\eta)\, d\eta \right)$$
$$\times \sin\frac{n\pi x}{2} \sin(m\pi y)$$

$$= -\frac{8}{\pi^2} \sum_{n=1}^{\infty} \sum_{m=1}^{\infty} \frac{1}{n^2 + 4m^2} \left[\frac{4}{n\pi} (-1)^{n+1} \right] \left[\frac{1}{m\pi} (-1)^{m+1} \right]$$
$$\times \sin\frac{n\pi x}{2} \sin(m\pi y)$$

$$= \sum_{n=1}^{\infty} \sum_{m=1}^{\infty} \frac{32}{\pi^4 nm(n^2 + 4m^2)} (-1)^{n+m+1} \sin\frac{n\pi x}{2} \sin(m\pi y).$$

VERIFICATION WITH MATHEMATICA®. Since the right-hand side above is the double Fourier sine series representation of u, we check that the $\{n,m\}$ term satisfies the PDE with xy replaced by the corresponding term in its own expansion. The BCs are also tested. Specifically, the input

```
q = x * y;
unm = (32 / (Pi^4 * n * m * (n^2 + 4 * m^2))) * (-1)^(n + m + 1) *
  Sin[n * Pi * x / 2] * Sin[m * Pi * y];
qnm = Simplify[2 * Integrate[q * Sin[n * Pi * x / 2] * Sin[m * Pi * y],
  {x,0,2}, {y,0,1}], {n,m}∈Integers] * Sin[n * Pi * x / 2] *
  Sin[m * Pi * y];
Simplify[{D[unm,x,x] + D[unm,y,y] - qnm, {unm /. x -> 0, unm /. x -> 2},
  {unm /. y -> 0, unm /. y -> 1}}]
```

generates the output $\{0, \{0, 0\}, \{0, 0\}\}$.

Exercises

Construct a series representation of the appropriate Green's function and solve the BVP for the nonhomogeneous Laplace equation

$$u_{xx}(x,y) + u_{yy}(x,y) = q(x,y), \quad 0 < x < L, \ 0 < y < K$$

with constants L, K, function q, and BCs as indicated.

1 $L = 2$, $K = 1$, $q(x,y) = 5\pi^2\big[-4\sin(\pi x)\sin(3\pi y) + \sin(2\pi x)\sin(\pi y)\big]$,
   $u(0,y) = 0$, $u(2,y) = 0$, $u(x,0) = 0$, $u(x,1) = 0$.

2 $L = 1$, $K = 2$, $q(x,y) = \pi^2\big[-15\sin(\pi x/2) + 13\sin(3\pi x/2)\big]\sin(\pi y)$,
   $u(0,y) = 0$, $u_x(1,y) = 0$, $u(x,0) = 0$, $u(x,2) = 0$.

3 $L = 1$, $K = 2$, $q(x,y) = -\pi^2\cos(3\pi x/2)\big[\sin(\pi y/4) - 3\sin(3\pi y/4)\big]$,
   $u_x(0,y) = 0$, $u(1,y) = 0$, $u(x,0) = 0$, $u_y(x,2) = 0$.

4 $L = 2$, $K = 1$, $q(x,y) = \pi^2\big[6\sin(\pi x)\cos(\pi y) - 25\sin(3\pi x/2)\cos(2\pi y)\big]$,
   $u(0,y) = 0$, $u(2,y) = 0$, $u_y(x,0) = 0$, $u_y(x,1) = 0$.

5 $L = 1$, $K = 2$, $q(x,y) = \sin(\pi y)$,
   $u(0,y) = 0$, $u(1,y) = 0$, $u(x,0) = 0$, $u(x,2) = 0$.

6 $L = 1$, $K = 1$, $q(x,y) = (y-1)\cos(\pi x)$,
   $u_x(0,y) = 0$, $u_x(1,y) = 0$, $u(x,0) = 0$, $u(x,1) = 0$.

7 $L = 2$, $K = 1$, $q(x,y) = \cos(\pi x/4)$,
   $u_x(0,y) = 0$, $u(2,y) = 0$, $u(x,0) = 0$, $u(x,1) = 0$.

8 $L = 1$, $K = 1$, $q(x,y) = x\sin(3\pi y/2)$,
   $u(0,y) = 0$, $u(1,y) = 0$, $u(x,0) = 0$, $u_y(x,1) = 0$.

9 $L = 1$, $K = 1$, $q(x,y) = 1$,
   $u(0,y) = 0$, $u(1,y) = 0$, $u(x,0) = 0$, $u(x,1) = 0$.

10 $L = 1$, $K = 1$, $q(x,y) = x$,
   $u(0,y) = 0$, $u(1,y) = 0$, $u_y(x,0) = 0$, $u(x,1) = 0$.

11 $L = 2$, $K = 1$, $q(x,y) = xy$,
   $u(0,y) = 0$, $u(2,y) = 0$, $u(x,0) = 0$, $u(x,1) = 0$.

12 $L = 1$, $K = 2$, $q(x,y) = y$,
   $u(0,y) = 0$, $u_x(1,y) = 0$, $u_y(x,0) = 0$, $u(x,2) = 0$.

Answers to Odd-Numbered Exercises

1 $G(x,y;\xi,\eta) = \sum\limits_{n=1}^{\infty}\sum\limits_{m=1}^{\infty} -\big\{8/\big[(n^2 + 4m^2)\pi^2\big]\big\}\sin(n\pi x/2)\sin(m\pi y)$
$$\times \sin(n\pi\xi/2)\sin(m\pi\eta),$$

$u(x,y) = 2\sin(\pi x)\sin(3\pi y) - \sin(2\pi x)\sin(\pi y)$.

3 $G(x, y; \xi, \eta) = \sum\limits_{n=1}^{\infty} \sum\limits_{m=1}^{\infty} -\{32/[(16n^2 - 16n + 4m^2 - 4m + 5)\pi^2]\}$

$$\times \cos\left((2n-1)\pi x/2\right) \sin\left((2m-1)\pi y/4\right)$$
$$\times \cos\left((2n-1)\pi \xi/2\right) \sin\left((2m-1)\pi \eta/4\right),$$

$u(x, y) = 16 \cos(3\pi x/2)\left[(1/37)\sin(\pi y/4) + (1/15)\sin(3\pi y/4)\right].$

5 $G(x, y; \xi, \eta) = \sum\limits_{n=1}^{\infty} \sum\limits_{m=1}^{\infty} -\left[8/\left((4n^2 + m^2)\pi^2\right)\right] \sin(n\pi x) \sin(m\pi y/2)$

$$\times \sin(n\pi \xi) \sin(m\pi \eta/2),$$

$u(x, y) = \sum\limits_{n=1}^{\infty} \left\{2\left[(-1)^n - 1\right]/\left[n(n^2+1)\pi^3\right]\right\} \sin(n\pi x) \sin(\pi y).$

7 $G(x, y; \xi, \eta) = \sum\limits_{n=1}^{\infty} \sum\limits_{m=1}^{\infty} -\{32/[(4n^2 - 4n + 16m^2 + 1)\pi^2]\}$

$$\times \cos\left((2n-1)\pi x/4\right) \sin(m\pi y)$$
$$\times \cos\left((2n-1)\pi \xi/4\right) \sin(m\pi \eta),$$

$u(x, y) = \sum\limits_{n=1}^{\infty} \left\{32\left[(-1)^m - 1\right]/\left[(16m^3 + m)\pi^3\right]\right\} \cos(\pi x/4) \sin(m\pi y).$

9 $G(x, y; \xi, \eta) = \sum\limits_{n=1}^{\infty} \sum\limits_{m=1}^{\infty} -\left[4/\left((n^2 + m^2)\pi^2\right)\right]$

$$\times \sin(n\pi x) \sin(m\pi y) \sin(n\pi \xi) \sin(m\pi \eta),$$

$u(x, y) = \sum\limits_{n=1}^{\infty} \sum\limits_{m=1}^{\infty} -\{4\left[(-1)^n - 1\right]\left[(-1)^m - 1\right]/\left[nm(n^2 + m^2)\pi^4\right]\}$

$$\times \sin(n\pi x) \sin(m\pi y).$$

11 $G(x, y; \xi, \eta) = \sum\limits_{m=1}^{\infty} \sum\limits_{n=1}^{\infty} -\left[8/\left((n^2 + 4m^2)\pi^2\right)\right] \sin(n\pi x/2) \sin(m\pi y)$

$$\times \sin(n\pi \xi/2) \sin(m\pi \eta),$$

$u(x, y) = \sum\limits_{n=1}^{\infty} \sum\limits_{m=1}^{\infty} -\{32(-1)^{n+m}/\left[nm(n^2 + 4m^2)\pi^4\right]\}$

$$\times \sin(n\pi x/2) \sin(m\pi y).$$

10.3   The Wave Equation

The vibrations of an infinite string are described by the IVP

$$u_{tt}(x, t) = c^2 u_{xx}(x, t) + q(x, t), \quad -\infty < x < \infty,\ t > 0,$$
$$u(x, t),\ u_x(x, t) \to 0 \quad \text{as } x \to \pm\infty,\ t > 0,$$
$$u(x, 0) = f(x), \quad u_t(x, 0) = g(x), \quad -\infty < x < \infty.$$

If we have a unit force acting at a point ξ at time $\tau > 0$, then its influence $G(x, t; \xi, \tau)$ on the vertical vibration of a point x at time t is the solution of the IVP

$$G_{tt}(x, t; \xi, \tau) = c^2 G_{xx}(x, t; \xi, \tau) + \delta(x - \xi, t - \tau),$$
$$-\infty < x < \infty, \ t > 0,$$
$$G(x, t; \xi, \tau), \ G_x(x, t; \xi, \tau) \to 0 \quad \text{as } x \to \pm\infty, \ t > 0,$$
$$G(x, t; \xi, \tau) = 0, \quad -\infty < x < \infty, \ t < \tau,$$

where

$$\delta(x - \xi, t - \tau) = \delta(x - \xi)\delta(t - \tau)$$

and the IC reflects the physical reality that the displacement of the point x is not affected by the unit force at ξ until this force has acted at time τ.

As we did in Chapter 8 in the case of the Cauchy problem for the heat equation, we find the function G by means of the full Fourier transformation. First, we note that, by (9.1),

$$\mathcal{F}[\delta(x - \xi)] = \frac{1}{\sqrt{2\pi}} \int\limits_{-\infty}^{\infty} \delta(x - \xi)e^{i\omega x} \, dx = \frac{1}{\sqrt{2\pi}} e^{i\omega \xi}.$$

Therefore, if we write $\mathcal{F}[G](\omega, t; \xi, \tau) = \tilde{G}(\omega, t; \xi, \tau)$ and apply $\mathcal{F}$ to the PDE and IC satisfied by G, we arrive at the transformed problem

$$\tilde{G}_{tt}(\omega, t; \xi, \tau) + c^2 \omega^2 \tilde{G}(\omega, t; \xi, \tau) = \frac{1}{\sqrt{2\pi}} e^{i\omega \xi} \delta(t - \tau), \quad t > 0,$$
$$\tilde{G}(\omega, t; \xi, \tau) = 0, \quad t < \tau. \tag{10.18}$$

Since $\delta(t - \tau) = 0$ for $t \neq \tau$, the solution of (10.18) is

$$\tilde{G}(\omega, t; \xi, \tau) = \begin{cases} 0, & t < \tau, \\ C_1 \cos\big(c\omega(t - \tau)\big) + C_2 \sin\big(c\omega(t - \tau)\big), & t > \tau, \end{cases} \tag{10.19}$$

where C_1 and C_2 are arbitrary functions of ω, ξ, and τ. Requiring $\tilde{G}$ to be continuous at $t = \tau$ yields $C_1 = 0$. To find C_2, we consider an interval $[\tau_1, \tau_2]$ such that $0 < \tau_1 < \tau < \tau_2$, and integrate (10.18) with respect to t over this interval:

$$\tilde{G}_t(\omega, \tau_2; \xi, \tau) - \tilde{G}_t(\omega, \tau_1; \xi, \tau) + c^2 \omega^2 \int\limits_{\tau_1}^{\tau_2} \tilde{G}(\omega, t; \xi, \tau) \, dt$$

$$= \frac{1}{\sqrt{2\pi}} e^{i\omega \xi} \int\limits_{\tau_1}^{\tau_2} \delta(t - \tau) \, dt = \frac{1}{\sqrt{2\pi}} e^{i\omega \xi} \int\limits_{-\infty}^{\infty} \delta(t - \tau) \, dt = \frac{1}{\sqrt{2\pi}} e^{i\omega \xi}.$$

By (10.19),

$$\tilde{G}_t(\omega, \tau_1; \xi, \tau) = 0,$$

$$\tilde{G}_t(\omega, \tau_2; \xi, \tau) = c\omega C_2 \cos\left(c\omega(\tau_2 - \tau)\right).$$

If we now let $\tau_1, \tau_2 \to \tau$, from the continuity of G at $t = \tau$ it follows that $C_2 = e^{i\omega\xi}/(\sqrt{2\pi}\, c\omega)$; hence,

$$\tilde{G}(\omega, t; \xi, \tau) = \begin{cases} 0, & t < \tau, \\ \dfrac{1}{\sqrt{2\pi}\, c}\, e^{i\omega\xi}\, \dfrac{\sin\left(c\omega(t - \tau)\right)}{\omega}, & t > \tau. \end{cases}$$

According to formulas 12 and 3 in Appendix A.3,

$$\mathcal{F}^{-1}\left[\sqrt{\frac{2}{\pi}}\, \frac{\sin(a\omega)}{\omega}\right] = H(a - |x|),$$

$$\mathcal{F}^{-1}\left[e^{i\omega a} \mathcal{F}[f](\omega)\right] = f(x - a);$$

so, setting $a = c(t - \tau)$ and $a = \xi$, respectively, we obtain

$$G(x, t; \xi, \tau) = \frac{1}{2c}\, H\left(c(t - \tau) - |x - \xi|\right). \tag{10.20}$$

The diagram in Fig. 10.1 shows the values of G in the upper half $(t > 0)$ of the (x, t)-plane, computed from (10.20). Using a similar diagram, it is easy to see that (10.20) can also be written in the form

$$G(x, t; \xi, \tau) = \frac{1}{2c}\left[H\left((x - \xi) + c(t - \tau)\right) - H\left((x - \xi) - c(t - \tau)\right)\right]. \tag{10.21}$$

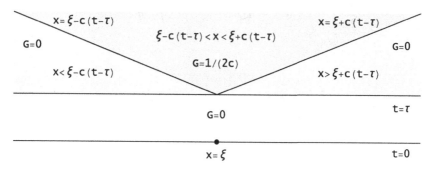

Figure 10.1: The values of $G(x, t; \xi, \tau)$ in the half-plane $t > 0$.

A procedure analogous to, but more involved than, that followed in the case of the Laplace equation can also be devised for the wave equation to obtain a symmetry relation for G and a representation formula for a solution u in terms of G. Thus, in its most general form, for $a < x < b$ the latter is

$$u(x,t) = \int_0^t \int_a^b G(x,t;\xi,\tau)q(\xi,\tau)\,d\xi\,d\tau$$

$$+ \int_a^b \left[G(x,t;\xi,0)u_\tau(\xi,0) - G_\tau(x,t;\xi,0)u(\xi,0)\right]d\xi$$

$$- c^2 \int_0^t \left[G_\xi(x,t;\xi,\tau)u(\xi,\tau) - G(x,t;\xi,\tau)u_\xi(\xi,\tau)\right]_{\xi=a}^{\xi=b} d\tau. \quad (10.22)$$

$G(x,t;\xi,\tau)$ is called the *Green's function* for the wave equation.

10.10 Remark. When $-\infty < x < \infty$, as in our problem, the corresponding formula is derived from the one above by letting $a \to -\infty$ and $b \to \infty$ and taking into account that $G(x,t;\xi,\tau) = 0$ for $|x|$ sufficiently large. If, additionally, $f(x) = g(x) = 0$, then (10.22) assumes a simpler form, established by means of the expression (10.21) for the Green's function:

$$u(x,t) = \int_0^t \int_{-\infty}^\infty G(x,t;\xi,\tau)q(\xi,\tau)\,d\xi\,d\tau$$

$$= \int_0^t \int_{-\infty}^\infty \frac{1}{2c}\left[H(x-\xi+t-\tau) - H(x-\xi-t+\tau)\right]q(\xi,\tau)\,d\xi\,d\tau$$

$$= \frac{1}{2c}\int_0^t \int_{-\infty}^{x+t-\tau} q(\xi,\tau)\,d\xi\,d\tau - \frac{1}{2c}\int_0^t \int_{-\infty}^{x-t+\tau} q(\xi,\tau)\,d\xi\,d\tau;$$

that is,

$$u(x,t) = \frac{1}{2c}\int_0^t \int_{x-t+\tau}^{x+t-\tau} q(\xi,\tau)\,d\xi\,d\tau. \quad (10.23)$$

10.11 Example. Consider the IBVP

$$u_{tt}(x,t) = u_{xx}(x,t) + \left[(6t^2+2)x - 4t^2x^3\right]e^{-x^2}, \quad -\infty < x < \infty, \ t > 0,$$

$$u(x,t), u_x(x,t) \to 0 \ \text{ as } x \to \pm\infty, \ t > 0,$$

$$u(x,0) = 0, \ u_t(x,0) = 0, \quad -\infty < x < \infty.$$

By (10.23) with $c = 1$,

$$u(x,t) = \frac{1}{2} \int_0^t \int_{x-t+\tau}^{x+t-\tau} \left[(6\tau^2 + 2)\xi - 4\tau^2\xi^3\right]e^{-\xi^2}\, d\xi\, d\tau$$

$$= \frac{1}{2} \int_0^t (2\tau^2\xi^2 - \tau^2 - 1)e^{-\xi^2}\Big|_{\xi=x+t-\tau}^{\xi=x-t+\tau}\, d\tau$$

$$= \frac{1}{2} \int_0^t \left\{\left[(2(x+t-\tau)^2 - 1)\tau^2 - 1\right]e^{-(x+t-\tau)^2}\right.$$
$$\left. - \left[(2(x-t+\tau)^2 - 1)\tau^2 - 1\right]e^{-(x-t+\tau)^2}\right\} d\tau$$

$$= \frac{1}{2}\left\{\left[(x+t-\tau)\tau^2 - \tau\right]e^{-(x+t-\tau)^2}\right.$$
$$\left. + \left[(x-t+\tau)\tau^2 + \tau\right]e^{-(x-t+\tau)^2}\right\}_{\tau=0}^{\tau=t}$$

$$= t^2 x e^{-x^2}.$$

VERIFICATION WITH MATHEMATICA®. The input

```
u = t^2 * x * E^(-x^2);
q = ((6*t^2+2)*x - 4*t^2*x^3) * E^(-x^2);
{Simplify[D[u,t,t] - D[u,x,x] - q], {u/.t->0, D[u,t] /.t->0},
Limit[u, x -> {-∞, ∞}]]}
```

generates the output $\{0, \{0,0\}, \{0,0\}\}$.

10.12 Example. Consider the IBVP

$$u_{tt}(x,t) = u_{xx}(x,t) + q(x,t), \quad -\infty < x < \infty,\ t > 0,$$
$$u(x,t),\ u_x(x,t) \to 0 \quad \text{as } x \to \pm\infty,\ t > 0,$$
$$u(x,0) = 0, \quad u_t(x,0) = 0, \quad -\infty < x < \infty,$$

where

$$q(x,t) = \begin{cases} t, & -1 < x < 1,\ t > 0, \\ 0 & \text{otherwise.} \end{cases}$$

By (10.23) with $c = 1$, the solution of this problem at, say, $(x,t) = (3,2)$, is computed as

$$u(3,2) = \frac{1}{2} \int_0^2 \int_{1+\tau}^{5-\tau} q(\xi,\tau)\, d\xi\, d\tau.$$

To calculate the above integral, we sketch the lines $\xi = 1 + \tau$, $\xi = 5 - \tau$, $\xi = -1$, and $\xi = 1$ in the (ξ,τ) system of axes and identify the domain of integration and the domain where q is nonzero (the darker and lighter shaded regions, respectively, in Fig. 10.2). Since, as the diagram shows, q is zero in the domain of integration, it follows that $u(3, 2) = 0$.

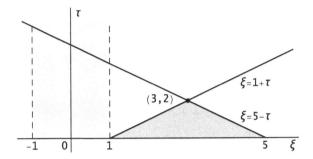

Figure 10.2: The relevant domains for $(x, t) = (3, 2)$.

Similarly, for $(x, t) = (2, 2)$,

$$u(2, 2) = \frac{1}{2} \int_0^2 \int_\tau^{4-\tau} q(\xi, \tau) \, d\xi \, \delta\tau.$$

Following the same procedure and taking into account the intersection of the domain of integration and the semi-infinite strip where q is nonzero (see Fig. 10.3), we have

$$u(2, 2) = \frac{1}{2} \int_0^1 \int_\tau^1 \tau \, d\xi \, d\tau = \frac{1}{2} \int_0^1 \tau\xi\big|_{\xi=\tau}^{\xi=1} \, d\tau = \frac{1}{2} \int_0^1 \tau(1 - \tau) \, d\tau = \frac{1}{12}.$$

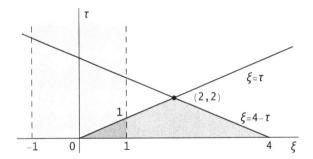

Figure 10.3: The relevant domains for $(x, t) = (2, 2)$.

Finally, to compute

$$u(1, 3) = \frac{1}{2} \int_0^3 \int_{-2+\tau}^{4-\tau} q(\xi, \tau) \, d\xi \, d\tau,$$

we make use of the sketch in Fig. 10.4 and obtain

$$u(1,3) = \frac{1}{2} \int\limits_{0}^{1} \int\limits_{-1}^{1} \tau \, d\xi \, d\tau + \frac{1}{2} \int\limits_{1}^{3} \int\limits_{-2+\tau}^{1} \tau \, d\xi \, d\tau = \frac{13}{6}.$$

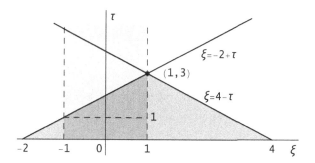

Figure 10.4: The relevant domains for $(x,t) = (1,3)$.

Alternatively, by changing the order of integration,

$$u(1,3) = \frac{1}{2} \int\limits_{-1}^{1} \int\limits_{0}^{\xi+2} \tau \, d\tau \, d\xi = \frac{1}{4} \int\limits_{-1}^{1} (\xi+2)^2 \, d\xi = \frac{13}{6}.$$

VERIFICATION WITH MATHEMATICA®. To confirm the values of the integrals computed above, we use the input

```
q = Piecewise[{{0, t<-1}, {t, -1<x<1}, {0, t>1}}];
(1/2) * Assuming[t ∈ Reals, {Integrate[Integrate[q, {x,1+t, 5-t}],
  {t,0,2}], Integrate[Integrate[q, {x,t,1}], {t,0,1}],
  (Integrate[Integrate[q, {x, -1,1}], {t,0,1}] + Integrate
  [Integrate[q, {x, -2+t,1}], {t,1,3}])}]
```

which generates the output $\left\{0, \frac{1}{12}, \frac{13}{6}\right\}$.

10.13 Remark. If there is no forcing term $(q = 0)$ in the general IVP with $-\infty < x < \infty$ and nonzero ICs, then the representation formula (10.22) reduces to

$$u(x,t) = \int\limits_{-\infty}^{\infty} \left[G(x,t;\xi,0)u_\tau(\xi,0) - G_\tau(x,t;\xi,0)u(\xi,0) \right] d\xi. \qquad (10.24)$$

This formula can be further simplified by using the explicit form of G. By (10.21) and Remark 9.4(ii), according to which

$$H'(\tau - a) = \delta(\tau - a),$$

we have

$$G_\tau(x, t; \xi, \tau) = -\frac{1}{2} \left[\delta\big((x - \xi) + c(t - \tau)\big) + \delta\big((x - \xi) - c(t - \tau)\big) \right];$$

so, by the definition of δ and H (see Section 9.1), (10.24) with $u(x, 0) = f(x)$ and $u_t(x, 0) = g(x)$ becomes

$$u(x, t) = \frac{1}{2} \int_{-\infty}^{\infty} \left[\delta(x - \xi - ct) + \delta(x - \xi + ct) \right] f(\xi) \, d\xi$$

$$+ \frac{1}{2c} \int_{-\infty}^{\infty} \left[H(x - \xi + ct) - H(x - \xi - ct) \right] g(\xi) \, d\xi$$

$$= \frac{1}{2} \left[f(x - ct) + f(x + ct) \right] + \frac{1}{2c} \int_{x-ct}^{x+ct} g(\xi) \, d\xi. \tag{10.25}$$

This formula, called *d'Alembert's solution*, is revisited in Chapter 12, where it is established by another method.

10.14 Example. Consider the IVP

$$u_{tt}(x, t) = u_{xx}(x, t), \quad -\infty < x < \infty, \ t > 0,$$

$$u(x, t), \ u_x(x, t) \to 0 \quad \text{as } x \to \pm\infty, \ t > 0,$$

$$u(x, 0) = f(x), \quad u_t(x, 0) = g(x), \quad -\infty < x < \infty,$$

where

$$f(x) = \begin{cases} 2x - 1, & |x| \le 3, \\ 0, & |x| > 3, \end{cases} \quad g(x) = \begin{cases} 4x, & |x| \le 3, \\ 0, & |x| > 3. \end{cases}$$

Then, choosing, say, (x, t) to be in turn $(2, 2)$, $(-2, 4)$, $(1, 1)$, and $(0, 5)$, by (10.25) we have

$$u(2, 2) = \frac{1}{2} \left[f(0) + f(4) \right] + \frac{1}{2} \int_0^4 g(\xi) \, d\xi$$

$$= \frac{1}{2} (-1 + 0) + \int_0^3 2\xi \, d\xi = -\frac{1}{2} + 9 = \frac{17}{2},$$

$$u(-2, 4) = \frac{1}{2} \left[f(-6) + f(2) \right] + \frac{1}{2} \int_{-6}^2 g(\xi) \, d\xi$$

$$= \frac{1}{2} (0 + 3) + \int_{-3}^2 2\xi \, d\xi = \frac{3}{2} - 5 = -\frac{7}{2},$$

$$u(1,1) = \tfrac{1}{2}\left[f(0) + f(2)\right] + \tfrac{1}{2}\int_0^2 g(\xi)\,d\xi$$

$$= \tfrac{1}{2}(-1+3) + \int_0^2 2\xi\,d\xi = 1 + 4 = 5,$$

$$u(0,5) = \tfrac{1}{2}\left[f(-5) + f(5)\right] + \tfrac{1}{2}\int_{-5}^5 g(\xi)\,d\xi$$

$$= \tfrac{1}{2}(0-0) + \int_{-3}^3 2\xi\,d\xi = 0 + 0 = 0.$$

VERIFICATION WITH MATHEMATICA®. The input

```
c = 1;
{{x1,t1},{x2,t2},{x3,t3},{x4,t4}} = {{2,2},{-2,4},{1,1},{0,5}};
f = Piecewise[{{0,ξ<-3},{2*ξ-1,-3<ξ<3},{0,ξ>3}}];
g = Piecewise[{{0,ξ<-3},{4*ξ,-3<ξ<3},{0,ξ>3}}];
{{l11,ul1},{l12,ul2},{l13,ul3},{l14,ul4}} =
  {{x1-c*t1,x1+c*t1},{x2-c*t2,x2+c*t2},
  {x3-c*t3,x3+c*t3},{x4-c*t4,x4+c*t4}};
{fTerm1,fTerm2,fTerm3,fTerm4} = {((f /.ξ->l11) + (f /.ξ->ul1))/2,
  ((f /.ξ->l12) + (f /.ξ->ul2))/2,((f /.ξ->l13) +
  (f /.ξ->ul3))/2,((f /.ξ->l14) + (f /.ξ->ul4))/2};
{gTerm1,gTerm2,gTerm3,gTerm4} = Assuming[{t ∈ Reals,x ∈ Reals},
  (1/(2*c)) * {Integrate[g,{ξ,l11,ul1}],Integrate[g,{ξ,l12,ul2}],
  Integrate[g,{ξ,l13,ul3}],Integrate[g,{ξ,l14,ul4}]}];
{u1,u2,u3,u4} = {fTerm1+gTerm1,fTerm2+gTerm2,fTerm3+gTerm3,
  fTerm4+gTerm4}
```

generates the output $\left\{\tfrac{17}{2}, -\tfrac{7}{2}, 5, 0\right\}$, which confirms the values computed above.

Exercises

In **1–10**, construct an integral representation of the solution in terms of the appropriate Green's function for the IVP

$$u_{tt}(x,t) = c^2 u_{xx}(x,t) + q(x,t), \quad -\infty < x < \infty, \ t > 0,$$
$$u(x,t),\ u_x(x,t) \to 0 \text{ as } x \to \pm\infty,$$
$$u(x,0) = 0, \quad u_t(x,0) = 0, \quad -\infty < x < \infty$$

with constant c and function q as indicated, then compute the solution at the given point x and time t.

1 $c = 1$, $\quad q(x,t) = \begin{cases} x, & -1 < x < 2, \ t > 0, \\ 0 & \text{otherwise,} \end{cases}$ $\quad (x,t) = (-1,2).$

2 $c = 1$, $\quad q(x,t) = \begin{cases} xt, & -1 < x < 3, \ t > 0, \\ 0 & \text{otherwise,} \end{cases}$ $\quad (x,t) = (2,3).$

3 $c = 2$, $\quad q(x,t) = \begin{cases} x - t, & -2 < x < 1, \ t > 0, \\ 0 & \text{otherwise,} \end{cases}$ $\quad (x,t) = (1,2).$

4 $c = 2$, $\quad q(x,t) = \begin{cases} 2x + t, & -3 < x < -1, \ t > 0, \\ 0 & \text{otherwise,} \end{cases}$ $\quad (x,t) = (-2,1).$

5 $c = 1$, $\quad q(x,t) = \begin{cases} 1 - x + 2t, & 0 < x < 4, \ t > 1, \\ 0 & \text{otherwise,} \end{cases}$ $\quad (x,t) = (3,2).$

6 $c = 1$, $\quad q(x,t) = \begin{cases} xt + 2, & -5 < x < 0, \ 0 < t < 1, \\ 0 & \text{otherwise,} \end{cases}$ $\quad (x,t) = (-4,2).$

7 $c = 2$, $\quad q(x,t) = \begin{cases} 2 - t, & -8 < x < 0, \ t > 0, \\ 0 & \text{otherwise,} \end{cases}$ $\quad (x,t) = (-3,4).$

8 $c = 2$, $\quad q(x,t) = \begin{cases} x - 1, & 3 < x < 10, \ t > 0, \\ 0 & \text{otherwise,} \end{cases}$ $\quad (x,t) = (5,3).$

9 $c = 1$, $\quad q(x,t) = \begin{cases} 2x - t - 1, & 0 < x < 5, \ 1/2 < t < 3/2, \\ 0 & \text{otherwise,} \end{cases}$ $\quad (x,t) = (4,2).$

10 $c = 1$, $\quad q(x,t) = \begin{cases} x(t - 1), & -3 < x < 2, \ t > 0, \\ 0 & \text{otherwise,} \end{cases}$ $\quad (x,t) = (-1,4).$

In **11** and **12**, use formula (10.25) to compute the solution of the IVP

$$u_{tt}(x,t) = c^2 u_{xx}(x,t), \quad -\infty < x < \infty, \ t > 0,$$
$$u(x,t), \ u_x(x,t) \to 0 \text{ as } x \to \pm\infty,$$
$$u(x,0) = f(x), \quad u_t(x,0) = g(x), \quad -\infty < x < \infty$$

with constant c and functions f and g as indicated, at the given point x and time t.

11 $c = 1,$ $f(x) = \begin{cases} 3 - 4x, & |x| \le 2, \\ 0, & |x| > 2, \end{cases}$ $g(x) = \begin{cases} 2x + 1, & |x| \le 2, \\ 0, & |x| > 2, \end{cases}$

$(x, t) = (4, 3),$ $(x, t) = (-2, 1),$ $(x, t) = (0, 1),$ $(x, t) = (7, 4).$

12 $c = 2,$ $f(x) = \begin{cases} x - x^2, & |x| \le 4, \\ 0, & |x| > 4, \end{cases}$ $g(x) = \begin{cases} 2 - x, & |x| \le 4, \\ 0, & |x| > 4, \end{cases}$

$(x, t) = (3, 1),$ $(x, t) = (-2, 3),$ $(x, t) = (1, 1),$ $(x, t) = (-6, 1/2).$

Answers to Odd-Numbered Exercises

1 $u(-1, 2) = -1/3.$ **3** $u(1, 2) = -27/32.$ **5** $u(3, 2) = 1/3.$

7 $u(-3, 4) = 8/3.$ **9** $u(4, 2) = 133/24.$

11 $u(4, 3) = 3/2,$ $u(-2, 1) = 5/2,$ $u(0, 1) = 4,$ $u(7, 4) = 0.$

CHAPTER 11

General Second-Order Linear Equations

Having studied several solution procedures for the heat, wave, and Laplace equations, we need to explain why we have chosen these particular models in preference to others. In Chapter 4, we mentioned that these were typical examples of what we called parabolic, hyperbolic, and elliptic equations, respectively. Below, we present a systematic discussion of the general second-order linear PDE in two independent variables and show how such an equation can be reduced to its simplest form. It will be seen that if the equation has constant coefficients, then its dominant part—that is, the sum of the terms containing the highest-order derivatives with respect to each of the variables—consists of the same terms as one of the above three equations. This gives us a good indication of what solution technique we should use, and what kind of behavior to expect from the solution.

11.1   The Canonical Form

We start with a preliminary examination of these PDEs in terms of their dominant coefficients. For simplicity, unless absolutely necessary, the variables on which the functions in this chapter depend will not be mentioned explicitly in the notation in the formulas where they occur.

11.1.1   Classification

The general form of a second-order linear PDE in two independent variables is

$$Au_{xx} + Bu_{xy} + Cu_{yy} + Du_x + Eu_y + Fu = G, \qquad (11.1)$$

where $u = u(x, y)$ is the unknown function and $A, \dots, G$ are given coefficients that may depend on x and y.

309

11.1 Definition. (i) If $B^2 - 4AC > 0$, (11.1) is called a *hyperbolic* equation.

(ii) If $B^2 - 4AC = 0$, (11.1) is called a *parabolic* equation.

(iii) If $B^2 - 4AC < 0$, (11.1) is called an *elliptic* equation.

11.2 Example. For the one-dimensional wave equation

$$u_{tt} - c^2 u_{xx} = 0$$

we have (considering t in place of y)

$$A = -c^2, \ C = 1, \ B = D = E = F = G = 0;$$

hence, $B^2 - 4AC = 4c^2 > 0$, which means that the equation is hyperbolic at all points in the (x, t)-plane.

11.3 Example. In the case of the one-dimensional heat equation

$$u_t - k u_{xx} = 0,$$

we have

$$A = -k, \quad E = 1, \quad B = C = D = F = G = 0,$$

so $B^2 - 4AC = 0$: the equation is parabolic in the entire (x, t)-plane.

11.4 Example. The (two-dimensional) Laplace equation

$$\Delta u = u_{xx} + u_{yy} = 0$$

is obtained for

$$A = 1, \ C = 1, \ B = D = E = F = G = 0;$$

therefore, $B^2 - 4AC = -4 < 0$, which means that this equation is elliptic throughout the (x, y)-plane.

11.5 Example. The equation

$$u_{xx} - \sqrt{y}\, u_{xy} + x u_{yy} + (2x + y)u_x - 3y u_y + 4u = \sin(x^2 - 2y), \quad y > 0,$$

fits the general template with

$$A = 1, \quad B = -\sqrt{y}, \quad C = x,$$
$$D = 2x + y, \quad E = -3y, \quad F = 4, \quad G = \sin(x^2 - 2y),$$

so $B^2 - 4AC = y - 4x$. Consequently,

(i) if $y > 4x$, the equation is hyperbolic;

(ii) if $y = 4x$, the equation is parabolic;

(iii) if $y < 4x$, the equation is elliptic.

In other words, the type of this equation at (x, y) depends on where the point lies in the half-plane $y > 0$ (see Fig. 11.1).

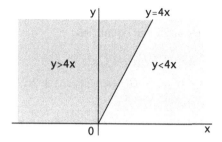

Figure 11.1: The type of the equation by region.

11.1.2 Reduction to Canonical Form

We introduce new coordinates

$$r = r(x, y), \quad s = s(x, y),$$

or, conversely,

$$x = x(r, s), \quad y = y(r, s),$$

and write

$$u\big(x(r, s), y(r, s)\big) = v(r, s).$$

By the chain rule,

$$u_x = v_r r_x + v_s s_x, \quad u_y = v_r r_y + v_s s_y,$$

and

$$
\begin{aligned}
u_{xx} &= (u_x)_x = (v_r r_x + v_s s_x)_x = (v_r)_x r_x + v_r (r_x)_x + (v_s)_x s_x + v_s (s_x)_x \\
&= \big[(v_r)_r r_x + (v_r)_s s_x\big] r_x + v_r r_{xx} + \big[(v_s)_r r_x + (v_s)_s s_x\big] s_x + v_s s_{xx} \\
&= v_{rr} r_x^2 + 2 v_{rs} r_x s_x + v_{ss} s_x^2 + v_r r_{xx} + v_s s_{xx}, \\
u_{yy} &= (u_y)_y = (v_r r_y + v_s s_y)_y = (v_r)_y r_y + v_r (r_y)_y + (v_s)_y s_y + v_s (s_y)_y \\
&= \big[(v_r)_r r_y + (v_r)_s s_y\big] r_y + v_r r_{yy} + \big[(v_s)_r r_y + (v_s)_s s_y\big] s_y + v_s s_{yy} \\
&= v_{rr} r_y^2 + 2 v_{rs} r_y s_y + v_{ss} s_y^2 + v_r r_{yy} + v_s s_{yy}, \\
u_{xy} &= (u_x)_y = (v_r r_x + v_s s_x)_y = (v_r)_y r_x + v_r (r_x)_y + (v_s)_y s_x + v_s (s_x)_y \\
&= \big[(v_r)_r r_y + (v_r)_s s_y\big] r_x + v_r r_{xy} + \big[(v_s)_r r_y + (v_s)_s s_y\big] s_x + v_s s_{xy} \\
&= v_{rr} r_x r_y + v_{rs}(r_x s_y + r_y s_x) + v_{ss} s_x s_y + v_r r_{xy} + v_s s_{xy}.
\end{aligned}
$$

Replacing all the derivatives in (11.1) and gathering the like terms, we arrive at an equation of the form

$$\bar{A} v_{rr} + \bar{B} v_{rs} + \bar{C} v_{ss} + \bar{D} v_r + \bar{E} v_s + \bar{F} v = \bar{G}, \tag{11.2}$$

where the new coefficients $\bar{A},\ldots,\bar{G}$ depend on r and s and are connected to the old ones $A,\ldots,G$ through the equalities

$$
\begin{aligned}
\bar{A} &= Ar_x^2 + Br_xr_y + Cr_y^2, \\
\bar{B} &= 2Ar_xs_x + B(r_xs_y + r_ys_x) + 2Cr_ys_y, \\
\bar{C} &= As_x^2 + Bs_xs_y + Cs_y^2, \\
\bar{D} &= Ar_{xx} + Br_{xy} + Cr_{yy} + Dr_x + Er_y, \\
\bar{E} &= As_{xx} + Bs_{xy} + Cs_{yy} + Ds_x + Es_y, \\
\bar{F} &= F, \\
\bar{G} &= G.
\end{aligned}
\tag{11.3}
$$

We now choose r and s so that $\bar{A} = \bar{C} = 0$. Since neither $r = r(x,y)$ nor $s = s(x,y)$ can be a constant, it follows that at least one of r_x, r_y and at least one of s_x, s_y must be nonzero. Suppose, for definiteness, that r_y and s_y are nonzero. Setting the expression of $\bar{A}$ in (11.3) equal to zero and dividing through by r_y^2, we find that

$$
A\left(\frac{r_x}{r_y}\right)^2 + B\left(\frac{r_x}{r_y}\right) + C = 0;
\tag{11.4}
$$

a similar procedure applied to the expression of $\bar{C}$ in (11.3) leads to

$$
A\left(\frac{s_x}{s_y}\right)^2 + B\left(\frac{s_x}{s_y}\right) + C = 0.
\tag{11.5}
$$

From (11.4) and (11.5) it follows that r_x/r_y and s_x/s_y are the roots of the quadratic equation with coefficients A, B, C, so

$$
\begin{aligned}
\frac{r_x}{r_y} &= \frac{-B + \sqrt{B^2 - 4AC}}{2A}, \\
\frac{s_x}{s_y} &= \frac{-B - \sqrt{B^2 - 4AC}}{2A}.
\end{aligned}
\tag{11.6}
$$

Obviously, the nature of the solutions of (11.6) depends on whether the given equation is hyperbolic, parabolic, or elliptic.

Exercises

Discuss the type (hyperbolic, parabolic, or elliptic) of the given PDE and sketch a graph in the (x,y)-plane to illustrate your findings.

1 $u_{xx} + (x - 1)u_{xy} + u_{yy} - 2x^2u_x + 3xyu_y + 2u = \sin x.$

2 $u_{xx} + \sqrt{y}\,u_{xy} - (x - 2)u_{yy} + 2u_y - (x - y)u = e^x \sin y, \quad y > 0.$

3 $yu_{xx} - xu_{xy} + yu_{yy} - 3x^2u_y = xe^{-xy}$.

4 $2xu_{xx} - u_{xy} + (y+1)u_{yy} - xu_x + (x-2y)u = x + 2y^2$.

5 $(x+2)u_{xx} + 2(x+y)u_{xy} + 2(y-1)u_{yy} - 3x^2u_x = x^3y^3$.

6 $4xu_{xx} + 4yu_{xy} + (4-x)u_{yy} - 2xyu_y + 2u = x(2y+1)$.

Answers to Odd-Numbered Exercises

1 $\{(x,y): -1 < x < 3\}$: elliptic; $\{(x,y): x < -1 \text{ or } x > 3\}$: hyperbolic; $\{(x,y): x = -1 \text{ or } x = 3\}$: parabolic.

3 $\{(x,y): |x| < 2|y|\}$: elliptic; $\{(x,y): |x| > 2|y|\}$: hyperbolic; $\{(x,y): |x| = 2|y|\}$: parabolic.

5 $\{(x,y): (x+1)^2 + (y-2)^2 < 1\}$: elliptic; $\{(x,y): (x+1)^2 + (y-2)^2 > 1\}$: hyperbolic; $\{(x,y): (x+1)^2 + (y-2)^2 = 1\}$: parabolic.

11.2   Hyperbolic Equations

Since here $B^2 - 4AC > 0$, there are two distinct equations (11.6). Consider the ODEs

$$\frac{dy}{dx} = \frac{B - \sqrt{B^2 - 4AC}}{2A},$$

$$\frac{dy}{dx} = \frac{B + \sqrt{B^2 - 4AC}}{2A},$$

(11.7)

called the *characteristic equations* for (11.1), and let

$$\varphi(x,y) = c_1, \quad \psi(x,y) = c_2, \quad c_1, c_2 \text{ arbitrary constants,}$$

be the families of their solution curves, called *characteristics*. Along these curves we have, respectively,

$$d\varphi = \varphi_x dx + \varphi_y dy = 0, \quad d\psi = \psi_x dx + \psi_y dy = 0,$$

which yields

$$\frac{\varphi_x}{\varphi_y} = -\frac{dy}{dx} = \frac{-B + \sqrt{B^2 - 4AC}}{2A},$$

$$\frac{\psi_x}{\psi_y} = -\frac{dy}{dx} = \frac{-B - \sqrt{B^2 - 4AC}}{2A}.$$

Consequently, the functions

$$r = \varphi(x, y), \quad s = \psi(x, y) \tag{11.8}$$

are solutions of (11.6); that is, they define the change of variables that produces $\bar{A} = \bar{C} = 0$. Using (11.3) and (11.8), we now compute the new coefficients $\bar{A}, \ldots, \bar{G}$ and write out the canonical form (11.2) as

$$\bar{B}(r, s)v_{rs} + \bar{D}(r, s)v_r + \bar{E}(r, s)v_s + \bar{F}(r, s)v = \bar{G}(r, s). \tag{11.9}$$

11.6 Remark. The hyperbolic equation has an alternative canonical form. If we introduce new variables α, β by means of the formulas

$$\alpha = \tfrac{1}{2}(r + s), \quad \beta = \tfrac{1}{2}(r - s)$$

and write $v\big(r(\alpha, \beta), s(\alpha, \beta)\big) = w(\alpha, \beta)$, then

$$v_r = w_\alpha \alpha_r + w_\beta \beta_r = \tfrac{1}{2}w_\alpha + \tfrac{1}{2}w_\beta,$$

$$v_s = w_\alpha \alpha_s + w_\beta \beta_s = \tfrac{1}{2}w_\alpha - \tfrac{1}{2}w_\beta,$$

$$v_{rs} = (v_r)_s = \tfrac{1}{2}\big[(w_\alpha + w_\beta)_\alpha \alpha_s + (w_\alpha + w_\beta)_\beta \beta_s\big]$$

$$= \tfrac{1}{2}\big[\tfrac{1}{2}(w_{\alpha\alpha} + w_{\beta\alpha}) - \tfrac{1}{2}(w_{\alpha\beta} + w_{\beta\beta})\big] = \tfrac{1}{4}w_{\alpha\alpha} - \tfrac{1}{4}w_{\beta\beta},$$

and (11.9) becomes

$$\bar{B}(w_{\alpha\alpha} - w_{\beta\beta}) + 2(\bar{D} + \bar{E})w_\alpha + 2(\bar{D} - \bar{E})w_\beta + 4\bar{F}w = 4\bar{G}.$$

The one-dimensional wave equation $u_{tt} - c^2 u_{xx} = 0$ can be brought to this form with new coefficients $\bar{D} = \bar{E} = \bar{F} = \bar{G} = 0$ by means of the substitution

$$\tau = ct.$$

11.7 Example. The coefficients of the PDE

$$u_{xx} - 7u_{xy} + 6u_{yy} = 0$$

are

$$A = 1, \quad B = -7, \quad C = 6, \quad D = E = F = G = 0.$$

Hence, $B^2 - 4AC = 25 > 0$, so the equation is hyperbolic at all points in the (x, y)-plane. The characteristic equations (11.7) are

$$y'(x) = -6, \quad y'(x) = -1,$$

with general solutions

$$y = -6x + c_1, \quad y = -x + c_2, \quad c_1, c_2 = \text{const},$$

respectively. Consequently, we choose the coordinate transformation

$$r = y + 6x, \quad s = y + x,$$

which, by (11.3), leads to $\bar{B} = -25$ and $\bar{D} = \bar{E} = \bar{F} = \bar{G} = 0$. (We already know that $\bar{A} = \bar{C} = 0$.) Replacing in (11.9), we obtain the canonical form

$$v_{rs} = 0.$$

Integrating first in terms of s, we find that

$$v_r(r, s) = \lambda(r),$$

where $\lambda(r)$ is an arbitrary function. A second integration, this time with respect to r, yields the general solution

$$v(r, s) = \varphi(r) + \psi(s),$$

or, in terms of the original variables x and y,

$$u(x, y) = v\big(r(x, y), s(x, y)\big) = \varphi(y + 6x) + \psi(y + x),$$

where φ and ψ are arbitrary one-variable functions.

VERIFICATION WITH MATHEMATICA®. The input

```
u = φ [y + 6 * x] + ψ [y + x];
Simplify[D[u,x,x] - 7 * D[u,x,y] + 6 * D[u,y,y]]
```

generates the output 0.

11.8 Example. For the PDE

$$u_{xx} + u_{xy} - 2u_{yy} - 3u_x - 6u_y = 18x - 9y$$

we have

$$A = 1, \quad B = 1, \quad C = -2, \quad D = -3, \quad E = -6, \quad F = 0, \quad G = 18x - 9y.$$

Since $B^2 - 4AC = 9 > 0$, this equation is hyperbolic at all points in the (x, y)-plane. By (11.7), the characteristic equations are

$$y'(x) = 2, \quad y'(x) = -1,$$

with solutions $y = 2x + c_1$ and $y = -x + c_2$, respectively, $c_1, c_2 = \text{const}$; hence, the coordinate transformation is

$$r = y - 2x, \quad s = y + x.$$

From (11.3) it follows that

$$\bar{B} = -9, \quad \bar{D} = 0, \quad \bar{E} = -9, \quad \bar{F} = 0, \quad \bar{G} = -9r;$$

therefore, the canonical form (11.9) of the given PDE is

$$v_{rs} + v_s = r,$$

or $(v_s)_r + v_s = r$, which, treated as a first-order equation for v_s, yields

$$v_s(r, s) = \lambda(s)e^{-r} + r - 1.$$

Then, after integration with respect to s,

$$v(r, s) = \varphi(s)e^{-r} + s(r - 1) + \psi(r),$$

so the general solution of the PDE is

$$u(x, y) = v(r(x, y), s(x, y))$$
$$= \varphi(x + y)e^{2x-y} + \psi(y - 2x) + (x + y)(y - 2x - 1),$$

where φ and ψ are arbitrary one-variable functions.

VERIFICATION WITH MATHEMATICA®. The input

```
u = φ [x+y] * E^(2*x-y) + ψ [y-2*x] + (x+y) * (y-2*x-1);
Simplify[D[u,x,x] +D[u,x,y] -2*D[u,y,y] -3*D[u,x] -6*D[u,y]
  - 18*x+9*y]
```

generates the output 0.

11.9 Example. For the PDE

$$4u_{xx} + 14u_{xy} + 6u_{yy} + 5u_x + 15u_y = -50 - 50x - 150y$$

we have

$$A = 4, \quad B = 14, \quad C = 6,$$
$$D = 5, \quad E = 15, \quad F = 0, \quad G = -50(1 + x + 3y).$$

Since $B^2 - 4AC = 100 > 0$, the equation is hyperbolic, and the characteristic equations (11.7) are

$$y'(x) = \tfrac{1}{2}, \quad y'(x) = 3,$$

with solutions $y = x/2 + c_1$ and $y = 3x + c_2$; hence, we make the transformation

$$r = 2y - x, \quad s = y - 3x,$$

or, conversely,

$$x = \tfrac{1}{5}(r - 2s), \quad y = \tfrac{1}{5}(3r - s).$$

Then, by (11.3) and (11.9), after some simplification we arrive at the canonical form

$$2v_{rs} - v_r = 4r - 2s + 2,$$

which, handled as the one in Example 11.8, yields the general solution

$$v(r, s) = \varphi(r)e^{s/2} + 2rs - 2r^2 + 2r + \psi(s).$$

In terms of the original variables x and y, this is

$$u(x,y) = \varphi(2y - x)e^{(y-3x)/2} + \psi(y - 3x) + 2(2y - x)(1 - 2x - y),$$

where φ and ψ are arbitrary one-variable functions.

VERIFICATION WITH MATHEMATICA®. The input

```
u = φ [2*y-x] *E^((y-3*x)/2)
   + ψ [y-3*x] +2* (2*y-x) * (1-2*x-y);
Simplify[4*D[u,x,x] +14*D[u,x,y] +6*D[u,y,y] +5*D[u,x]
   + 15*D[u,y] +50+50*x+150*y]
```

generates the output 0.

Exercises

Verify that the given PDE is hyperbolic everywhere in the (x, y)-plane, reduce it to its canonical form, and find its general solution.

1 $2u_{xx} - 7u_{xy} + 3u_{yy} = -150x - 50y.$

2 $u_{xx} + u_{xy} - 2u_{yy} = 72(2x^2 + xy - y^2) - 9.$

3 $3u_{xx} + 2u_{xy} - u_{yy} = -32e^{-2x+2y}.$

4 $2u_{xx} + u_{xy} - 6u_{yy} = 98e^{7x}.$

5 $6u_{xx} + u_{xy} - 2u_{yy} + 42u_x - 21u_y = 0.$

6 $u_{xx} + 2u_{xy} - 3u_{yy} + 4u_x - 4u_y = 32(3x - y).$

7 $3u_{xx} + u_{xy} - 2u_{yy} - 30u_x + 20u_y = 25(2x + 3y).$

8 $2u_{xx} - 3u_{xy} - 2u_{yy} + 10u_x - 20u_y = -25(4x + 2y + 5).$

9 $4u_{xx} + 14u_{xy} + 6u_{yy} - 10u_x - 5u_y = 25(4y - 7x - 2).$

10 $2u_{xx} - 2u_{xy} - 4u_{yy} - 9u_x + 18u_y = 9(24x - 6y - 1).$

Answers to Odd-Numbered Exercises

1 $v_{rs} = 2r,\quad u(x,y) = \varphi(3x + y) + \psi(x + 2y) + (3x + y)^2(x + 2y).$

3 $v_{rs} = 2e^{2s},\quad u(x,y) = \varphi(x + 3y) + \psi(y - x) + (x + 3y)e^{-2x+2y}.$

5 $v_{rs} + 3v_s = 0,\quad u(x,y) = \varphi(3y - 2x)e^{-3(x+2y)} + \psi(x + 2y).$

7 $v_{rs} - 2v_s = -r,\quad u(x,y) = \varphi(2x + 3y) + \psi(y - x)e^{4x+6y}$
$$+ (1/4)(y - x) + (1/2)(2x + 3y)(y - x).$$

9 $2v_{rs} - v_s = 2 - r - 2s$, $u(x, y) = \varphi(2y - x) + \psi(y - 3x)e^{y-x/2}$

$$+ (4x - 3y)(3x - y).$$

11.3 Parabolic Equations

Since here $B^2 - 4AC = 0$, from (11.6) we see that there is only one characteristic equation, which means that we can make only one of $\bar{A}$ and $\bar{C}$ zero. Let $\bar{A} = 0$. Then (11.6) reduces to

$$\frac{r_x}{r_y} = -\frac{B}{2A},$$

from which

$$\frac{dy}{dx} = -\frac{r_x}{r_y} = \frac{B}{2A}. \tag{11.10}$$

The general solution of this equation provides us with the function $r(x, y)$.

Now $B^2 - 4AC = 0$ implies that $AC \geq 0$ and $B = 2\sqrt{AC}$. Without loss of generality, we may also assume that $A, C \geq 0$. Then, by (11.3),

$$\bar{B} = 2Ar_xs_x + B(r_xs_y + r_ys_x) + 2Cr_ys_y$$

$$= 2\left[Ar_xs_x + \sqrt{A}\sqrt{C}(r_xs_y + r_ys_x) + Cr_ys_y\right]$$

$$= 2\left[\sqrt{A}\,r_x(\sqrt{A}\,s_x + \sqrt{C}\,s_y) + \sqrt{C}\,r_y(\sqrt{A}\,s_x + \sqrt{C}\,s_y)\right]$$

$$= 2(\sqrt{A}\,r_x + \sqrt{C}\,r_y)(\sqrt{A}\,s_x + \sqrt{C}\,s_y).$$

But, according to (11.10),

$$\frac{r_x}{r_y} = -\frac{B}{2A} = -\frac{2\sqrt{A}\sqrt{C}}{2A} = -\frac{\sqrt{C}}{\sqrt{A}},$$

so $\bar{B} = 0$. Consequently, s can be chosen arbitrarily, in any manner that does not 'clash' with r given by (11.10) (more precisely, so that the Jacobian of the transformation is nonzero). The canonical form in this case is

$$\bar{C}(r, s)v_{ss} + \bar{D}(r, s)v_r + \bar{E}(r, s)v_s + \bar{F}(r, s)v = \bar{G}(r, s). \tag{11.11}$$

11.10 Remark. As commented earlier, the one-dimensional heat equation $u_t - ku_{xx} = 0$ is of the form (11.11) with $\bar{C} = -k$, $\bar{D} = 1$, and $\bar{E} = \bar{F} = \bar{G} = 0$.

11.11 Example. The coefficients of the PDE

$$u_{xx} + 2u_{xy} + u_{yy} = 0$$

are

$$A = 1, \quad B = 2, \quad C = 1, \quad D = E = F = G = 0.$$

Since $B^2 - 4AC = 0$, the equation is parabolic, and its characteristic equation (11.10) is

$$y'(x) = 1,$$

with general solution $y = x + c$, $c = \text{const}$. Hence, we take $r = y - x$; for the function s we can make any suitable choice, for example, $s = y$. Then

$$\bar{C} = 1, \quad \bar{D} = \bar{E} = \bar{F} = \bar{G} = 0$$

(we already know that $\bar{A} = \bar{B} = 0$), which, by (11.11), leads to the canonical form

$$v_{ss} = 0.$$

Integrating twice with respect to s, we obtain the general solution

$$v(r, s) = s\varphi(r) + \psi(r),$$

or, in terms of x and y,

$$u(x, y) = v\big(r(x, y), s(x, y)\big) = y\varphi(y - x) + \psi(y - x),$$

where φ and ψ are arbitrary one-variable functions.

VERIFICATION WITH MATHEMATICA®. The input

```
u = y * φ [y - x] + ψ [y - x] ;
Simplify[D[u,x,x] + 2 * D[u,x,y] + D[u,y,y]]
```

generates the output 0.

11.12 Example. For the equation

$$4u_{xx} + 12u_{xy} + 9u_{yy} - 9u = 9$$

we have

$$A = 4, \quad B = 12, \quad C = 9, \quad D = E = 0, \quad F = -9, \quad G = 9,$$

so $B^2 - 4AC = 0$; that is, the PDE is parabolic. By (11.10), the characteristic equation is

$$y'(x) = \tfrac{3}{2},$$

with general solution

$$y = \tfrac{3}{2} x + c, \quad c = \text{const},$$

or

$$2y - 3x = c', \quad c' = \text{const};$$

consequently, we can take $r = 2y - 3x$ and, say, $s = y$, as above. Then

$$\bar{C} = 9, \quad \bar{F} = -9, \quad \bar{G} = 9,$$

which, replaced in (11.11), yields the canonical form

$$v_{ss} - v = 1.$$

The general solution of this equation may be written in the form

$$v(r, s) = \varphi(r) \cosh s + \psi(r) \sinh s - 1.$$

Other equivalent expressions in terms of hyperbolic functions or straight exponentials may also be used. The choice depends on whether the ranges of the variables x and y and BCs or ICs are specified in addition to the equation itself. For our choice, the general solution of the given PDE is written as

$$u(x, y) = v(r(x, y), s(x, y))$$
$$= \varphi(2y - 3x) \cosh y + \psi(2y - 3x) \sinh y - 1,$$

where φ and ψ are arbitrary one-variable functions.

VERIFICATION WITH MATHEMATICA®. The input

```
u = φ [2 * y - 3 * x] * Cosh[y] + ψ [2 * y - 3 * x] * Sinh[y] - 1;
Simplify[4 * D[u,x,x] + 12 * D[u,x,y] + 9 * D[u,y,y] - 9 * u - 9]
```

generates the output 0.

11.13 Example. The coefficients of the PDE

$$u_{xx} + 4u_{xy} + 4u_{yy} + 6u_x + 12u_y = 16 + 24x + 36y$$

show that $B^2 - 4AC = 0$, so the equation is parabolic. In this case, (11.10) becomes

$$y'(x) = 2,$$

with solution $y = 2x + c$, $c = $ const. Then we perform, say, the transformation

$$r = y - 2x, \quad s = y,$$

or

$$x = \tfrac{1}{2}(s - r), \quad y = s,$$

and, from (11.3) and (11.11), obtain the canonical form

$$v_{ss} + 3v_s = 4 - 3r + 12s.$$

The general solution of this equation is

$$v(r, s) = \varphi(r)e^{-3s} + \psi(r) + 2s^2 - rs,$$

which means that

$$u(x, y) = \varphi(y - 2x)e^{-3y} + \psi(y - 2x) + y(2x + y),$$

where φ and ψ are arbitrary one-variable functions.

VERIFICATION WITH MATHEMATICA®. The input

```
u = φ [y - 2 * x] * E^ ( - 3 * y) + ψ [y - 2 * x] + y * (2 * x + y);
Simplify [D [u,x,x] + 4 * D [u,x,y] + 4 * D [u,y,y] + 6 * D [u,x]
   + 12 * D [u,y] - 16 - 24 * x - 36 * y]
```

generates the output 0.

11.14 Example. The PDE

$$u_{xx} - 6u_{xy} + 9u_{yy} + 6u_x - 18u_y - 27u = 9(9xy + 6x - 2y - 4)$$

is parabolic since $B^2 - 4AC = 0$. By (11.10),

$$y'(x) = -3,$$

so $y = -3x + c$, $c = $ const. Implementing the transformation

$$r = y + 3x, \quad s = y,$$

or, the other way around,

$$x = \frac{1}{3}(r - s), \quad y = s,$$

we use (11.3) and (11.11) to obtain the canonical form

$$v_{ss} - 2v_s - 3v = 3rs - 3s^2 + 2r - 4s - 4.$$

The GS of this ODE (with respect to the variable s) is

$$v(r, s) = \varphi(r)e^{-s} + \psi(r)e^{3s} + s^2 - rs + 2,$$

so

$$u(x, y) = \varphi(y + 3x)e^{-y} + \psi(y + 3x)e^{3y} + 2 - 3xy,$$

where φ and ψ are arbitrary one-variable functions.
VERIFICATION WITH MATHEMATICA®. The input

```
u = φ [y + 3 * x] * E^ ( - y) + ψ [y + 3 * x] * E^ (3 * y) + 2 - 3 * x * y;
Simplify [D [u,x,x] - 6 * D [u,x,y] + 9 * D [u,y,y] + 6 * D [u,x] -
   18 * D [u,y] - 27 * u - 9 * (9 * x * y + 6 * x - 2 * y - 4)]
```

generates the output 0.

Exercises

Verify that the given PDE is parabolic everywhere in the (x, y)-plane, reduce it to its canonical form, and find its general solution.

1 $u_{xx} + 8u_{xy} + 16u_{yy} + 64u = 16.$

2 $16u_{xx} - 24u_{xy} + 9u_{yy} + 36u_x - 27u_y = 9.$

3 $25u_{xx} + 30u_{xy} + 9u_{yy} - 45u_x - 27u_y + 18u = 18(3xy - 5y^2).$

4 $9u_{xx} - 6u_{xy} + u_{yy} + 12u_x - 4u_y + 3u = 9x + 21y + 8.$

5 $4u_{xx} + 4u_{xy} + u_{yy} - 2u_x - u_y - 2u = 3y - x.$

6 $u_{xx} + 6u_{xy} + 9u_{yy} + 9u_x + 27u_y + 18u = 27(4x - 2y - 1).$

Answers to Odd-Numbered Exercises

1 $v_{ss} + 4v = 1,$ $u(x, y) = \varphi(y - 4x)\cos(2y) + \psi(y - 4x)\sin(2y) + 1/4.$

3 $v_{ss} - 3v_s + 2v = -2rs,$ $u(x, y) = \varphi(5y - 3x)e^y + \psi(5y - 3x)e^{2y}$
$$+ (1/2)(2y + 3)(3x - 5y).$$

5 $v_{ss} - v_s - 2v = r + s,$ $u(x, y) = \varphi(2y - x)e^{-y} + \psi(2y - x)e^{2y}$
$$+ (1/4)(1 + 2x - 6y).$$

11.4   Elliptic Equations

The procedure in this case is the same as for hyperbolic equations, but, since this time $B^2 - 4AC < 0$, the characteristic curves are complex (see Section 14.1). However, a real canonical form can still be obtained.

11.15 Example. The coefficients of the PDE

$$u_{xx} + 2u_{xy} + 5u_{yy} + u_x = 0$$

are

$$A = 1, \quad B = 2, \quad C = 5, \quad D = 1, \quad E = F = G = 0.$$

Thus, $B^2 - 4AC = -16 < 0$, so the equation is elliptic. By (11.7), the characteristic equations are

$$y'(x) = 1 - 2i, \quad y'(x) = 1 + 2i,$$

with general solutions

$$y = (1 - 2i)x + c_1, \quad y = (1 + 2i)x + c_2,$$

respectively. Therefore, the coordinate transformation is

$$r = y - (1 - 2i)x, \quad s = y - (1 + 2i)x.$$

Then
$$\bar{B} = 16, \quad \bar{D} = -(1 - 2i), \quad \bar{E} = -(1 + 2i), \quad \bar{F} = \bar{G} = 0$$
($\bar{A} = \bar{C} = 0$ because of the transformation), which yields the complex canonical form
$$16v_{rs} - (1 - 2i)v_r - (1 + 2i)v_s = 0,$$
where $v(r, s) = u(x(r, s), y(r, s))$.
    Performing the second transformation
$$\alpha = \tfrac{1}{2}(r + s), \quad \beta = -\tfrac{1}{2}i(r - s)$$
and writing $w(\alpha, \beta) = v(r(\alpha, \beta), s(\alpha, \beta))$, we have
$$v_r = w_\alpha \alpha_r + w_\beta \beta_r = \tfrac{1}{2}w_\alpha - \tfrac{1}{2}iw_\beta,$$
$$v_s = w_\alpha \alpha_s + w_\beta \beta_s = \tfrac{1}{2}w_\alpha + \tfrac{1}{2}iw_\beta,$$
$$v_{rs} = (v_r)_s = \left(\tfrac{1}{2}w_\alpha - \tfrac{1}{2}iw_\beta\right)_\alpha \alpha_s + \left(\tfrac{1}{2}w_\alpha - \tfrac{1}{2}iw_\beta\right)_\beta \beta_s$$
$$= \tfrac{1}{2}\left(\tfrac{1}{2}w_{\alpha\alpha} - \tfrac{1}{2}iw_{\alpha\beta}\right) + \tfrac{1}{2}i\left(\tfrac{1}{2}w_{\alpha\beta} - \tfrac{1}{2}iw_{\beta\beta}\right)$$
$$= \tfrac{1}{4}(w_{\alpha\alpha} + w_{\beta\beta}),$$

which, replaced in the PDE satisfied by v, produce the new (real) canonical form
$$4(w_{\alpha\alpha} + w_{\beta\beta}) - w_\alpha + 2w_\beta = 0.$$

11.16 Remark. The Laplace equation $u_{xx} + u_{yy} = 0$ is already in this real canonical form.

11.17 Example. The PDE
$$4u_{xx} + 8u_{xy} + 5u_{yy} - 32u_x + 8u_y + 32u = 32(x - 2y)$$
is elliptic since its coefficients satisfy $B^2 - 4AC = -16 < 0$. According to (11.7), the general solutions of the characteristic equations
$$y'(x) = 1 - \tfrac{1}{2}i, \quad y'(x) = 1 + \tfrac{1}{2}i$$
are, respectively,
$$y = \left(1 - \tfrac{1}{2}i\right)x + c_1, \quad y = \left(1 + \tfrac{1}{2}i\right)x + c_2,$$
which suggests the coordinate transformation
$$r = 2y - (2 - i)x, \quad s = 2y - (2 + i)x,$$
with inverse
$$x = -\tfrac{1}{2}i(r - s), \quad y = \tfrac{1}{4}[(1 - 2i)r + (1 + 2i)s].$$

In this case, the computed coefficients of the complex canonical form are

$$\bar{A} = \bar{C} = 0, \quad \bar{B} = 16, \quad \bar{D} = 16(5 - 2i), \quad \bar{E} = 16(5 + 2i),$$
$$\bar{F} = 32, \quad \bar{G} = -16\big[(1 - i)r + (1 + i)s\big],$$

so

$$v_{rs} + (5 - 2i)v_r + (5 + 2i)v_s + 2v = -(1 - i)r - (1 + i)s,$$

where $v(r, s) = u(x(r, s), y(r, s))$. The second transformation of variables introduced in Example 11.15 now yields the real canonical form

$$w_{\alpha\alpha} + w_{\beta\beta} + 20w_\alpha - 8w_\beta + 8w = -8(\alpha + \beta),$$

where $w(\alpha, \beta) = v\big(r(\alpha, \beta), s(\alpha, \beta)\big)$.

Exercises

Verify that the given PDE is elliptic everywhere in the (x, y)-plane and reduce it to its real canonical form.

1 $u_{xx} + 4u_{xy} + 5u_{yy} - 2u_x + u_y = 3.$

2 $u_{xx} + 2u_{xy} + 10u_{yy} + 3u_x - 2u_y + 2u = x + y.$

3 $u_{xx} + 6u_{xy} + 10u_{yy} - u_x + u_y - 3u = 2x - y.$

4 $u_{xx} + 4u_{xy} + 13u_{yy} + 2u_x - u_y + u = 3x + 2y.$

Answers to Odd-Numbered Exercises

1 $w_{\alpha\alpha} + w_{\beta\beta} + 5w_\alpha - 2w_\beta = 3.$

3 $w_{\alpha\alpha} + w_{\beta\beta} + 4w_\alpha - w_\beta - 3w = -\alpha - \beta.$

11.5 Other Problems

In this section, we consider equations with variable coefficients and equations accompanied by initial or boundary conditions.

11.18 Example. The coefficients of the PDE

$$yu_{xx} + 3yu_{xy} + 3u_x = 0, \quad y > 0$$

are

$$A = y, \quad B = 3y, \quad C = 0, \quad D = 3, \quad E = F = G = 0.$$

They yield $B^2 - 4AC = 9y^2 > 0$ for $y > 0$, so the equation is hyperbolic. Here, the characteristic equations (11.7) are

$$y'(x) = 0, \quad y'(x) = 3,$$

with general solutions

$$y = c_1, \quad y = 3x + c_2, \quad c_1, c_2 = \text{const},$$

respectively. Hence, we can take our coordinate transformation to be

$$r = y, \quad s = y - 3x,$$

which, by (11.3), leads to

$$\bar{B} = -9y = -9r, \quad \bar{E} = -9, \quad \bar{D} = \bar{F} = \bar{G} = 0$$

(we already know that $\bar{A} = \bar{C} = 0$). Replacing in (11.9), we obtain the canonical form

$$r v_{rs} + v_s = 0.$$

Writing this equation as

$$r(v_s)_r + v_s = 0$$

and using, for example, the integrating factor method or noting that the left-hand side is $(rv_s)_r$, we find that

$$v_s(r, s) = \frac{1}{r} c(s),$$

where $c(s)$ is an arbitrary function. Integration with respect to s now produces the solution

$$v(r, s) = \frac{1}{r} \varphi(s) + \psi(r),$$

or, in terms of x and y,

$$u(x, y) = v(r(x, y), s(x, y)) = \frac{1}{y} \varphi(y - 3x) + \psi(y),$$

where φ and ψ are arbitrary one-variable functions.

VERIFICATION WITH MATHEMATICA®. The input

```
u = (1/y) * φ [y - 3 * x] + ψ [y];
Simplify[y * D[u,x,x] + 3 * y * D[u,x,y] + 3 * D[u,x]]
```

generates the output 0.

11.19 Example. The PDE

$$x^2 u_{xx} + 2xy u_{xy} + y^2 u_{yy} + (x + y)u_x = x^2 - 2y,$$

considered at all points (x, y) except the origin, has coefficients

$$A = x^2, \quad B = 2xy, \quad C = y^2,$$
$$D = x + y, \quad E = F = 0, \quad G = x^2 - 2y.$$

Since $B^2 - 4AC = 4x^2y^2 - 4x^2y^2 = 0$, the equation is parabolic and its characteristic equation is

$$y'(x) = \frac{y}{x},$$

with solution $y = cx$, $c = \text{const}$. Hence, we may use the transformation

$$r = \frac{y}{x}, \quad s = y,$$

under which the new coefficients, given by (11.3), are

$$\bar{C} = s^2, \quad \bar{D} = -r - r^2, \quad \bar{E} = \bar{F} = 0, \quad \bar{G} = \frac{s^2}{r^2} - 2s.$$

In view of (11.11), we arrive at the canonical form

$$r^2 s^2 v_{ss} - (r^4 + r^3)v_r = s^2 - 2r^2 s.$$

11.20 Example. Consider the IVP (with the variable t replaced by y to facilitate the use of the general formulas already derived in this chapter)

$$2u_{xx} - 5u_{xy} + 2u_{yy} = -36x - 18y, \quad -\infty < x < \infty, \ y > 0,$$
$$u(x,0) = 4x^3 + 3x, \quad u_y(x,0) = 12x^2 + 4, \quad -\infty < x < \infty.$$

Here,

$$A = 2, \quad B = -5, \quad C = 2, \quad D = E = F = 0, \quad G = -36x - 18y,$$

so $B^2 - 4AC = 9 > 0$, which means that the equation is hyperbolic. From (11.7) it follows that the characteristic equations are

$$y'(x) = -2, \quad y'(x) = -\tfrac{1}{2},$$

with solutions $y = -2x + c_1$ and $y = -x/2 + c_2$, where $c_1, c_2 = \text{const}$. Therefore, we operate the coordinate transformation

$$r = 2x + y, \quad s = x + 2y$$

and, using (11.3), find that

$$\bar{B} = -9, \quad \bar{D} = \bar{E} = \bar{F} = 0, \quad \bar{G} = -18r.$$

This yields the canonical form (see (11.9))

$$v_{rs} = 2r,$$

with general solution

$$v(r, s) = r^2 s + \varphi(r) + \psi(s).$$

Hence, the general solution of the given PDE is

$$u(x, y) = \varphi(2x + y) + \psi(x + 2y) + (2x + y)^2(x + 2y), \qquad (11.12)$$

where, as before, φ and ψ are arbitrary one-variable functions.

To apply the ICs, we first differentiate u with respect to y to obtain

$$u_y(x, y) = \varphi'(2x + y) + 2\psi'(x + 2y) + 2(2x + y)(x + 2y) + 2(2x + y)^2,$$

and then set $y = 0$ in u and u_y. Canceling out the like terms, we arrive at the system of equations

$$\varphi(2x) + \psi(x) = 3x,$$
$$\varphi'(2x) + 2\psi'(x) = 4.$$

Differentiation of the first equation with respect to x yields

$$2\varphi'(2x) + \psi'(x) = 3,$$

which, combined with the second equation, leads to

$$\psi'(x) = \tfrac{5}{3};$$

consequently,

$$\psi(x) = \tfrac{5}{3}x + c,$$

where c is an arbitrary constant.

Next,

$$\varphi(2x) = 3x - \psi(x) = 3x - \tfrac{5}{3}x - c = \tfrac{2}{3} \cdot 2x - c,$$

so $\varphi(x) = 2x/3 - c$. Then the solution of the IVP, given by (11.12), is

$$u(x, y) = \tfrac{2}{3}(2x + y) + \tfrac{5}{3}(x + 2y) + (2x + y)^2(x + 2y)$$
$$= 3x + 4y + (2x + y)^2(x + 2y).$$

VERIFICATION WITH MATHEMATICA®. The input

```
u = 3 * x + 4 * y + (2 * x + y)^2 * (x + 2 * y);
{f, g} = {4 * x^3 + 3 * x, 12 * x^2 + 4};
Simplify[{2 * D[u,x,x] - 5 * D[u,x,y] + 2 * D[u,y,y] + 36 * x + 18 * y,
  {(u /. y -> 0) - f, (D[u,y] /. y -> 0) - g}}]
```

generates the output $\{0, \{0, 0\}\}$.

11.21 Example. Checking the type of the PDE in the IVP

$$u_{xx}(x,y) - 4u_{xy}(x,y) + 4u_{yy}(x,y) - 10u_x(x,y) + 20u_y(x,y) + 24u(x,y)$$
$$= 4(5 + 12x + 12y), \quad -\infty < x < \infty, \ y > 0,$$
$$u(x,0) = 8x, \quad u_y(x,0) = 4 - 14x, \quad -\infty < x < \infty,$$

we see that $B^2 - 4AC = 0$, so the equation is parabolic. Then the single characteristic equation is

$$y' = -2,$$

with solution $y = -2x + c$, $c = $ const. Consequently, as before, we may choose the change of variables

$$r = y + 2x, \quad s = y.$$

Formulas (11.3) now yield the coefficients

$$\bar{C} = 4, \quad \bar{D} = 0, \quad \bar{E} = 20, \quad \bar{F} = 24, \quad \bar{G} = 4(5 + 6r + 6s),$$

which produce the canonical form

$$v_{ss} + 5v_s + 6v = 5 + 6r + 6s$$

with general solution

$$v(r,s) = \varphi(r)e^{-2s} + \psi(r)e^{-3s} + r + s.$$

Consequently, the general solution of our PDE is

$$u(x,y) = \varphi(y + 2x)e^{-2y} + \psi(y + 2x)e^{-3y} + 2x + 2y,$$

where φ and ψ are arbitrary one-variable functions.

Next, we have

$$u_y(x,y) = \varphi'(y + 2x)e^{-2y} - 2\varphi(y + 2x)e^{-2y}$$
$$+ \psi'(y + 2x)e^{-3y} - 3\psi(y + 2x)e^{-y} + 2.$$

Using the ICs and performing some simple algebra, we arrive at the system

$$\varphi(2x) + \psi(2x) = 6x,$$
$$\varphi'(2x) + \psi'(2x) - 2\varphi(2x) - 3\psi(2x) = 2 - 14x,$$

or, with $2x$ replaced by x,

$$\varphi(x) + \psi(x) = 3x,$$
$$\varphi'(x) + \psi'(x) - 2\varphi(x) - 3\psi(x) = 2 - 7x.$$

Differentiating the first equation term by term, we see that

$$\varphi'(x) + \psi'(x) = 3,$$

which, combined with the second equation, leads to

$$2\varphi(x) + 3\psi(x) = 1 + 7x.$$

Recalling that $\varphi(x) + \psi(x) = 3x$, the two unknown functions are now easily obtained as $\varphi(x) = 2x - 1$ and $\psi(x) = x + 1$. Therefore, the solution to the given IVP is

$$u(x, y) = (4x + 2y - 1)e^{-2y} + (2x + y + 1)e^{-3y} + 2x + 2y.$$

VERIFICATION WITH MATHEMATICA®. The input

```
u = (4 * x + 2 * y - 1) * E^(-2 * y) + (2 * x + y + 1) * E^(-3 * y) + 2 * x + 2 * y;
{q,f,g} = {4 * (5 + 12 * x + 12 * y), 8 * x, 4 - 14 * x};
Simplify[{D[u,x,x] - 4 * D[u,x,y] + 4 * D[u,y,y] - 10 * D[u,x] +
  20 * D[u,y] + 24 * u - q, (u /. y -> 0) - f, (D[u,y] /. y -> 0) - g}]
```

generates the output $\{0, 0, 0\}$.

Exercises

In **1–6**, identify the type of the PDE and reduce the equation to its (real) canonical form.

1 $\quad u_{xx} + (x + 1)u_{xy} + xu_{yy} + (2y - x^2)u = y + x - x^2, \quad x \neq 1.$

2 $\quad u_{xx} + xu_{xy} + (2x - 4)u_{yy} + (4y - x^2)u = (y - 2x)(2y + 4x - x^2), \quad x \neq 4.$

3 $\quad u_{xx} + 2yu_{xy} + y^2u_{yy} + 2u_y = x + y, \quad y > 0.$

4 $\quad u_{xx} + 2\sqrt{x}\,u_{xy} + xu_{yy} - yu_y = 2y + 1, \quad x > 0.$

5 $\quad u_{xx} + 2xu_{xy} + (x^2 + 1)u_{yy} = x + y.$

6 $\quad u_{xx} + 2u_{xy} + (x^2 + 1)u_{yy} = 4(y - x), \quad x \neq 0.$

In **7–12**, identify the type of the PDE, reduce the equation to its canonical form, and compute the solution of the given IVP.

7 $\quad u_{xx} + 6u_{xy} + 8u_{yy} = 8(10x - 3y), \quad -\infty < x < \infty, \ y > 0,$
$\quad u(x, 0) = -10(8x^3 + x), \quad u_y(x, 0) = 84x^2 + 3, \quad -\infty < x < \infty.$

8 $\quad u_{xx} - u_{xy} - 6u_{yy} - 10u_x + 30u_y = 50, \quad -\infty < x < \infty, \ y > 0,$
$\quad u(x, 0) = 4x + e^{6x}, \quad u_y(x, 0) = 3 + 2e^{6x}, \quad -\infty < x < \infty.$

9 $\quad 4u_{xx} + 12u_{xy} + 9u_{yy} - 6u_x - 9u_y = 27(3x - 2y), \quad -\infty < x < \infty, \ y > 0,$
$\quad u(x, 0) = -3x, \quad u_y(x, 0) = 1, \quad -\infty < x < \infty.$

10 $u_{xx} - 2u_{xy} + u_{yy} + 2u_x - 2u_y + u = x + 2y - 4,\quad -\infty < x < \infty,\ y > 0,$

$\quad u(x,0) = 3x - 2,\quad u_y(x,0) = 3x + 2,\quad -\infty < x < \infty.$

11 $9u_{xx} + 6u_{xy} + u_{yy} + 3u_x + u_y = 2x - 6y + 1,\quad x > 0,\ -\infty < y < \infty,$

$\quad u(0,y) = y,\quad u_x(0,y) = 2y,\quad -\infty < y < \infty.$

12 $u_{xx} - 6u_{xy} + 9u_{yy} - 12u_x + 36u_y - 45u = 9(13 + 15x - 5y),$

$$x > 0,\quad -\infty < y < \infty,$$

$\quad u(0,y) = y - 1 + 2e^y + ye^{-5y},\quad u_x(0,y) = -3 + 3e^{-5y},\quad -\infty < y < \infty.$

Answers to Odd-Numbered Exercises

1 Hyperbolic; $2(s - 2r - 1)v_{rs} - 2v_s + sv = s - r.$

3 Parabolic; $s^2 v_{ss} + (2r/s - r)v_r + 2v_s = s - \ln(r/s).$

5 Elliptic; $512v_{rs} - 64v_r - 64v_s = 8(1 - i)r + 8(1 + i)s - (r - s)^2;$

$\quad 32w_{\alpha\alpha} + 32w_{\beta\beta} - 16w_\alpha = 4\alpha + 4\beta + \beta^2.$

7 Hyperbolic; $v_{rs} = 2r + 4s,$

$\quad u(x,y) = (3y - 10x)(8x^2 - 6xy + y^2 + 1) + 3y - 10x.$

9 Parabolic; $v_{ss} - v_s = -3r,$

$\quad u(x,y) = (9x - 6y - 1)e^y + 1 - 12x + 8y - 9xy + 6y^2.$

11 Parabolic; $v_{ss} + v_s = 1 - 2r,$

$\quad u(x,y) = 4(x - 3y)e^{-x/3} + (2/3)x^2 - 2xy - 4x + 13y.$

CHAPTER 12

The Method of Characteristics

The equations in the problems we have investigated so far are all linear, and the terms containing the unknown function and its derivatives have, in general, constant coefficients. The only exception is the type of problem where we need to make use of polar coordinates, but in such problems the polar radius is present in some of the coefficients in a very specific way, which does not disturb the solution scheme. Below, we discuss a procedure for solving first-order linear PDEs with more general variable coefficients, and first-order nonlinear PDEs of a particular form. We also re-examine the one-dimensional wave equation from the perspective of this new technique.

12.1 First-Order Linear Equations

Consider the IVP

$$u_t(x,t) + cu_x(x,t) = 0, \quad -\infty < x < \infty, \ t > 0, \qquad \text{(PDE)}$$

$$u(x,0) = f(x), \quad -\infty < x < \infty, \qquad \text{(IC)}$$

where $c = \text{const}$. If we measure the rate of change of u from a moving position given by $x = x(t)$, then, by the chain rule,

$$\frac{d}{dt} u(x(t),t) = u_t\big(x(t),t\big) + u_x\big(x(t),t\big)x'(t).$$

The first term on the right-hand side above is the change in u at a fixed point x, while the second one is the change in u resulting from the movement of the observation position.

Assuming that $x'(t) = c$, from the PDE we see that

$$\frac{d}{dt} u\big(x(t),t\big) = u_t\big(x(t),t\big) + cu_x\big(x(t),t\big) = 0;$$

that is, $u = $ const as perceived from the moving observation point. The position of this point is obtained by integrating its velocity $x'(t) = c$:

$$x = ct + x_0, \quad x_0 = x(0). \tag{12.1}$$

This formula defines a family of lines in the upper half of the (x,t)-plane, which are called *characteristics* (see Fig. 12.1). As mentioned above, the characteristics have the property that $u(x,t)$ takes a constant value along each one of them (but, in general, different constant values on different characteristics).

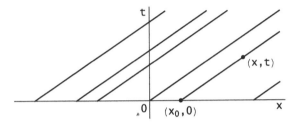

Figure 12.1: Characteristic lines.

Hence, to find the value of the solution u at (x,t), we consider the characteristic through (x,t), of equation $x = ct + x_0$, which intersects the x-axis at $(x_0, 0)$. Since u is constant on this line, its value at (x,t) is the same as at $(x_0, 0)$. But the latter is known from the IC, so

$$u(x,t) = u(x_0, 0) = f(x_0). \tag{12.2}$$

The parameter x_0 is now replaced from the equation (12.1) of the characteristic line: $x_0 = x - ct$. So, by (12.2), the solution of the given IVP is

$$u(x,t) = f(x - ct). \tag{12.3}$$

This formula shows that at a fixed time t, the shape of the solution is the same as at $t = 0$, but is shifted by ct along the x-axis. In other words, the shape of the initial data function travels in the positive (negative) x-direction with velocity c if $c > 0$ ($c < 0$), which means that the solution is a wave.

The above solution procedure can also be applied when c is replaced by a nonconstant function. What changes in this case is the shape of the characteristics and that of the wave.

12.1 Example. In the IVP

$$u_t(x,t) + \tfrac{1}{2} u_x(x,t) = 0, \quad -\infty < x < \infty, \ t > 0,$$

$$u(x,0) = \begin{cases} \sin x, & 0 \leq x \leq \pi, \\ 0 & \text{otherwise,} \end{cases}$$

the ODE of the characteristics is $x'(t) = 1/2$, so the characteristic passing through $x = x_0$ at $t = 0$ has equation $x = t/2 + x_0$. The dotted lines in Fig. 12.2 are the characteristics passing through $x = 0$ and $x = \pi$ at $t = 0$; that is, the lines of equations $x = t/2$ and $x = t/2 + \pi$, respectively.

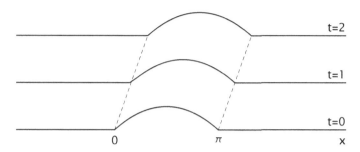

Figure 12.2: The characteristic lines through $(0,0)$ and $(\pi, 0)$.

Figure 12.3 offers a visualization of the moving wave represented by the solution of the given IVP, constructed by the superposition of the curves at $t = 0, 1, 2$ in Fig. 12.2.

Figure 12.3: Diagram of the moving wave.

Since

$$\frac{du}{dt} = u_t + u_x x' = u_t + \frac{1}{2} u_x = 0,$$

the solution u is constant along the characteristics:

$$u(x,t) = u(x_0, 0) = \begin{cases} \sin x_0, & 0 \leq x_0 \leq \pi, \\ 0 & \text{otherwise}; \end{cases}$$

therefore, since $x_0 = x - t/2$ on the characteristic through (x, t), it follows that

$$u(x,t) = \begin{cases} \sin(x - t/2), & 0 \leq x - t/2 \leq \pi, \\ 0 & \text{otherwise}. \end{cases}$$

This can also be written as

$$u(x,t) = \begin{cases} \sin(x - t/2), & t/2 \leq x \leq t/2 + \pi, \\ 0 & \text{otherwise}. \end{cases}$$

VERIFICATION WITH MATHEMATICA®. The input

```
{u1,u2}={Sin[x-t/2],0};
{f1,f2}={Sin[x],0};
Simplify[{D[{u1,u2},t]+(1/2)*D[{u1,u2},x],({u1,u2}/.t->0)
  -{f1,f2}}]
```

generates the output $\{\{0,0\},\{0,0\}\}$.

12.2 Example. The velocity of the observation point in the IVP

$$u_t(x,t) + 3tu_x(x,t) = u, \quad -\infty < x < \infty, \ t > 0,$$
$$u(x,0) = \cos x, \quad -\infty < x < \infty$$

is $x'(t) = 3t$, so the family of characteristics is given by

$$x = \frac{3}{2}t^2 + x_0,$$

where, obviously, $x_0 = x(0)$. A few of these curves are sketched in Fig. 12.4 for various values of x_0, after t has been expressed as a function of x.

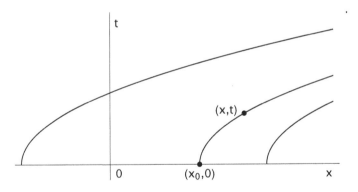

Figure 12.4: Characteristic curves.

Along the characteristic through the point (x,t) we have

$$\frac{du}{dt} = u_t + u_x x' = u_t + 3tu_x = u,$$

with solution

$$u(x,t) = Ce^t, \quad C = \text{const.}$$

Since the characteristic through (x,t) also passes through $(x_0,0)$ and $ue^{-t} = C$ is constant on this curve, we use the IC to write

$$C = u(x,t)e^{-t} = u(x_0,0)e^0 = u(x_0,0) = \cos x_0.$$

But

$$x_0 = x - \frac{3}{2}t^2$$

on the characteristic; consequently,

$$u(x,t) = Ce^t = e^t \cos x_0 = e^t \cos\left(x - \frac{3}{2}t^2\right).$$

VERIFICATION WITH MATHEMATICA®. The input

```
u = E^t * Cos [x - 3 * t^2/2];
f = Cos [x];
Simplify [{D[u,t] + 3 * t * D[u,x] - u, (u/. t -> 0) - f}]
```

generates the output $\{0,0\}$.

12.3 Example. In the IVP

$$u_t(x,t) + xu_x(x,t) = 1, \quad -\infty < x < \infty, \ t > 0,$$
$$u(x,0) = x^2, \quad -\infty < x < \infty,$$

the velocity of the observation point satisfies the ODE $x'(t) = x$, with general solution $x = ce^t$, $c = \text{const}$. Thus, the characteristic through (x,t) that also passes through $(x_0, 0)$ (see Fig. 12.5) has equation

$$x = x_0 e^t.$$

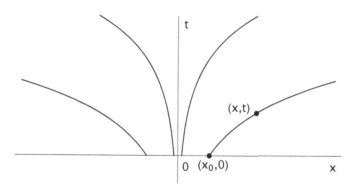

Figure 12.5: Characteristic curves.

On this characteristic,

$$\frac{du}{dt} = u_t + u_x x' = u_t + xu_x = 1;$$

that is, $u(x,t) = t + C$, $C = \text{const}$. Using the IC, we see that

$$C = u(x,t) - t = u(x_0, 0) - 0 = u(x_0, 0) = x_0^2.$$

The solution of the IVP is now obtained by replacing $x_0 = xe^{-t}$ from the equation of the characteristic:

$$u(x,t) = t + C = t + x_0^2 = t + x^2 e^{-2t}.$$

VERIFICATION WITH MATHEMATICA®. The input

```
u = t + x^2 * E^(-2*t);
f = x^2;
Simplify[{D[u,t] + x*D[u,x] - 1, (u/. t -> 0) - f}]
```

generates the output $\{0, 0\}$.

12.4 Example. Consider the IVP

$$u_t(x,t) + (3t^2 - 1)u_x(x,t) = u + 2x - 2t^3, \quad -\infty < x < \infty, \ t > 0,$$
$$u(x,0) = -2x - 3, \quad -\infty < x < \infty.$$

Here, $x' = 3t^2 - 1$ yields the characteristics $x = t^3 - t + x_0$, with $x_0 = x(0)$. Since t cannot be expressed easily as a function of x, we are not sketching these curves and go directly to the solution procedure.

On the characteristic passing through the generic point (x,t) we have

$$\frac{du}{dt} = u_t + u_x x' = u_t + (3t^2 - 1)u_x = u + 2x - 2t^3$$
$$= u + 2(t^3 - t + x_0) - 2t^3 = u - 2t + 2x_0,$$

with general solution $u(x,t) = Ce^t + 2t + 2 - 2x_0$, $C = $ const, obtained as the sum of the complementary function and a particular integral of the form $at + b$. The IC then leads to

$$C = \left[u(x,t) - 2t - 2 + 2x_0\right]e^{-t} = u(x_0,0) - 2 + 2x_0 = -5,$$

which, replaced above together with $x_0 = x - t^3 + t$ from the equation of the characteristic, shows that the solution of the given IVP is

$$u(x,t) = 2t^3 - 2x + 2 - 5e^t.$$

VERIFICATION WITH MATHEMATICA®. The input

```
u = 2*t^3 - 2*x + 2 - 5*E^t;
{q,f} = {u + 2*x - 2*t^3, -2*x - 3};
Simplify[{D[u,t] + (3*t^2 - 1)*D[u,x] - q, (u/.t -> 0) - f}]
```

generates the output $\{0, 0\}$.

12.5 Example. The IBVP

$$u_t(x,t) + u_x(x,t) = x, \quad x > 0, \ t > 0,$$
$$u(0,t) = t, \quad t > 0,$$
$$u(x,0) = \sin x, \quad x > 0$$

needs slightly different handling since here x is restricted to nonnegative values and we also have a BC at $x = 0$. First, using the standard argument, we see that the velocity of the moving observation point satisfies $x'(t) = 1$; therefore, the equation of the family of characteristics is $x = t + c$, $c = \text{const}$. Since this problem is defined in the first quadrant of the (x,t)-plane, we notice (see Fig. 12.6) that if the point (x,t) is above the line $x = t$, then the characteristic passing through this point never reaches the x-axis, so the IC cannot be used for it. However, this characteristic reaches the t-axis, and we can use the BC instead. We split the discussion into three parts.

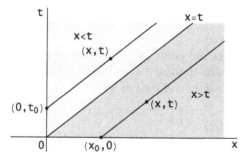

Figure 12.6: Characteristic lines in the first quadrant.

(i) Let $x > t$. Since the characteristic through (x,t) also passes through $(x_0, 0)$, its equation is written in the form $x = t + c = t + x_0$, and the PDE shows that on this line we have

$$\frac{du}{dt} = u_t + u_x x' = u_t + u_x = x = t + x_0;$$

hence,

$$u(x,t) = \tfrac{1}{2} t^2 + x_0 t + C, \quad C = \text{const}.$$

Consequently,

$$C = u(x,t) - \tfrac{1}{2} t^2 - x_0 t = u(x_0, 0) - 0 - 0 = \sin x_0,$$

which, on substituting $x_0 = x - t$ from the equation of the characteristic, yields

$$u(x,t) = \tfrac{1}{2} t^2 + x_0 t + \sin x_0$$
$$= \tfrac{1}{2} t^2 + t(x - t) + \sin(x - t) = tx - \tfrac{1}{2} t^2 + \sin(x - t).$$

(ii) Let $x < t$. Then, since the characteristic through (x, t) also passes through $(0, t_0)$, its equation is written as $x = t + c = t - t_0$, and on it,

$$\frac{du}{dt} = x = t - t_0,$$

with solution
$$u(x, t) = \frac{1}{2} t^2 - t_0 t + C, \quad C = \text{const};$$

therefore, using the BC, we deduce that

$$C = u(x, t) - \frac{1}{2} t^2 + t_0 t = u(0, t_0) - \frac{1}{2} t_0^2 + t_0^2 = t_0 + \frac{1}{2} t_0^2.$$

To find the solution at (x, t), we now need to replace the parameter $t_0 = t - x$ from the equation of the characteristic:

$$u(x, t) = \frac{1}{2} t^2 - t_0 t + t_0 + \frac{1}{2} t_0^2$$
$$= \frac{1}{2} t^2 - t(t - x) + t - x + \frac{1}{2} (t - x)^2$$
$$= \frac{1}{2} x^2 - x + t.$$

(iii) We see that as the point (x, t) approaches the line $x = t$ from either side, we obtain the same limiting value $u(x, t) = x^2/2$. Hence, the solution is continuous across this line, and we can write

$$u(x, t) = \begin{cases} \frac{1}{2} x^2 - x + t, & x \le t, \\ xt - \frac{1}{2} t^2 + \sin(x - t), & x > t. \end{cases}$$

VERIFICATION WITH MATHEMATICA®. Given the split form of the solution, we use the input

```
{u1, u2} = {x^2/2 - x + t, x * t - t^2/2 + Sin[x - t]};
{f, g} = {t, Sin[x]};
Simplify[{D[{u1, u2}, t] + D[{u1, u2}, x] - x, (u1 /. x -> 0) - f,
  (u2 /. t -> 0) - g}]
```

which generates the output $\{\{0, 0\}, 0, 0\}$.

12.6 Remark. The continuity of u across the line $x = t$ is due to the continuity of the data at $(0, 0)$; that is,

$$\lim_{x \to 0} \sin x = \lim_{t \to 0} t = 0.$$

When this condition is not satisfied in an IBVP of this type, then the solution u is discontinuous across the corresponding dividing line in the (x, t)-plane. Discontinuities—and, in general, any perturbations—in the solution always propagate along characteristic lines.

12.7 Example. The problem

$$u_x(x, y) + u_y(x, y) + 2u(x, y) = 0, \quad -\infty < x, \, y < \infty,$$
$$u(x, y) = x + 1 \quad \text{on the line } 2x + y + 1 = 0$$

is neither an IVP nor a BVP. However, the method of characteristics works in this case as well. Thus, assuming that $x = x(y)$, we can write

$$\frac{d}{dy} u(x(y), y) = u_y(x(y), y) + u_x(x(y), y)x'(y);$$

so, if $x'(y) = 1$—that is, $x = y + c$, $c = \text{const}$—then the PDE becomes

$$\frac{du}{dy} + 2u = 0,$$

with general solution

$$u(x, y) = Ce^{-2y}, \quad C = \text{const}.$$

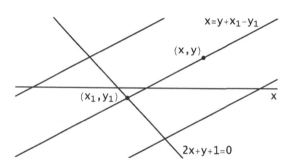

Figure 12.7: Characteristics and the data line.

The equation of the characteristic through (x, y) and (x_1, y_1) (see Fig. 12.7) is $x = y + x_1 - y_1$, and, in view of the prescribed data, on this line we have

$$u(x, y)e^{2y} = C = u(x_1, y_1)e^{2y_1} = (x_1 + 1)e^{2y_1}.$$

Since the point (x_1, y_1) lies on both the characteristic and the data lines, its coordinates satisfy the system

$$x_1 - y_1 = x - y,$$
$$2x_1 + y_1 = -1,$$

with solution

$$x_1 = \frac{1}{3}(x - y - 1), \quad y_1 = \frac{1}{3}(-2x + 2y - 1).$$

Consequently, the solution of the given problem is

$$u(x, y) = Ce^{-2y} = (x_1 + 1)e^{2(y_1 - y)}$$
$$= \tfrac{1}{3}(x - y + 2)e^{-2(2x + y + 1)/3}.$$

VERIFICATION WITH MATHEMATICA®. Given that $y = -2x - 1$ on the data line, the input

```
u = (1/3) * (x - y + 2) * E^( - 2 * (2 * x + y + 1)/3);
f = x + 1;
Simplify[{D[u,x] + D[u,y] + 2 * u, (u /. y -> ( - 2 * x - 1)) - f}]
```

generates the output $\{0, 0\}$.

Exercises

In **1–10**, use the method of characteristics to solve the IVP

$$u_t(x, t) + c(x, t)u_x(x, t) = q(x, t, u), \quad -\infty < x < \infty, \ t > 0,$$
$$u(x, 0) = f(x), \quad -\infty < x < \infty$$

with functions c, q, and f as indicated. In each case, sketch the family of characteristics in the (x, t)-plane.

1 $c(x, t) = 2$, $q(x, t, u) = t$, $f(x) = 1 - x$.

2 $c(x, t) = -2t$, $q(x, t, u) = 2$, $f(x) = x^3$.

3 $c(x, t) = 2t + 1$, $q(x, t, u) = 2u$, $f(x) = \sin x$.

4 $c(x, t) = t + 2$, $q(x, t, u) = u + 2t$, $f(x) = x + 2$.

5 $c(x, t) = x$, $q(x, t, u) = e^{-t}$, $f(x) = x^2 + 1$.

6 $c(x, t) = x - 1$, $q(x, t, u) = 3t - 2$, $f(x) = 2x - 1$.

7 $c(x, t) = -3$, $q(x, t, u) = 1 - x$, $f(x) = e^{-2x}$.

8 $c(x, t) = x + 2$, $q(x, t, u) = 2x + t^2$, $f(x) = 2x + 3$.

9 $c(x, t) = 4$, $q(x, t, u) = u + x + t$, $f(x) = \cos(2x)$.

10 $c(x, t) = -2x$, $q(x, t, u) = u - 2x$, $f(x) = 2 - x$.

In **11–16**, use the method of characteristics to solve the IBVP

$$u_t(x, t) + cu_x(x, t) = q(x, t, u), \quad x > 0, \ t > 0,$$
$$u(0, t) = g(t), \quad t > 0,$$
$$u(x, 0) = f(x), \quad x > 0$$

with constant c and functions q, f, and g as indicated. In each case, sketch the family of characteristics in the (x, t)-plane.

11 $c = 1/2$, $q(x, t, u) = u$, $g(t) = 1$, $f(x) = e^x$.

12 $c = 1$, $q(x, t, u) = u - 1$, $g(t) = t - 1$, $f(x) = x^2$.

13 $c = 2$, $q(x, t, u) = x^2$, $g(t) = t^2 + 1$, $f(x) = x$.

14 $c = 3/2$, $q(x, t, u) = x + t$, $g(t) = e^t$, $f(x) = 1 - x$.

15 $c = 1$, $q(x, t, u) = u + x$, $g(t) = 2t + 1$, $f(x) = \cos x$.

16 $c = 2$, $q(x, t, u) = 2u - t$, $g(t) = t^2$, $f(x) = 2x - 1$.

In **17–22**, use the method of characteristics to solve the problem

$$u_y(x, y) + cu_x(x, y) = q(x, y, u), \quad -\infty < x, \ y < \infty,$$
$$u(x, y) = f(x, y) \quad \text{on the line } ax + by + d = 0$$

with constant c, functions q and f, and line $ax + by + d = 0$ as indicated. In each case, sketch the family of characteristics and the data line in the (x, y)-plane.

17 $c = 2$, $q(x, y, u) = u$, $f(x, y) = 2x - y$, $x + y - 1 = 0$.

18 $c = -1$, $q(x, y, u) = 2y$, $f(x, y) = xy$, $x - 2y - 1 = 0$.

19 $c = 1/2$, $q(x, y, u) = 2u - y$, $f(x, y) = x + y$, $2x + 3y - 2 = 0$.

20 $c = 1$, $q(x, y, u) = u + x$, $f(x, y) = 2x - y - 1$, $2x + y - 3 = 0$.

21 $c = -2$, $q(x, y, u) = u - x - y$, $f(x, y) = x - 2xy$, $x - y - 2 = 0$.

22 $c = -1/2$, $q(x, y, u) = 2x + y$, $f(x, y) = e^{x-y}$, $3x - y - 3 = 0$.

Answers to Odd-Numbered Exercises

1 $x = 2t + x_0$, $u(x, t) = 1 - x + 2t + t^2/2$.

3 $x = t^2 + t + x_0$, $u(x, t) = e^{2t} \sin(x - t^2 - t)$.

5 $x = x_0 e^t$, $u(x, t) = x^2 e^{-2t} - e^{-t} + 2$.

7 $x = -3t + x_0$, $u(x, t) = -xt - 3t^2/2 + t + e^{-2(x+3t)}$.

9 $x = 4t + x_0$, $u(x, t) = [x - 4t + 5 + \cos(2x - 8t)]e^t - x - t - 5$.

11 $x = \begin{cases} t/2 + x_0, & x \geq t/2, \\ (t - t_0)/2, & x < t/2, \end{cases}$ $\quad u(x, t) = \begin{cases} e^{x+t/2}, & x \geq t/2, \\ e^{2x}, & x < t/2. \end{cases}$

13 $\quad x = \begin{cases} 2t + x_0, & x \geq 2t, \\ 2(t - t_0), & x < 2t, \end{cases}$

$u(x,t) = \begin{cases} x^2 t - 2xt^2 + 4t^3/3 + x - 2t, & x > 2t, \\ x^3/6 + x^2/4 - xt + t^2 + 1, & x < 2t. \end{cases}$

15 $\quad x = \begin{cases} t + x_0, & x \geq t, \\ t - t_0, & x < t, \end{cases}$

$u(x,t) = \begin{cases} [x - t + 1 + \cos(x - t)]e^t - x - 1, & x \geq t, \\ 2(t - x + 1)e^x - x - 1, & x < t. \end{cases}$

17 $\quad x = 2y + x_0, \quad u(x,y) = (x - 2y + 1)e^{(x+y-1)/3}$.

19 $\quad x = y/2 + x_0, \quad u(x,y) = (1/4)[(2x - y + 1)e^{(2x+3y-2)/2} + 2y + 1]$.

21 $\quad x = -2y + x_0, \quad u(x,y) = -(1/9)(2x^2 + 8xy + 8y^2 + 7x + 14y - 31)e^{(y-x+2)/3}$
$$+ x + y - 1.$$

12.2 First-Order Quasilinear Equations

A PDE of the form

$$u_t(x,t) + c(x,t,u)u_x(x,t) = q(x,t,u)$$

is called *quasilinear*. Although technically nonlinear, it is linear in the first-order derivatives of u. Such equations arise in the modeling of a variety of phenomena (for example, traffic flow) and can be solved by the method of characteristics.

12.8 Example. Consider the IVP

$$u_t(x,t) + u^3(x,t)u_x(x,t) = 0, \quad -\infty < x < \infty, \ t > 0,$$
$$u(x,0) = x^{1/3}, \quad -\infty < x < \infty.$$

If $x = x(t)$, then the usual procedure leads to the conclusion that the characteristic line through (x,t) and $(x_0,0)$ satisfies

$$x'(t) = u^3(x(t),t), \quad x(0) = x_0.$$

On this line,

$$\frac{du}{dt} = u_t + u_x x' = u_t + u^3 u_x = 0,$$

which, in view of the IC, yields

$$u(x,t) = C = u(x_0,0) = x_0^{1/3}.$$

Hence, the ODE problem for the characteristic line becomes

$$x'(t) = x_0, \quad x(0) = x_0,$$

with solution $x = x_0 t + x_0 = x_0(t+1)$. Since on this line we have $x_0 = x/(t+1)$, we can now write the solution of the given IVP as

$$u(x,t) = x_0^{1/3} = \left(\frac{x}{t+1}\right)^{1/3}.$$

VERIFICATION WITH MATHEMATICA®. The input

```
u = (x/(t+1))^(1/3);
f = x^(1/3);
Simplify[{D[u,t] + u^3*D[u,x], (u/. t->0) - f}]
```

generates the output $\{0,0\}$.

12.9 Example. A similar procedure is used to solve the IVP

$$u_t(x,t) + u(x,t)u_x(x,t) = 2t, \quad -\infty < x < \infty, \ t > 0,$$
$$u(x,0) = x, \quad -\infty < x < \infty.$$

If $x = x(t)$ satisfies

$$x'(t) = u(x(t), t), \quad x(0) = x_0,$$

then on the characteristic curve through (x,t) and $(x_0, 0)$ we have

$$\frac{du}{dt} = u_t + u_x x' = u_t + u u_x = 2t.$$

Therefore, $u(x,t) = t^2 + C$, or

$$u(x,t) - t^2 = C = u(x_0, 0) - 0 = x_0,$$

so $u(x,t) = t^2 + x_0$. This means that the characteristic curve satisfies

$$x'(t) = t^2 + x_0, \quad x(0) = x_0,$$

with solution

$$x = \tfrac{1}{3}t^3 + x_0 t + x_0 = \tfrac{1}{3}t^3 + x_0(t+1).$$

Since on this curve

$$x_0 = \frac{x - \tfrac{1}{3}t^3}{t+1} = \frac{3x - t^3}{3(t+1)},$$

the solution of the IBVP is

$$u(x,t) = t^2 + \frac{3x - t^3}{3(t+1)}.$$

A few characteristic curves are shown in Figure 12.8.

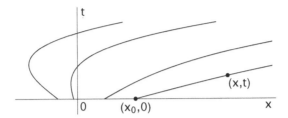

Figure 12.8: Characteristics curves.

VERIFICATION WITH MATHEMATICA®. The input

```
u = t^2 + (3*x-t^3)/(3*(t+1));
f = x;
Simplify[{D[u,t] +u*D[u,x] -2*t, (u/.t->0)-f}]
```

generates the output $\{0,0\}$.

12.10 Remark. If the IC in Example 12.9 is changed to $u(x,0) = -x$, then the method yields the solution

$$u(x,t) = t^2 + \frac{3x - t^3}{3(t-1)},$$

which is defined only for $0 < t < 1$.

Exercises

Use the method of characteristics to solve the quasilinear IVP

$$u_t(x,t) + c(x,t,u)u_x(x,t) = q(x,t,u), \quad -\infty < x < \infty, \ t > 0,$$
$$u(x,0) = f(x), \quad -\infty < x < \infty$$

with functions c, q, and f as indicated.

1 $c(x,t,u) = -2u - 1$, $q(x,t,u) = 3$, $f(x) = 1 - x$.

2 $c(x,t,u) = u - 1$, $q(x,t,u) = t + 1$, $f(x) = 2x$.

3 $c(x,t,u) = u + t$, $q(x,t,u) = 1$, $f(x) = x + 1$.

4 $c(x,t,u) = 1 + 2t - u$, $q(x,t,u) = 4t - 1$, $f(x) = 2 - 3x$.

5 $c(x,t,u) = 1$, $q(x,t,u) = 2tu^2$, $f(x) = -e^{-x}$.

6 $c(x,t,u) = (t+2)u$, $q(x,t,u) = u + 1$, $f(x) = x$.

Answers to Odd-Numbered Exercises

1 $x = (2t + 1)x_0 - 3t^2 - 3t$, $u(x, t) = (3t^2 + 2t - x + 1)/(2t + 1)$.

3 $x = t^2 + t + (t + 1)x_0$, $u(x, t) = (x + t + 1)/(t + 1)$.

5 $x = t + x_0$, $u(x, t) = -1/(e^{x-t} + t^2)$.

12.3   The One-Dimensional Wave Equation

We now reconsider the wave equation in terms of details provided by the method of characteristics.

12.3.1   The d'Alembert Solution

Using operational notation, we can write the one-dimensional wave equation formally as

$$
\begin{aligned}
u_{tt}(x, t) - c^2 u_{xx}(x, t) &= \left(\frac{\partial}{\partial t} + c \frac{\partial}{\partial x} \right)\left(\frac{\partial}{\partial t} - c \frac{\partial}{\partial x} \right) u(x, t) \\
&= \left(\frac{\partial}{\partial t} + c \frac{\partial}{\partial x} \right) v(x, t) \\
&= v_t(x, t) + c v_x(x, t) = 0.
\end{aligned}
\tag{12.4}
$$

We know from Section 12.1 that the general solution of the equation satisfied by v in (12.4) is

$$
v(x, t) = u_t(x, t) - c u_x(x, t) = P(x - ct),
\tag{12.5}
$$

where P is an arbitrary one-variable function. On the other hand, if we write the wave equation in the alternative form

$$
\begin{aligned}
u_{tt}(x, t) - c^2 u_{xx}(x, t) &= \left(\frac{\partial}{\partial t} - c \frac{\partial}{\partial x} \right)\left(\frac{\partial}{\partial t} + c \frac{\partial}{\partial x} \right) u(x, t) \\
&= \left(\frac{\partial}{\partial t} - c \frac{\partial}{\partial x} \right) w(x, t) \\
&= w_t(x, t) - c w_x(x, t) = 0,
\end{aligned}
$$

then, as above,

$$
w(x, t) = u_t(x, t) + c u_x(x, t) = Q(x + ct),
\tag{12.6}
$$

where Q is another arbitrary one-variable function. Adding (12.5) and (12.6) side by side, we find that

$$
u_t(x, t) = \tfrac{1}{2} \left[P(x - ct) + Q(x + ct) \right].
$$

Direct integration now yields

$$u(x,t) = F(x - ct) + G(x + ct), \qquad (12.7)$$

where F and G are arbitrary one-variable functions.

12.11 Remark. According to the explanation given in Section 12.1, $F(x-ct)$ is a fixed-shape wave traveling to the right with velocity c, and is constant on the characteristics $x - ct = $ const. Similarly, $G(x + ct)$ is a fixed-shape wave traveling to the left with velocity $-c$, and is constant on the characteristics $x + ct = $ const (see Fig. 12.9). Through every point (x,t) in the half-plane $t > 0$ there pass two characteristics, one from each family.

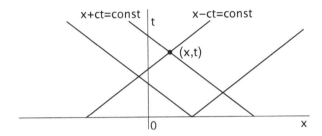

Figure 12.9: The two families of characteristics.

Consider an infinite vibrating string, which is modeled by

$$\begin{aligned} u_{tt}(x,t) &= c^2 u_{xx}(x,t), \quad -\infty < x < \infty, \ t > 0, \\ u(x,0) &= f(x), \quad u_t(x,0) = g(x), \quad -\infty < x < \infty. \end{aligned} \qquad \text{(IVP)}$$

Differentiating the solution (12.7) of the PDE with respect to t and recalling that F and G are one-variable functions, we obtain

$$u_t(x,t) = -c\, F'(x - ct) + c\, G'(x + ct).$$

From the ICs it now follows that

$$\begin{aligned} f(x) &= u(x,0) = F(x) + G(x), \\ g(x) &= u_t(x,0) = -c\, F'(x) + c\, G'(x). \end{aligned}$$

We solve the above equations for F and G. Thus,

$$\frac{1}{c} \int_0^x g(y)\, dy = -F(x) + G(x) + C, \quad C = \text{const},$$

so

$$F(x) = \frac{1}{2} \left\{ f(x) - \frac{1}{c} \int_0^x g(y)\, dy - C \right\},$$

$$\qquad (12.8)$$

$$G(x) = \frac{1}{2} \left\{ f(x) + \frac{1}{c} \int_0^x g(y)\, dy + C \right\}.$$

From these expressions and (12.7) it follows that

$$u(x,t) = \frac{1}{2}\left[f(x-ct) + f(x+ct)\right] + \frac{1}{2c}\int_{x-ct}^{x+ct} g(y)\,dy. \tag{12.9}$$

This is *d'Alembert's solution*, which was derived earlier in Section 10.3 by means of the Green's function for the wave equation.

12.12 Remarks.   (i) Technically, (12.7), which is obtained by integrating $u_t(x,t)$ with respect to t, should be written as

$$u(x,t) = F(x-ct) + G(x+ct) + \varphi(x),$$

where φ is an arbitrary function of x. Replaced in the equation $u_{tt} = c^2 u_{xx}$, this leads to $\varphi''(x) = 0$, so

$$\varphi(x) = ax + b, \quad a, b = \text{const},$$

or, what is the same,

$$\varphi(x) = a\left[\frac{1}{2}(x-ct) + \frac{1}{2}(x+ct)\right] + b.$$

Since the terms above can be incorporated in the arbitrary functions $F(x-ct)$ and $G(x+ct)$, it follows that there is no need for φ to be specifically included in (12.7).

(ii) If the wave equation contains a nonhomogeneous term of the form $(ax + b)p(t)$, where $a, b = \text{const}$ and p is an integrable function, then our procedure leads to the more general d'Alembert formula

$$u(x,t) = \frac{1}{2}\left[f(x-ct) + f(x+ct)\right] + \frac{1}{2c}\int_{x-ct}^{x+ct} g(y)\,dy$$

$$+ (ax+b)\int_0^t (t-\tau)p(\tau)\,d\tau; \tag{12.10}$$

here, the last term on the right-hand side is easily verified to be a particular integral of the full equation.

12.13 Example. Suppose that in the above IVP we have

$$u(x,0) = f(x) = \begin{cases} 1, & |x| < h, \\ 0, & |x| > h, \end{cases} \quad u_t(x,0) = g(x) = 0.$$

By (12.9), the solution is

$$u(x,t) = \frac{1}{2}\left[f(x-ct) + f(x+ct)\right],$$

where

$$\tfrac{1}{2} f(x - ct) = \begin{cases} 1/2, & |x - ct| < h, \text{ or } -h + ct < x < h + ct, \\ 0 & \text{otherwise}, \end{cases}$$

$$\tfrac{1}{2} f(x + ct) = \begin{cases} 1/2, & |x + ct| < h, \text{ or } -h - ct < x < h - ct, \\ 0 & \text{otherwise}. \end{cases}$$

Thus, the solution is the sum of two pulses of amplitude $1/2$, which move away from each other with velocity $2c$. Since their (common) endpoints are initially $2h$ apart, they separate after $t = h/c$.

12.14 Example. In the case of the IVP

$$u_{tt}(x, t) = c^2 u_{xx}(x, t), \quad -\infty < x < \infty, \ t > 0,$$
$$u(x, 0) = \sin x, \quad u_t(x, 0) = 0, \quad -\infty < x < \infty,$$

d'Alembert's solution (12.9) yields

$$u(x, t) = \tfrac{1}{2} \left[\sin(x - ct) + \sin(x + ct) \right] = \sin x \cos(ct).$$

Waves represented by solutions like this, where the variables separate, are called *standing waves*.

VERIFICATION WITH MATHEMATICA®. The input

```
u = Sin[x] * Cos[c * t];
{f, g} = {Sin[x], 0};
Simplify[{D[u,t,t] - c^2 * D[u,x,x], {(u/.t->0) - f,
   (D[u,t] /.t -> 0) - g}}]
```

generates the output $\{0, \{0, 0\}\}$.

12.15 Example. Again by (12.9), the solution of the IVP

$$u_{tt}(x, t) = c^2 u_{xx}(x, t), \quad -\infty < x < \infty, \ t > 0,$$
$$u(x, 0) = 0, \quad u_t(x, 0) = \sin x, \quad -\infty < x < \infty$$

is

$$u(x, t) = \frac{1}{2c} \int_{x-ct}^{x+ct} \sin y \, dy$$

$$= \frac{1}{2c} \left[\cos(x - ct) - \cos(x + ct) \right] = \frac{1}{c} \sin x \sin(ct).$$

This solution also represents standing waves.

VERIFICATION WITH MATHEMATICA®. The input

```
u = (1/c) * Sin[x] * Sin[c * t];
{f, g} = {0, Sin[x]};
Simplify[{D[u,t,t] - c^2 * D[u,x,x], {(u/.t -> 0) - f,
   (D[u,t] /.t -> 0) - g}}]
```

generates the output $\{0, \{0, 0\}\}$.

12.16 Example. For the IVP

$$u_{tt}(x,t) = 4u_{xx}(x,t), \quad -\infty < x < \infty, \ t > 0,$$
$$u(x,0) = x, \quad u_t(x,0) = 2x^2, \quad -\infty < x < \infty,$$

d'Alembert's formula (12.9) with $c = 2$ produces the solution

$$u(x,t) = \tfrac{1}{2}\left[(x - 2t) + (x + 2t)\right] + \frac{1}{4}\int_{x-2t}^{x+2t} 2y^2\, dy = x + 2x^2 t + \tfrac{8}{3}t^3.$$

VERIFICATION WITH MATHEMATICA®. The input

```
u = x + 2 * x^2 * t + (8/3) * t^3;
{f, g} = {x, 2 * x^2};
Simplify[{D[u,t,t] - 4 * D[u,x,x], {(u/.t -> 0) - f,
   (D[u,t] /.t -> 0) - g}}]
```

generates the output $\{0, \{0, 0\}\}$.

12.17 Example. In the case of the IVP

$$u_{tt}(x,t) = u_{xx}(x,t) + 2 - 12t, \quad -\infty < x < \infty, \ t > 0,$$
$$u(x,0) = -2x, \quad u_t(x,0) = 3x^2, \quad -\infty < x < \infty,$$

we apply the generalized d'Alembert's formula (12.10) with $c = 1$ and compute the solution as

$$u(x,t) = \tfrac{1}{2}\left[(-2(x-t) - 2(x+t)\right] + \frac{1}{2}\int_{x-t}^{x+t} 3y^2\, dy + \int_0^t (t - \tau)(2 - 12\tau)\, d\tau$$
$$= 3x^2 t - t^3 + t^2 - 2x.$$

VERIFICATION WITH MATHEMATICA®. The input

```
u = 3 * x^2 * t - t^3 + t^2 - 2 * x;
{f, g, p} = {-2 * x, 3 * x^2, 2 - 12 * t};
```

```
Simplify[{D[u,t,t] - D[u,x,x] - p, {(u/.t -> 0) - f,
    (D[u,t] /.t -> 0) - g}}]
```

generates the output $\{0, \{0,0\}\}$.

Exercises

Use the method of operator decomposition to solve the IVP

$$u_{tt}(x,t) = c^2 u_{xx}(x,t) + q(x,t), \quad -\infty < x < \infty, \ t > 0,$$
$$u(x,0) = f(x), \quad u_t(x,0) = g(x), \quad -\infty < x < \infty$$

with constant c and functions q, f, and g as indicated. (Do not use the formula for d'Alembert's solution.)

1   $c = 1$, $q(x,t) = 0$, $f(x) = 2x + 3$, $g(x) = x^2 + 1$.
2   $c = 3$, $q(x,t) = 0$, $f(x) = x^2 + x$, $g(x) = 2 - x$.
3   $c = 1/2$, $q(x,t) = 0$, $f(x) = e^{-2x}$, $g(x) = x - x^2$.
4   $c = 2$, $q(x,t) = 0$, $f(x) = x^2 - 3x + 1$, $g(x) = \sin x$.
5   $c = 1$, $q(x,t) = 6t + 4$, $f(x) = 6x^2$, $g(x) = 4x$.
6   $c = 2$, $q(x,t) = (2x - 1)(18t - 6)$, $f(x) = 3x - 2$, $g(x) = 2x - 6x^2$.

Answers to Odd-Numbered Exercises

1   $u(x,t) = x^2 t + t^3/3 + 2x + t + 3$.
3   $u(x,t) = e^{t-2x}/2 + e^{-t-2x}/2 - x^2 t - t^3/12 + xt$.
5   $u(x,t) = 6x^2 + 4xt + 8t^2 + t^3$.

12.3.2   The Semi-Infinite Vibrating String

Consider the IBVP

$$u_{tt}(x,t) = c^2 u_{xx}(x,t), \quad x > 0, \ t > 0,$$
$$u(0,t) = 0, \quad t > 0,$$
$$u(x,0) = f(x), \quad u_t(x,0) = g(x), \quad x > 0.$$

As in the preceding case, from the PDE we obtain

$$u(x,t) = F(x - ct) + G(x + ct),$$

where $F(x)$ and $G(x)$ are computed by means of (12.8). Since $x > 0$, these functions are determined only for positive values of their arguments. This does not affect $G(x+ct)$, since $t > 0$. But the argument of $F(x-ct)$ is negative if $0 < x < ct$. To obtain $F(x-ct)$ for $x - ct < 0$, we use the BC. Thus,

$$u(0,t) = 0 = F(-ct) + G(ct), \quad t > 0,$$

so for $\xi < 0$ we have $F(\xi) = -G(-\xi)$, which means that the solution for $0 < x < ct$ is

$$u(x,t) = F(x-ct) + G(x+ct) = -G(ct-x) + G(x+ct)$$

$$= \tfrac{1}{2}\left[-f(ct-x) + f(x+ct)\right] + \frac{1}{2c}\left[-\int_0^{ct-x} g(y)\,dy + \int_0^{x+ct} g(y)\,dy\right]$$

$$= \tfrac{1}{2}\left[f(ct+x) - f(ct-x)\right] + \frac{1}{2c}\int_{ct-x}^{ct+x} g(y)\,dy. \tag{12.11}$$

The term $-G(ct-x)$ is a fixed-shape wave that travels to the right and is called the *reflected wave*.

For $x > ct$, the solution of the problem is given by d'Alembert's formula (12.9), as before.

12.18 Example. The solution of the IBVP

$$u_{tt}(x,t) = 4u_{xx}(x,t), \quad x > 0, \ t > 0,$$

$$u(0,t) = 0, \quad t > 0,$$

$$u(x,0) = 4x, \quad u_t(x,0) = 2x + 6, \quad x > 0$$

is computed in two stages. First, for $0 < x < 2t$ we use (12.11) with $c = 2$ to find that

$$u(x,t) = \tfrac{1}{2}\left[4(2t+x) - 4(2t-x)\right] + \tfrac{1}{4}\int_{2t-x}^{2t+x} (2y+6)\,dy = 2xt + 7x;$$

then, by (12.9), for $x > 2t$ we have

$$u(x,t) = \tfrac{1}{2}\left[4(x+2t) + 4(x-2t)\right] + \tfrac{1}{4}\int_{x-2t}^{x+2t} (2y+6)\,dy$$

$$= 2xt + 4x + 6t.$$

The solution is continuous across the characteristic line $x = 2t$ since

$$\lim_{x\to 0} u(x,0) = \lim_{t\to 0} u(0,t) = 0.$$

Thus, we can write

$$u(x,t) = \begin{cases} 2xt + 7x, & x \le 2t, \\ 2xt + 4x + 6t, & x > 2t. \end{cases}$$

VERIFICATION WITH MATHEMATICA®. The input, which reflects the split definition of u,

```
{u1,u2} = {2*x*t+7*x, 2*x*t+4*x+6*t};
{f1,f2} = {4*x, 2*x+6};
g = 0;
Simplify[{D[{u1,u2},t,t] - 4*D[{u1,u2},x,x], (u1/.x->0) - g,
    {(u2/.t->0) - f1, (D[u2,t]/.t->0) - f2}}]
```

generates the output $\{\{0,0\},0,\{0,0\}\}$.

Exercises

Use the method of operator decomposition to solve the IBVP

$$u_{tt}(x,t) = c^2 u_{xx}(x,t), \quad x > 0, \ t > 0,$$
$$u(0,t) = 0, \quad t > 0,$$
$$u(x,0) = f(x), \quad u_t(x,0) = g(x), \quad x > 0$$

with constant c and functions f and g as indicated. (Do not use the formula for d'Alembert's solution.)

1 $c = 1$, $f(x) = 2x^2 - x$, $g(x) = 4x + 1$.

2 $c = 1/2$, $f(x) = x + 2$, $g(x) = 3x^2$.

3 $c = 3$, $f(x) = 1 - x^2$, $g(x) = \cos x$.

4 $c = 2$, $f(x) = e^{-x} + 2$, $g(x) = 3x^2 - 2x$.

Answers to Odd-Numbered Exercises

1 $u(x,t) = \begin{cases} 8xt, & x < t, \\ 2x^2 + 4xt + 2t^2 - x + t, & x \geq t. \end{cases}$

3 $u(x,t) = \begin{cases} -6xt + (1/3)\sin x \cos(3t), & x < 3t, \\ 1 - x^2 - 9t^2 + (1/3)\cos x \sin(3t), & x > 3t. \end{cases}$

12.3.3 The Finite String

Consider the IBVP

$$u_{tt}(x,t) = c^2 u_{xx}(x,t), \quad 0 < x < L, \ t > 0,$$
$$u(0,t) = 0, \quad u(L,t) = 0, \quad t > 0,$$
$$u(x,0) = f(x), \quad u_t(x,0) = g(x), \quad 0 < x < L.$$

Using separation of variables, in Section 5.2 we obtained the solution

$$u(x,t) = \sum_{n=1}^{\infty} \sin \frac{n\pi x}{L} \left(b_{1n} \cos \frac{n\pi ct}{L} + b_{2n} \sin \frac{n\pi ct}{L} \right),$$

where the coefficients b_{1n} and b_{2n} are determined from the expansions

$$f(x) = \sum_{n=1}^{\infty} b_{1n} \sin \frac{n\pi x}{L},$$

$$g(x) = \sum_{n=1}^{\infty} b_{2n} \frac{n\pi c}{L} \sin \frac{n\pi x}{L}.$$

Suppose that $f \neq 0$ and $g = 0$; then

$$b_{2n} = 0, \quad n = 1, 2, \ldots.$$

Using the formula

$$\sin\alpha \cos\beta = \tfrac{1}{2} \left[\sin(\alpha + \beta) + \sin(\alpha - \beta) \right],$$

we write the solution in the form

$$u(x,t) = \tfrac{1}{2} \sum_{n=1}^{\infty} b_{1n} \left[\sin \frac{n\pi(x + ct)}{L} + \sin \frac{n\pi(x - ct)}{L} \right]$$

$$= \tfrac{1}{2} \left[f(x + ct) + f(x - ct) \right].$$

Suppose now that $f = 0$ and $g \neq 0$; then $b_{1n} = 0$, $n = 1, 2, \ldots$. Since

$$\sin\alpha \sin\beta = \tfrac{1}{2} \left[\cos(\alpha - \beta) - \cos(\alpha + \beta) \right],$$

we rewrite the solution as

$$u(x,t) = \tfrac{1}{2} \sum_{n=1}^{\infty} b_{2n} \left[\cos \frac{n\pi(x - ct)}{L} - \cos \frac{n\pi(x + ct)}{L} \right].$$

On the other hand,

$$\int_{x-ct}^{x+ct} g(y)\, dy = \int_{x-ct}^{x+ct} u_t(y,0)\, dy = \sum_{n=1}^{\infty} cb_{2n} \int_{x-ct}^{x+ct} \frac{n\pi}{L} \sin \frac{n\pi y}{L}\, dy$$

$$= c \sum_{n=1}^{\infty} b_{2n} \left[\cos \frac{n\pi(x - ct)}{L} - \cos \frac{n\pi(x + ct)}{L} \right];$$

hence,

$$u(x,t) = \frac{1}{2c} \int_{x-ct}^{x+ct} g(y)\, dy.$$

Combining the two separate solutions and using the superposition principle, we now regain d'Alembert's formula (12.9).

12.4 Other Hyperbolic Equations

The method set out in the preceding section can be extended to more general hyperbolic equations. We illustrate how this is done in two particular cases.

12.4.1 One-Dimensional Waves

The solution of the type of problem discussed here is based on a decomposition of the second-order linear differential operator which is similar to that performed for the wave equation.

Suppose that the PDE can be written in the alternative forms

$$\left(\frac{\partial}{\partial t} + a\frac{\partial}{\partial x}\right)\left(\frac{\partial}{\partial t} + b\frac{\partial}{\partial x}\right)u(x,t) = 0,$$

$$\left(\frac{\partial}{\partial t} + b\frac{\partial}{\partial x}\right)\left(\frac{\partial}{\partial t} + a\frac{\partial}{\partial x}\right)u(x,t) = 0,$$

(12.12)

where $-\infty < x < \infty$, $t > 0$, and a and b are distinct real numbers. As in Section 12.3.1, we make the notation

$$v = u_t + bu_x, \quad w = u_t + au_x$$

and arrive at the pair of equations

$$v_t + av_x = 0, \quad w_t + bw_x = 0,$$

with solutions

$$v = P(x - at), \quad w = Q(x - bt),$$

where, as before, P and Q are arbitrary one-variable functions. Then

$$v(x,t) = u_t(x,t) + bu_x(x,t) = P(x - at),$$
$$w(x,t) = u_t(x,t) + au_x(x,t) = Q(x - bt),$$

from which

$$u_t(x,t) = \frac{1}{a - b}\left[aP(x - at) - bQ(x - bt)\right].$$

Integration followed by differentiation with respect to t now yield

$$u(x,t) = F(x - at) + G(x - bt),$$
$$u_t(x,t) = -aF'(x - at) - bG'(x - bt),$$

where F and G are also arbitrary one-variable functions. Setting $t = 0$ in both these equalities and using the ICs, we arrive at

$$F(x) + G(x) = f(x), \quad -aF'(x) - bG'(x) = g(x),$$

so, as in the case of the wave equation, we find that

$$F(x) = \frac{1}{a-b}\left\{-bf(x) - \int_0^x g(y)\,dy - C\right\},$$

$$(12.13)$$

$$G(x) = \frac{1}{a-b}\left\{af(x) + \int_0^x g(y)\,dy + C\right\},$$

where $C = $ const. This means that the solution of the IVP is given by

$$u(x,t) = \frac{1}{a-b}\left\{af(x-bt) - bf(x-at) + \int_{x-at}^{x-bt} g(y)\,dy\right\},\qquad (12.14)$$

which may be regarded as an extension of the d'Alembert formula (12.9).

12.19 Example. The IVP

$$u_{tt}(x,t) + 5u_{xt}(x,t) + 6u_{xx}(x,t) = 0, \quad -\infty < x < \infty, \ t > 0,$$
$$u(x,0) = x + 2, \quad u_t(x,0) = 2x, \quad -\infty < x < \infty$$

is hyperbolic because $B^2 - 4AC = 25 - 24 = 1 > 0$. It is easily verified that the PDE can be rewritten in the alternative forms (12.12) with $a = 3$ and $b = 2$. Consequently, by (12.14),

$$u(x,t) = 3(x - 2t + 2) - 2(x - 3t + 2) + \int_{x-3t}^{x-2t} 2y\,dy$$

$$= 2xt - 5t^2 + x + 2.$$

VERIFICATION WITH MATHEMATICA®. The input

```
u = 2 * x * t - 5 * t^2 + x + 2;
{f1, f2} = {x + 2, 2 * x};
Simplify[{D[u,t,t] + 5 * D[u,x,t] + 6 * D[u,x,x], {(u /. t -> 0) - f1,
    (D[u,t] /. t -> 0) - f2}}]
```

generates the output $\{0, \{0, 0\}\}$.

We now consider an IBVP where $x > 0$ and $t > 0$ and the PDE admits a dual decomposition of the form (12.12). For such a problem to be fully solvable, the numbers a and b have to be of opposite signs, to allow us to use the condition prescribed at the boundary point $x = 0$. A formula like (12.14) can also be designed in this case, but it is too cumbersome and we omit it, preferring to compute the solution from first principles.

12.20 Example. Consider the IBVP

$$u_{tt}(x,t) - 3u_{xt}(x,t) - 4u_{xx}(x,t) = 0, \quad x > 0, \ t > 0,$$

$$u(0,t) = 0, \quad t > 0,$$

$$u(x,0) = 1 - x^2, \quad u_t(x,0) = x + 2, \quad x > 0.$$

Since $B^2 - 4AC = 9 + 16 = 25 > 0$, the PDE is hyperbolic. Rewriting the equation alternatively as

$$\left(\frac{\partial}{\partial t} + \frac{\partial}{\partial x}\right)\left(\frac{\partial}{\partial t} - 4\frac{\partial}{\partial x}\right)u(x,t) = 0,$$

$$\left(\frac{\partial}{\partial t} - 4\frac{\partial}{\partial x}\right)\left(\frac{\partial}{\partial t} + \frac{\partial}{\partial x}\right)u(x,t) = 0$$

and applying the earlier general method, we find that

$$u(x,t) = F(x - t) + G(x + 4t), \tag{12.15}$$

where the functions

$$F(x) = \tfrac{1}{10}\left(-9x^2 - 4x + 8 - C\right),$$
$$G(x) = \tfrac{1}{10}\left(-x^2 + 4x + 2 + C\right), \tag{12.16}$$

computed by means of (12.13) with $a = 1$ and $b = -4$, are defined for positive values of their arguments.

If $x > t$, then $x - t > 0$ and we replace (12.16) in (12.15):

$$u(x,t) = \tfrac{1}{10}\Big[-9(x - t)^2 - 4(x - t) + 8 - C$$
$$- (x + 4t)^2 + 4(x + 4t) + 2 + C\Big]$$
$$= -x^2 + xt - \tfrac{5}{2}t^2 + 2t + 1.$$

If $0 \le x < t$, then $x - t < 0$, and we make use of the BC:

$$0 = u(0,t) = F(-t) + G(4t) = F(-t) + G(-4(-t)),$$

which implies that

$$F(z) = -G(-4z), \quad z < 0,$$

so, by (12.15) and (12.16),

$$u(x,t) = -G(4t - 4x) + G(x + 4t)$$
$$= \tfrac{1}{10}\Big[(4t - 4x)^2 - 4(4t - 4x) - 2 - C$$
$$- (x + 4t)^2 + 4(x + 4t) + 2 + C\Big]$$
$$= \tfrac{3}{2}x^2 - 4xt + 2x.$$

The two cases can be written together in the form

$$u(x,t) = \begin{cases} \frac{3}{2}x^2 - 4xt + 2x, & x < t, \\ -x^2 + xt - \frac{5}{2}t^2 + 2t + 1, & x > t. \end{cases}$$

We notice that the solution is discontinuous across the line $x = t$.

VERIFICATION WITH MATHEMATICA®. The split form of u prompts us to use the input

```
{u1, u2} = {3*x^2/2-4*x*t+2*x, -x^2+x*t-5*t^2/2+2*t+1};
{f1, f2} = {1-x^2, x+2};
g = 0;
Simplify[{D[{u1, u2},t,t] -3*D[{u1, u2},x,t] -4*D[{u1, u2},x,x],
  (u1 /. x->0) -g, {(u2 /. t->0) -f1, (D[u2,t] /. t->0) -f2}}]
```

which generates the output $\{\{0,0\}, 0, \{0,0\}\}$.

Exercises

In **1–4**, use the method of operator decomposition to solve the IVP

$$au_{tt}(x,t) + bu_{xt}(x,t) + cu_{xx}(x,t) = 0, \quad -\infty < x < \infty, \ t > 0,$$
$$u(x,0) = f(x), \quad u_t(x,0) = g(x), \quad -\infty < x < \infty$$

with constants a, b, c and functions f, g as indicated.

1 $a = 1$, $b = 1$, $c = -2$, $f(x) = x^2$, $g(x) = 2x + 1$.

2 $a = 1$, $b = 1$, $c = -6$, $f(x) = 1 - 2x$, $g(x) = 3x^2$.

3 $a = 1$, $b = 4$, $c = 3$, $f(x) = x^2 - x$, $g(x) = 4x - 2$.

4 $a = 2$, $b = -7$, $c = -4$, $f(x) = 3x + 1$, $g(x) = 1 - 2x$.

In **5–8**, use the method of operator decomposition to solve the IBVP

$$au_{tt}(x,t) + bu_{xt}(x,t) + cu_{xx}(x,t) = 0, \quad x > 0, \ t > 0,$$
$$u(0,t) = 0, \quad t > 0,$$
$$u(x,0) = f(x), \quad u_t(x,0) = g(x), \quad x > 0$$

with constants a, b, c and functions f, g as indicated.

5 $a = 1$, $b = 2$, $c = -3$, $f(x) = x + 1$, $g(x) = 2x$.

6 $a = 1$, $b = 2$, $c = -8$, $f(x) = 2x - 1$, $g(x) = 4x + 1$.

7 $a = 2$, $b = 1$, $c = -1$, $f(x) = x^2$, $g(x) = 2x + 3$.

8 $a = 2$, $b = -3$, $c = -2$, $f(x) = x^2 + 1$, $g(x) = x - 3$.

Answers to Odd-Numbered Exercises

1 $u(x,t) = x^2 + 2xt + t^2 + t.$

3 $u(x,t) = x^2 + 4xt - 11t^2 - x - 2t.$

5 $u(x,t) = \begin{cases} 2x^2/9 + 2xt/3 + x, & x < 3t, \\ 2xt - 2t^2 + x + 1, & x > 3t. \end{cases}$

7 $u(x,t) = \begin{cases} x^2 + 2xt + 3x, & x < t, \\ x^2 + 2xt + 3t, & x \geq t. \end{cases}$

12.4.2 Moving Boundary Problems

We start by considering an IBVP for the wave equation, nominally set in the first quadrant of the (x,t)–plane; that is, for $x \geq 0$ and $t \geq 0$. If we now allow the boundary point $x = 0$ to move, as t increases, along a curve of equation $x = \varphi(t)$, $\varphi(0) = 0$, then the IBVP becomes

$$u_{tt}(x,t) = c^2 u_{xx}(x,t), \quad \varphi(t) < x < \infty, \ t > 0,$$
$$u(x,0) = f(x), \quad u_t(x,0) = g(x), \quad x > 0,$$
$$u(\varphi(t),t) = \psi(t), \quad t > 0$$

with given (continuous) functions f, g, and ψ.

Suppose that the speed of the boundary point satisfies

$$\left| \frac{dx}{dt} \right| = |\varphi'(t)| > c, \quad t > 0.$$

If $\varphi'(t) > 0$, then $\varphi'(t) > c$, and, looking upon t as a function of x, we are led to the geometric configuration sketched in Fig. 12.10. The problem is set in the shaded region, and its solution at a generic point (x,t) is computed by means of d'Alembert's formula (12.9) in terms of the IC functions f and g. This is possible because both characteristic lines $x - ct = \text{const}$ (shown in Fig. 12.10) and $x + ct = \text{const}$ passing through (x,t) intersect the border segment $t = 0$ of the solution domain. The same can be done at any moving boundary point $(\varphi(t),t)$, but there, the result computed with (12.9) must equal $\psi(t)$, otherwise the IBVP has no solution.

If, on the other hand, $\varphi'(t) < 0$, then $\varphi'(t) < -c$, and it can be shown that under this condition the solution is not unique.

Suppose now that $|\varphi'(t)| < c$. In this case, the IBVP has a unique solution, which we will prove by actually constructing it. For definiteness, let $\varphi'(t) > 0$, $t > 0$. The solution domain, with the same graphical convention as before, is illustrated in Fig. 12.11.

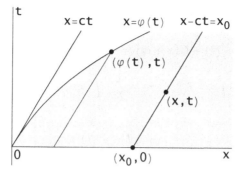

Figure 12.10: $\varphi'(t) > c$.

According to the earlier argument, the solution at every point (x, t) in the darker-shaded region $(x > ct)$ is once again obtained from d'Alembert's formula (12.9). However, this is not possible when (x, t) is in the lighter-shaded zone $(\varphi(t) < x < ct)$ since the characteristic $x - ct = $ const through (x, t) does not now reach the $t = 0$ boundary segment of the solution domain and, therefore, does not give us access to the ICs. To resolve this difficulty, we consider the point (ξ, τ) where that characteristic, of equation $x - ct = \xi - c\tau$, intersects the curve $x = \varphi(t)$.

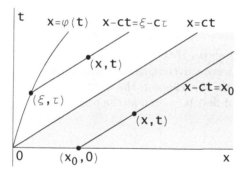

Figure 12.11: $0 < \varphi'(t) < c$.

By (12.7) and the fact that, as mentioned in Remark 12.11, $F(x - ct)$ is constant along this characteristic, we have

$$u(\xi, \tau) = F(\xi - c\tau) + G(\xi + c\tau)$$
$$= F(x - ct) + G(\xi + c\tau),$$

so

$$F(x - ct) = u(\xi, \tau) - G(\xi + c\tau);$$

hence, using (12.7) and (12.8), we now find that

$$u(x,t) = F(x - ct) + G(x + ct) = u(\xi, \tau) - G(\xi + c\tau) + G(x + ct)$$

$$= u(\xi, \tau) - \tfrac{1}{2} f(\xi + c\tau) - \frac{1}{2c} \int_0^{\xi+c\tau} g(y)\, dy$$

$$+ \tfrac{1}{2} f(x + ct) + \frac{1}{2c} \int_0^{x+ct} g(y)\, dy$$

$$= \psi(\tau) + \tfrac{1}{2} [f(x + ct) - f(\xi + c\tau)] + \frac{1}{2c} \int_{\xi+c\tau}^{x+ct} g(y)\, dy. \qquad (12.17)$$

The coordinates ξ and τ are computed from the system

$$x - ct = \xi - c\tau,$$

$$\xi = \varphi(\tau),$$

which leads to

$$\varphi(\tau) - c\tau = x - ct. \qquad (12.18)$$

Formula (12.18) shows that τ is a function of $x - ct$, which means that $u(\xi, \tau) = \psi(\tau)$ is a wave traveling to the right with velocity c.

As noted earlier in this chapter, the solution is continuous across the line $x = ct$ if the BC and first IC match at the origin. Here, if $(x, t) \to (0, 0)$, then $(\xi, \tau) \to (0, 0)$, so

$$\lim_{t \to 0} u(\varphi(t), t) = \lim_{t \to 0} \psi(t) = \psi(0),$$

$$\lim_{x \to 0} u(x, 0) = \lim_{x \to 0} f(x) = f(0),$$

so the continuity condition is $\psi(0) = f(0)$.

12.21 Remark. The above analysis yields the same results if $\varphi'(t) < 0$.

12.22 Example. Let

$$x = \varphi(t) = kt, \quad k = \text{const}, \quad 0 < |k| = |\varphi'(t)| < c,$$

$$f(x) = g(x) = 0,$$

$$\psi(t) = A\cos(\omega t), \quad A, \omega = \text{const} > 0,$$

which means that the boundary point moves with constant velocity k. By (12.18), $k\tau - c\tau = x - ct$, so

$$\tau = \frac{ct - x}{c - k} = \frac{c}{c - k} t - \frac{c}{c - k} \cdot \frac{x}{c} = \frac{c}{c - k}\left(t - \frac{x}{c}\right),$$

$$\psi(\tau) = A\cos(\omega\tau) = A\cos\left[\frac{c\omega}{c - k}\left(t - \frac{x}{c}\right)\right]. \qquad (12.19)$$

Therefore, following the conclusions of our discussion, for $x > ct$ we use d'Alembert's formula with $f = g = 0$ to obtain $u(x,t) = 0$, whereas for $\varphi(t) = kt < x < ct$, (12.17) and (12.19) lead to

$$u(x,t) = \psi(\tau) = A\cos\left[\frac{c\omega}{c - k}\left(t - \frac{x}{c}\right)\right]. \qquad (12.20)$$

If we regard the moving boundary point as an energy source, then (12.20) shows that the wave $u(x,t)$ produced by the boundary motion oscillates with frequency

$$\tilde{\omega} = \frac{c}{c - k}\,\omega.$$

Two cases arise in this context.

(i) If $k > 0$, then $0 < c - k < c$. Both the source and the wave are traveling to the right and $\tilde{\omega} > \omega$, with $\tilde{\omega}$ increasing as k increases.

(ii) If $k < 0$, then $c - k > c$. The wave is traveling to the right, but the source is traveling to the left, and $\tilde{\omega} < \omega$, with $\tilde{\omega}$ decreasing as k decreases.

In other words, if the boundary source moves in the same direction as the wave that it generates, its oscillation frequency is higher than that of the latter; if it moves in the opposite direction, its frequency is lower. This phenomenon is known as the *Doppler effect*.

12.4.3 Spherical Waves

The three-dimensional wave equation is

$$u_{tt} = c^2 \Delta u, \qquad (12.21)$$

where the Laplacian Δ is expressed in terms of a system of coordinates appropriate for the geometry of the problem. If the amplitude of the waves depends only on the distance r of the wave front from a point source, then we choose spherical coordinates, in which case $u = u(r,t)$ and, by Remark 4.12(v),

$$\Delta u = r^{-2}(r^2 u_r)_r = r^{-2}(2r u_r + r^2 u_{rr}) = u_{rr} + 2r^{-1}u_r.$$

Therefore, the equation governing the propagation of spherical waves is

$$u_{tt}(r,t) = c^2\left[u_{rr}(r,t) + 2r^{-1}u_r(r,t)\right].$$

Multiplying this equality by r and manipulating the right-hand side, we find that

$$ru_{tt} = c^2\left[(ru_{rr} + u_r) + u_r\right] = c^2\left[(ru_r)_r + u_r\right]$$
$$= c^2(ru_r + u)_r = c^2((ru)_r)_r = c^2(ru)_{rr}.$$

The substitution $ru(r,t) = v(r,t)$ now reduces the above equation to

$$v_{tt}(r,t) = c^2 v_{rr}(r,t),$$

so, by analogy with the argument developed in Section 12.1,

$$v(r,t) = F(r - ct) + G(r + ct).$$

Consequently, the general solution of (12.21) is

$$u(r,t) = \frac{1}{r}\left[F(r - ct) + G(r + ct)\right],$$

where F and G are arbitrary one-variable functions.

CHAPTER 13

Perturbation and Asymptotic Methods

Owing to the complexity of the PDEs involved in some mathematical models, it is not always possible to find an exact solution to an initial/boundary value problem. The next best thing in such situations is to compute an approximate solution instead. This is the idea behind the method of asymptotic expansion, which is applicable to problems that contain a small positive parameter, and relies on the expansion of the solution in a series of powers of the parameter. If the series converges, then the technique is called a *perturbation method;* when the series diverges but is asymptotic (in the sense explained below), we have an *asymptotic method.*

In what follows, we discuss only the formal construction of the series solution and obtain the first few terms, without considering the question of convergence.

13.1 Asymptotic Series

In order to state what an asymptotic series is, we need a mechanism to compare the 'magnitude' of functions.

13.1 Definition. Let f and g be two functions of a real variable x. We say that *f is of order g near* $x = a$, and write

$$f(x) = O(g(x)) \quad \text{as } x \to a,$$

if

$$\left| \frac{f(x)}{g(x)} \right| \text{ is bounded as } x \to a.$$

We also write

$$f(x) = o(g(x)) \quad \text{as } x \to a$$

if

$$\frac{f(x)}{g(x)} \to 0 \quad \text{as } x \to a.$$

13.2 Examples.   (i) The function $\sin x$ is of order x near $x = 0$ since

$$\lim_{x \to 0} \frac{\sin x}{x} = 1;$$

therefore, $(\sin x)/x$ is bounded for x close to 0.

(ii) We have $x^2 \ln|x| = o(x)$ near $x = 0$ since

$$\lim_{x \to 0} \frac{x^2 \ln|x|}{x} = 0.$$

(iii) Similarly, $e^{-1/|x|} = o(x^n)$ near $x = 0$ for any positive integer n because

$$\lim_{x \to 0} \frac{e^{-1/|x|}}{x^n} = 0.$$

13.3 Definition. A function $f(x, \varepsilon)$, where $0 < \varepsilon \ll 1$ is a small parameter, is said to have the *asymptotic (power) series*

$$f(x, \varepsilon) \approx \sum_{n=0}^{\infty} f_n(x)\varepsilon^n \quad \text{as } \varepsilon \to 0+$$

if for any positive integer N,

$$f(x, \varepsilon) = \sum_{n=0}^{N-1} f_n(x)\varepsilon^n + O(\varepsilon^N) \quad \text{as } \varepsilon \to 0+, \qquad (13.1)$$

uniformly for x in some interval.

13.4 Remarks.   (i) Equality (13.1) means that the remainder after N terms is of order ε^N as $\varepsilon \to 0+$. An alternative way to write this is

$$f(x, \varepsilon) = \sum_{n=0}^{N-1} f_n(x)\varepsilon^n + o(\varepsilon^{N-1}) \quad \text{as } \varepsilon \to 0+.$$

(ii) The right-hand side of (13.1) may diverge as $N \to \infty$, but it yields a good approximation of $f(x, \varepsilon)$ when N is fixed and $\varepsilon > 0$ is very small. We do not need convergence for the result to be acceptable. Also, the approximation may not get better if we take additional terms because, as mentioned above, the series may be divergent.

(iii) In an asymptotic series, it is assumed that the terms containing higher powers of ε are much smaller than those with lower powers.

(iv) In this chapter, we assume that asymptotic series may be differentiated and integrated term by term.

(v) If $k \gg 1$ is a large parameter, then we can reduce the problem to one with a small parameter by setting $k = 1/\varepsilon$.

13.5 Definition. Consider an initial/boundary value problem that depends (smoothly) on a small parameter $\varepsilon > 0$. The problem obtained by setting $\varepsilon = 0$ in the equation and data functions is called the *reduced (unperturbed) problem*. If the reduced problem is of the same type and order as the given one and both have unique solutions, then the given problem is called a *regular perturbation problem*; otherwise, it is called a *singular perturbation problem*.

13.6 Example. The IBVP

$$u_t(x,t) = ku_{xx}(x,t) + \varepsilon u_x(x,t), \quad 0 < x < L, \ t > 0,$$
$$u(0,t) = 0, \quad u(L,t) = 0, \quad t > 0,$$
$$u(x,0) = f(x), \quad 0 < x < L,$$

where $0 < \varepsilon \ll 1$, is a regular perturbation problem since both this IBVP and its reduced version (involving the heat equation) are second-order parabolic problems with unique solutions.

13.7 Example. The signalling (hyperbolic) problem

$$\varepsilon u_{tt}(x,t) + u_t(x,t) = c^2 u_{xx}(x,t), \quad x > 0, \ -\infty < t < \infty,$$
$$u(0,\ t) = f(t), \quad -\infty < t < \infty,$$

where $0 < \varepsilon \ll 1$, reduces to a parabolic one when we set $\varepsilon = 0$. Therefore, although both the given and reduced problems have unique solutions, the given BVP is a singular perturbation problem.

13.8 Example. Let D be a finite region bounded by a smooth, closed, simple curve ∂D in the (x,y)-plane, and let n be the unit outward normal to ∂D. The fourth-order BVP

$$(\Delta\Delta u)(x,y) = 0, \quad (x,y) \text{ in } D,$$
$$u(x,y) = f(x,y), \quad \varepsilon u_n(x,y) + u(x,y) = g(x,y), \quad (x,y) \text{ on } \partial D,$$

where $0 < \varepsilon \ll 1$ and $u_n = \partial u / \partial n$, reduces, when we set $\varepsilon = 0$, to the BVP

$$(\Delta\Delta u)(x,y) = 0, \quad (x,y) \text{ in } D,$$
$$u(x,y) = f(x,y), \quad u(x,y) = g(x,y), \quad (x,y) \text{ on } \partial D.$$

If $f \neq g$, then the reduced problem has no solution. On the other hand, if $f = g$, then the solution is not unique, since we have lost one of the BCs. Hence, the given BVP is a singular perturbation problem.

13.9 Definition. If the solution u of a perturbation problem has the asymptotic series $u \approx \sum_{n=0}^{\infty} u_n \varepsilon^n$, then $u - u_0$ is called a *perturbation* of the solution of the reduced problem.

13.2   Regular Perturbation Problems

The best way to see how the method works in this case is to examine a number of specific models.

13.2.1   Formal Solutions

We start with a couple of problems that reduce to ordinary differential equations.

13.10 Example. Let

$$D = \{(r, \theta) : 0 \leq r < 1, \ -\pi \leq \theta < \pi\},$$
$$\partial D = \{(r, \theta) : r = 1, \ -\pi \leq \theta < \pi\}$$

be the unit disc (with the center at the origin) and its circular boundary, and consider the Dirichlet problem for the two-dimensional Helmholtz equation

$$(\Delta u)(r, \theta) + \varepsilon u(r, \theta) = 0, \quad (r, \theta) \text{ in } D, \ r \neq 0,$$
$$u(r, \theta) = 1, \quad (r, \theta) \text{ on } \partial D,$$

where $0 < \varepsilon \ll 1$, which ensures that the above BVP has a unique solution. The reduced problem (for $\varepsilon = 0$) is the Dirichlet problem for the Laplace equation, also uniquely solvable. Both the reduced and the perturbed problems are elliptic and of second order, so the given BVP is a regular perturbation problem.

Assuming a perturbation series of the form $u(r, \theta) \approx \sum\limits_{n=0}^{\infty} u_n(r, \theta)\varepsilon^n$, from the PDE we find that

$$\Delta u + \varepsilon u \approx \sum_{n=0}^{\infty} (\Delta u_n)\varepsilon^n + \sum_{n=0}^{\infty} u_n \varepsilon^{n+1}$$

$$= \Delta u_0 + \sum_{n=1}^{\infty} (\Delta u_n + u_{n-1})\varepsilon^n = 0 \quad \text{in } D, \ r \neq 0.$$

At the same time, the BC yields

$$u \approx u_0 + \sum_{n=1}^{\infty} u_n \varepsilon^n = 1 \quad \text{on } \partial D.$$

Equating the coefficients of every power of ε on both sides, we then obtain the BVPs

$$\Delta u_0 = 0 \quad \text{in } D, \ r \neq 0,$$
$$u_0 = 1 \quad \text{on } \partial D$$

and, for $n \geq 1$,

$$\Delta u_n = -u_{n-1} \quad \text{in } D, \ r \neq 0,$$
$$u_n = 0 \quad \text{on } \partial D.$$

Since the region where the problem is posed and the boundary data are independent of the polar angle θ, we may assume that $u_n = u_n(r)$; consequently, by Remark 4.11(ii), the problem for u_0 can be written as

$$(\Delta u_0)(r) = u_0''(r) + r^{-1} u_0'(r) = 0, \quad 0 < r < 1,$$
$$u_0(1) = 1.$$

Multiplying the differential equation by r and noting that the left-hand side of the new ODE is, in fact, $(r u_0')'$, we find that

$$u_0(r) = C_1 \ln r + C_2, \quad C_1, C_2 = \text{const.}$$

We know (see Section 5.3) that for this type of problem we also have additional conditions, generated by physical considerations. Since we have assumed that the solution is independent of θ, the only other condition of this kind to be satisfied here is that the solution and its derivative be continuous (hence, bounded) in D. This implies that $C_1 = 0$; the value of C_2 is then found from the BC at $r = 1$, which yields $u_0(r) = 1$.

Next, applying the same argument to the BVP satisfied by u_1, namely

$$(\Delta u_1)(r) = u_1''(r) + r^{-1} u_1'(r) = -u_0(r) = -1, \quad 0 < r < 1,$$
$$u_1(1) = 0,$$

we find that $u_1(r) = \frac{1}{4}(1 - r^2)$; hence,

$$u(r) = 1 + \frac{1}{4} \varepsilon (1 - r^2) + O(\varepsilon^2). \tag{13.2}$$

For this problem we can actually see how good an approximation (13.2) is. The given (perturbed) equation can be written in the form

$$u''(r) + r^{-1} u'(r) + \varepsilon u(r) = 0,$$

which is Bessel's equation of order zero (see (3.14) with $m = 0$, $\lambda = \varepsilon$, and x replaced by r). Its bounded solution satisfying $u(1) = 1$ is

$$u(r) = J_0(\sqrt{\varepsilon} r)/J_0(\sqrt{\varepsilon}),$$

where J_0 is the Bessel function of the first kind and order zero. This formula makes sense because $0 < \varepsilon \ll 1$; that is, $\sqrt{\varepsilon}$ is smaller than the first zero ($\xi \cong 2.4$) of $J_0(\xi)$. Since, for small values of ξ,

$$J_0(\xi) = 1 - \frac{1}{4} \xi^2 + O(\xi^4),$$

we use the formula

$$1 + q + q^2 + \cdots + q^n + \cdots = \frac{1}{1-q}$$

for the sum of an infinite geometric progression with ratio q, $0 < |q| < 1$, to find that

$$\frac{J_0(\sqrt{\varepsilon}\, r)}{J_0(\sqrt{\varepsilon})} = \frac{1 - \frac{1}{4}\varepsilon r^2 + O(\varepsilon^2)}{1 - \frac{1}{4}\varepsilon + O(\varepsilon^2)} = \frac{1 - \frac{1}{4}\varepsilon r^2 + O(\varepsilon^2)}{1 - \left(\frac{1}{4}\varepsilon + O(\varepsilon^2)\right)}$$

$$= \left[1 - \tfrac{1}{4}\varepsilon r^2 + O(\varepsilon^2)\right]\left[1 + \left(\tfrac{1}{4}\varepsilon + O(\varepsilon^2)\right) + O(\varepsilon^2)\right]$$

$$= 1 + \tfrac{1}{4}\varepsilon(1 - r^2) + O(\varepsilon^2).$$

Thus, the perturbation solution (13.2) coincides with the exact solution to $O(\varepsilon)$-terms throughout D.

VERIFICATION WITH MATHEMATICA®. This is also easily seen to be the case if we replace the truncation of (13.2) after two terms in the PDE and the BC, and restrict the computation to the first two terms of the asymptotic series in the results. Specifically, the input

```
u = 1 + (1/4) * ε * (1 - r^2);
f = 1;
Simplify[Series[{D[u,r,r] + r^(-1) * D[u,r] + ε * u,
  (u /. r -> 1) - f}, {ε,0,1}] // Normal]
```

generates the output $\{0, 0\}$.

13.11 Example. The elliptic nonlinear BVP

$$\Delta u(r, \theta) + \varepsilon u^2(r, \theta) = 36r, \quad (r, \theta) \text{ in } D,$$

$$u(r, \theta) = 4, \quad (r, \theta) \text{ on } \partial D,$$

where $0 < \varepsilon \ll 1$ and D and ∂D are the same as in Example 13.10, is a regular perturbation problem. Since we again notice that the solution u is expected to depend only on r, we assume for the solution a series of the form $u(r) \approx \sum\limits_{n=0}^{\infty} u_n(r)\varepsilon^n$. Then, in the usual way, taking into account the asymptotic expansion

$$u^2 = \left[u_0 + \varepsilon u_1 + O(\varepsilon^2)\right]^2 = u_0^2 + 2\varepsilon u_0 u_1 + O(\varepsilon^2)$$

and recalling the boundedness condition mentioned in Example 13.10, we see that u_0 and u_1 are, respectively, the solutions of the linear BVPs

$$u_0''(r) + r^{-1}u_0'(r) = 36r, \quad 0 < r < 1,$$

$$u_0(r), \; u_0'(r) \text{ bounded as } r \to 0+, \quad u_0(1) = 4$$

and

$$u_1''(r) + r^{-1}u_1'(r) = -u_0^2(r), \quad 0 < r < 1,$$

$$u_1(r), \ u_1'(r) \text{ bounded as } r \to 0+, \quad u_1(1) = 0;$$

that is,

$$u_0(r) = 4r^3, \quad u_1(r) = \tfrac{1}{4}(1 - r^8).$$

Consequently,

$$u(r) = 4r^3 + \tfrac{1}{4}\varepsilon(1 - r^8) + O(\varepsilon^2).$$

This example shows how, in certain circumstances, the perturbation method can reduce a nonlinear problem to a sequence of linear ones.

VERIFICATION WITH MATHEMATICA®. The input

```
u = 4 * r^3 + (1/4) * ε * (1 - r^8);
{q,f} = {36 * r, 4};
Simplify[Series[{D[u,r,r] + r^(-1) * D[u,r] + ε * u^2 - q,
    (u/.r->1) - f}, {ε,0,1}] // Normal]
```

generates the output $\{0, 0\}$.

13.12 Remark. Sometimes, the small parameter may occur not only in the equation, but also in any accompanying ICs and/or BCs. The solution procedure in such cases remains the same.

13.13 Example. The IVP

$$u_{xx}(x, y) + (\varepsilon - 1)u_x(x, y) + 2\varepsilon u_y(x, y) - (2 + \varepsilon)u(x, y)$$

$$= 3e^{2x} + \varepsilon\big[2y - 2x - 7 + (3 + x + y)e^{2x}\big],$$

$$x > 0, \ -\infty < y < \infty,$$

$$u(0, y) = y + \varepsilon(3 - y), \quad u_x(0, y) = 1 + 2y + \varepsilon, \quad -\infty < y < \infty,$$

where $0 < \varepsilon \ll 1$, is a regular perturbation problem since the PDEs in both the given IVP and the reduced one are second-order parabolic equations and the two problems have unique solutions. To find a formal two-term asymptotic approximation of the solution, we replace $u(x, y) = u_0(x, y) + \varepsilon u_1(x, y) + O(\varepsilon^2)$ in the PDE and ICs and, equating the constant terms and the coefficients of ε on both sides, find that u_0 and u_1 are, respectively, the solutions of the IVPs

$$(u_0)_{xx}(x, y) - (u_0)_x(x, y) - 2u_0(x, y) = 3e^{2x},$$

$$x > 0, \ -\infty < y < \infty,$$

$$u_0(0, y) = y, \quad (u_0)_x(0, y) = 1 + 2y, \quad -\infty < y < \infty$$

and

$$(u_1)_{xx}(x, y) - (u_1)_x(x, y) - 2u_1(x, y)$$
$$= -(u_0)_x(x, y) - 2(u_0)_y(x, y) + u_0(x, y)$$
$$+ 2y - 2x - 7 + (3 + x + y)e^{2x}, \quad x > 0, \quad -\infty < y < \infty,$$
$$u_1(0, y) = 3 - y, \quad (u_1)_x(0, y) = 1, \quad -\infty < y < \infty.$$

Solving the first of these two IVPs, we arrive at

$$u_0(x, y) = C_1(y)e^{2x} + C_2(y)e^{-x} + A(y)xe^{2x} = (x + y)e^{2x},$$

which reduces the equation in the second IVP to

$$(u_1)_{xx}(x, y) - (u_1)_x(x, y) - 2u_1(x, y) = 2y - 2x - 7.$$

Using the ICs, we then find that

$$u_1(x, y) = C_1(y)e^{2x} + C_2(y)e^{-x} + A(y)x + B(y) = x - y + 3.$$

Consequently, the desired approximate solution is

$$u(x, y) = (x + y)e^{2x} + \varepsilon(x - y + 3) + O(\varepsilon^2).$$

VERIFICATION WITH MATHEMATICA®. The input

```
u = (x + y) * E^(2 * x) + ε * (x - y + 3);
{q, f1, f2} = {3 * E^(2 * x) + ε * (2 * y - 2 * x - 7) + (3 + x + y) * E^(2 * x)),
    y + ε * (3 - y), 1 + 2 * y + ε};
Simplify[Series{D[u,x,x] + (ε - 1) * D[u,x] + 2 * ε * D[u,y] -
    (2 + ε) * u - q, {(u /. x -> 0) - f1, (D[u,x] /. x -> 0) - f2},
    {ε, 0, 1}] // Normal]
```

generates the output $\{0, \{0, 0\}\}$.

13.14 Example. Consider the BVP

$$u_{xx}(x, y) + (2 - \varepsilon)u_x(x, y) - \varepsilon u_y(x, y) - 3u(x, y)$$
$$= 2 - 3x + 3y^2 + \varepsilon(2y - 1 - 8e^x \cos y),$$
$$0 < x < 1, \quad -\infty < y < \infty,$$
$$u(0, y) = -y^2 + \varepsilon, \quad u(1, y) = 1 - y^2 + \varepsilon e(1 - 2 \cos y), \quad -\infty < y < \infty,$$

where $0 < \varepsilon \ll 1$. This is a regular perturbation problem, so we consider for its solution an expansion of the form $u(x, y) = u_0(x, y) + \varepsilon u_1(x, y) + O(\varepsilon^2)$.

Following the standard technique explained earlier, we find that the PDE and BCs give rise to the BVPs

$$(u_0)_{xx}(x, y) + 2(u_0)_x(x, y) - 3u_0(x, y) = 2 - 3x + 3y^2,$$

$$0 < x < 1, \quad -\infty < y < \infty,$$

$$u_0(0, y) = -y^2, \quad u_0(1, y) = 1 - y^2, \quad -\infty < y < \infty$$

and

$$(u_1)_{xx}(x, y) + 2(u_1)_x(x, y) - 3u_1(x, y)$$

$$= (u_0)_x(x, y) + (u_0)_y(x, y) + 2y - 1 - 8e^x \cos y,$$

$$0 < x < 1, \quad -\infty < y < \infty,$$

$$u_1(0, y) = 1, \quad u_1(1, y) = e(1 - 2\cos y), \quad -\infty < y < \infty.$$

The solution of the former is

$$u_0(x, y) = C_1(y)e^x + C_2(y)e^{-3x} + A(y)x + B(y) = x - y^2.$$

Then the equation in the latter becomes

$$(u_1)_{xx}(x, y) + 2(u_1)_x(x, y) - 3u_1(x, y) = -8e^x \cos y,$$

which, together with the appropriate BCs, yields

$$u_1(x, y) = C_1(y)e^x + C_2(y)e^{-3x} + A(y)xe^x = (1 - 2x \cos y)e^x;$$

therefore,

$$u(x, y) = x - y^2 + \varepsilon(1 - 2x \cos y)e^x + O(\varepsilon^2).$$

VERIFICATION WITH MATHEMATICA®. The input

```
u = x - y^2 + ε * (1 - 2 * x * Cos [y]) * E^x;
{q,f1,f2} = {2 - 3 * x + 3 * y^2 + ε * (2 * y - 1 - 8 * E^x * Cos [y]),
   - y^2 + ε, 1 - y^2 + ε * E * (1 - 2 * Cos [y])};
Simplify [Series [{D [u,x,x] + (2 - ε) * D [u,x] - ε * D [u,y] - 3 * u - q,
   {(u /. x -> 0) - f1, (u /. x -> 1) - f2}}, {ε,0,1}] // Normal]
```

generates the output $\{0, \{0, 0\}\}$.

13.15 Example. The BVP

$$u_{yy}(x, y) - 2\varepsilon u_x(x, y) + \varepsilon u_y(x, y) - (4 + 2\varepsilon)u(x, y)$$

$$= 4y^2 - 2 + \varepsilon[8x - 6y + 2y^2 - 8xy^2 - (2x + 2)\cosh(2y)$$

$$+ 2x \sinh(2y)], \quad -\infty < x < \infty, \quad 0 < y < 1,$$

$$u_y(x, 0) = \varepsilon, \quad u(x, 1) = x \cosh 2 - 1 + \varepsilon(x + 1), \quad -\infty < x < \infty,$$

where $0 < \varepsilon \ll 1$, is a regular parabolic perturbation problem. Let

$$u(x, y) = u_0(x, y) + \varepsilon u_1(x, y) + O(\varepsilon^2).$$

Applying the usual procedure, we arrive at the BVPs

$$(u_0)_{yy}(x, y) - 4u_0(x, y) = 4y^2 - 2, \quad -\infty < x < \infty, \ 0 < y < 1,$$

$$(u_0)_y(x, 0) = 0, \quad u_0(x, 1) = x \cosh 2 - 1, \quad -\infty < x < \infty$$

and

$$(u_1)_{yy}(x, y) - 4u_1(x, y)$$
$$= 2(u_0)_x(x, y) - (u_0)_y(x, y) + 2u_0(x, y) + 8x - 6y + 2y^2 - 8xy^2$$
$$- (2x + 2) \cosh(2y) + 2x \sinh(2y), \quad -\infty < x < \infty, \ 0 < y < 1,$$

$$(u_1)_y(x, 0) = 1, \quad u_1(x, 1) = x + 1, \quad -\infty < x < \infty.$$

The first one has solution

$$u_0(x, y) = C_1(x) \cosh(2y) + C_2(x) \sinh \left(2(y - 1)\right)$$
$$+ A(x)y^2 + B(x)y + C(x) = x \cosh(2y) - y^2,$$

so the DE in the second one changes to

$$(u_1)_{yy}(x, y) - 4u_1(x, y) = 8x - 4y - 8xy^2.$$

Writing out the GS and then applying the BCs, we find that

$$u_1(x, y) = C_1(x) \cosh(2y) + C_2(x) \sinh \left(2(y - 1)\right)$$
$$+ A(x)y^2 + B(x)y + C(x) = 2xy^2 + y - x;$$

hence,
$$u(x, y) = x \cosh(2y) - y^2 + \varepsilon(2xy^2 + y - x) + O(\varepsilon^2).$$

VERIFICATION WITH MATHEMATICA®. The input

```
u = x * Cosh [2 * y] - y^2 + ε * (2 * x * y^2 + y - x);
{q, f1, f2} = {4 * y^2 - 2 + ε * (8 * x - 6 * y + 2 * y^2 - 8 * x * y^2 -
    (2 * x + 2) * Cosh [2 * y] + 2 * x * Sinh [2 * y]), ε, x * Cosh [2] -
    1 + ε * (x + 1)};
Simplify [Series [{D[u, y, y] - 2 ε * D[u, x] + ε * D[u, y] -
    (4 + 2 * ε) * u - q, {(D[u, y] /. y -> 0) - f1, (u /. y -> 1) - f2}},
    {ε, 0, 1}] // Normal]
```

generates the output $\{0, \{0, 0\}\}$.

13.16 Example. The IVP

$$u_{xx}(x, y) - (1 + \varepsilon)u_{yy}(x, y) + \varepsilon u_x(x, y) + 2\varepsilon u(x, y)$$

$$= 4y + 4\varepsilon(x^2 y + xy + 1), \quad -\infty < x < \infty, \ y > 0,$$

$$u(x, 0) = 1, \quad u_y(x, 0) = 2x^2 + 3\varepsilon x, \quad -\infty < x < \infty,$$

where $0 < \varepsilon \ll 1$, is a regular, second-order, hyperbolic perturbation problem. Writing, as usual,

$$u(x, y) = u_0(x, y) + \varepsilon u_1(x, y) + O(\varepsilon^2),$$

we generate the separate IVPs

$$(u_0)_{xx}(x, y) - (u_0)_{yy}(x, y) = 4y, \quad -\infty < x < \infty, \ y > 0,$$

$$u_0(x, 0) = 1, \quad (u_0)_y(x, 0) = 2x^2, \quad -\infty < x < \infty$$

and

$$(u_1)_{xx}(x, y) - (u_1)_{yy}(x, y)$$

$$= (u_0)_{yy}(x, y) - (u_0)_x(x, y) - 2u_0(x, y) + 4x^2 y + 4xy + 4,$$

$$-\infty < x < \infty, \ y > 0,$$

$$u_1(x, 0) = 0, \quad (u_1)_y(x, 0) = 3x, \quad -\infty < x < \infty.$$

The solution of the former is given by the generalized d'Alembert formula (12.10) with t replaced by y, $c = 1$, $f(x) = 1$, $g(x) = 2x^2$, and $p(y) = -4y$:

$$u_0(x, y) = 1 + 2x^2 y,$$

which, substituted in the IVP for u_1, yields, by the same procedure with $c = 1$, $f(x) = 0$, $g(x) = 3x$, and $p(y) = -2$,

$$u_1(x, y) = 3xy - y^2.$$

Thus, the solution of the given problem is

$$u(x, y) = 1 + 2x^2 y + \varepsilon(3xy - y^2) + O(\varepsilon^2).$$

VERIFICATION WITH MATHEMATICA®. The input

```
u = 1 + 2 * x^2 * y + ε * (3 * x * y - y^2);
{q, f1, f2} = {4 * y + 4 * ε * (x^2 * y + x * y + 1), 1, 2 * x^2 + 3 * ε * x};
Simplify [Series [{D[u,x,x] - (1 + ε) * D[u,y,y] + ε * D[u,x] +
   2 * ε * u - q, {(u /. y -> 0) - f1, (D[u,y] /. y -> 0) - f2}}, {ε, 0, 1}]
   // Normal]
```

generates the output $\{0, \{0, 0\}\}$.

13.17 Example. The BVP

$$u_{xy}(x, y) + (1 - \varepsilon)u_x(x, y) + \varepsilon u(x, y)$$
$$= 2x - 2y - 2 + \varepsilon(x^2 - 2xy + 3y^2 - 2x + 4y), \quad x > 0, \quad y > 0,$$
$$u(0, y) = 3y^2 + \varepsilon y, \quad y > 0,$$
$$u(x, 0) = x^2 - 2\varepsilon x, \quad x > 0,$$

where $0 < \varepsilon \ll 1$, is also a regular, second-order, hyperbolic perturbation problem. Seeking the solution as

$$u(x, t) = u_0(x, t) + \varepsilon u_1(x, t) + O(\varepsilon^2),$$

from the PDE and BCs we deduce that u_0 and u_1 are, respectively, the solutions of the BVPs

$$(u_0)_{xy}(x, y) + (u_0)_x(x, y) = 2x - 2y - 2, \quad x > 0, \quad y > 0,$$
$$u_0(0, y) = 3y^2, \quad y > 0,$$
$$u_0(x, 0) = x^2, \quad x > 0$$

and

$$(u_1)_{xy}(x, y) + (u_1)_x(x, y)$$
$$= (u_0)_x(x, y) - u_0(x, y) + x^2 - 2xy + 3y^2 - 2x + 4y,$$
$$x > 0, \quad y > 0,$$
$$u_1(0, y) = y, \quad y > 0,$$
$$u_1(x, 0) = -2x, \quad x > 0.$$

Solving first for $(u_0)_x$ as the unknown function in the former and using the condition $(u_0)_x(x, 0) = 2x$ obtained by differentiation from the second BC for u_0, we arrive at

$$(u_0)_x(x, y) = 2x - 2y,$$

which, in view of the first BC, yields

$$u_0(x, y) = x^2 - 2xy + 3y^2.$$

We now replace u_0 in the second BVP above and, following the same procedure, find that

$$u_1(x, y) = 2xy - 2x + y;$$

therefore,

$$u(x, y) = x^2 - 2xy + 3y^2 + \varepsilon(2xy - 2x + y) + O(\varepsilon^2).$$

VERIFICATION WITH MATHEMATICA®. The input

```
u=x^2-2*x*y+3*y^2+ ε * (2*x*y-2*x+y);
{q,f1,f2}={2*x-2*y-2+ ε * (x^2-2*x*y+3*y^2-2*x+4*y),
   3*y^2+ ε *y, x^2-2* ε *x};
Simplify[Series[{D[u,x,y] + (1- ε ) *D[u,x] + ε *u-q,
   (u/.x->0)-f1, (u/.y->0)-f2}, { ε ,0,1}]//Normal]
```

generates the output $\{0, 0, 0\}$.

Exercises

In **1–4**, use the method of asymptotic expansion to compute a formal approximate solution (to $O(\varepsilon)$–terms) for the regular perturbation problem

$$u''(r) + r^{-1}u'(r) + \varepsilon f(u) = q(r), \quad 0 < r < 1,$$
$$u(r), \ u'(r) \text{ bounded as } r \to 0+, \quad u(1) = g,$$

where $0 < \varepsilon \ll 1$, with functions f, q, and number g as indicated.

1 $f(u) = 4u$, $q(r) = 1$, $g = 1 + \varepsilon$.

2 $f(u) = -2u$, $q(r) = -4 + 2\varepsilon(r^2 + 9r)$, $g = 1 + 3\varepsilon$.

3 $f(u) = -u^3$, $q(r) = \varepsilon$, $g = -3$.

4 $f(u) = u - u^2$, $q(r) = 6(3r - 2) + \varepsilon(12 - 3r^2 + 2r^3 - 9r^4 + 12r^5 - 4r^6)$,
   $g = -1 + 3\varepsilon$.

In **5–14**, use the method of asymptotic expansion to compute a formal approximate solution (to $O(\varepsilon)$–terms) for the regular perturbation problem

$$u_t(x,t) + au_x(x,t) + bu(x,t) = q(x,t), \quad -\infty < x < \infty, \ t > 0,$$
$$u(x,0) = f(x), \quad -\infty < x < \infty,$$

where $0 < \varepsilon \ll 1$, with functions q, f, and coefficients a, b as indicated.

5 $a = \varepsilon$, $b = 1 + \varepsilon$, $q(x,t) = 2x + 2xt + \varepsilon(2t + 2xt)$, $f(x) = \varepsilon x$.

6 $a = -\varepsilon$, $b = 2 - \varepsilon$, $q(x,t) = \varepsilon[-1 - 2t - 2x + (1+x)e^{-2t}]$, $f(x) = -x - \varepsilon x$.

7 $a = 2\varepsilon$, $b = 3 - \varepsilon$, $q(x,t) = 1 + 3x^2 + 3t + \varepsilon(4x - x^2 - t - e^{-3t})$,
   $f(x) = x^2 + \varepsilon x$.

8 $a = -2\varepsilon$, $b = 2 + \varepsilon$, $q(x,t) = e^{-2t} + \varepsilon[2x + 4xt + (x + t - 2)e^{-2t}]$,
   $f(x) = x$.

9 $a = 3$, $b = -2\varepsilon$, $q(x,t) = 0$, $f(x) = x$.

10 $a = -1$, $b = 1 + \varepsilon$, $q(x,t) = 0$, $f(x) = 2x - 1$.

11 $a = 2 + \varepsilon$, $b = -1 + \varepsilon$, $q(x,t) = 2 + \varepsilon[(4t - 1)e^t - 2]$, $f(x) = -x - 2$.

12 $a = 1 - \varepsilon$, $b = 1 + 2\varepsilon$, $q(x,t) = x + \varepsilon(4x - 1 + 5e^{-t})$,

$f(x) = 1 + x + 2\varepsilon x$.

13 $a = 2 + \varepsilon$, $b = 0$, $q(x,t) = 2t$, $f(x) = e^x$.

14 $a = 1 - 3\varepsilon$, $b = 0$, $q(x,t) = 1 + 3x + t + \varepsilon(x - 8t - 2)$, $f(x) = x + \varepsilon x$.

In **15–36**, use the method of asymptotic expansion to compute a formal approximate solution (to $O(\varepsilon)$–terms) for the regular perturbation problem consisting of the PDE

$$au_{xx}(x,y) + bu_{xy}(x,y) + cu_{yy}(x,y)$$
$$+ fu_x(x,y) + gu_y(x,y) + hu(x,y) = q(x,y)$$

and ranges of the variables, coefficients, function q, and BCs/ICs as indicated, with $0 < \varepsilon \ll 1$.

15 $0 < x < 1$, $-\infty < y < \infty$, $a = 1$, $b = c = f = 0$, $g = -\varepsilon$, $h = -1$,

$q(x,y) = -xy^2$, $u(0,y) = \varepsilon y$, $u(1,y) = y^2 + \varepsilon(e - 2)y$.

16 $-\infty < x < \infty$, $0 < y < 1$, $a = b = 0$, $c = 1$, $f = -\varepsilon$, $g = 0$, $h = \varepsilon - 4$,

$q(x,y) = 4x^2 y + \varepsilon[2xy - x^2 y - 4(y+1)e^{-2x}]$, $u(x,0) = \varepsilon e^{-2x}$,

$u(x,1) = -x^2 + 2\varepsilon e^{-2x}$.

17 $-\infty < x < \infty$, $0 < y < 1$, $a = b = 0$, $c = 1$, $f = \varepsilon$, $g = \varepsilon - 1$,

$h = 3\varepsilon - 2$, $q(x,y) = 2(1 - y - y^2)e^{-x} + \varepsilon[1 + 2y - 2x^2 + 2(y^2 + y)e^{-x}]$,

$u(x,0) = \varepsilon x^2$, $u(x,1) = e^{-x} + \varepsilon(x^2 - 1)$.

18 $0 < x < 1$, $-\infty < y < \infty$, $a = 1$, $b = c = 0$, $f = -2(1 + \varepsilon)$,

$g = 2\varepsilon$, $h = 1$, $q(x,y) = 2 - 4x + 2y + x^2 + \varepsilon(3 - 4x - 2y^2 + xy^2)$,

$u(0,y) = 2y - \varepsilon$, $u(1,y) = 2y + 1 + \varepsilon(y^2 - 1)$.

19 $0 < x < 1$, $-\infty < y < \infty$, $a = 1$, $b = c = 0$, $f = 2 + \varepsilon$, $g = 2\varepsilon$,

$h = 1 - \varepsilon$, $q(x,y) = y - 8x - 4 - 2x^2 + \varepsilon(2 - 4x - y + 2x^2 - 2y \sin x)$,

$u(0,y) = y + \varepsilon y$, $u(1,y) = y - 2 + \varepsilon y \cos 1$.

20 $-\infty < x < \infty$, $0 < y < 1$, $a = b = 0$, $c = 1$, $f = -2\varepsilon$, $g = \varepsilon - 2$,

$h = -3 - \varepsilon$, $q(x,y) = -(4 + 6y)e^{3x} + \varepsilon[12 - 2x - 3xy + (2 - 14y)e^{3x}]$,

$u(x,0) = -4\varepsilon$, $u(x,1) = 2e^{3x} + \varepsilon(x - 4)$.

21 $x > 0$, $-\infty < y < \infty$, $a = 1$, $b = c = 0$, $f = 3$, $g = 2\varepsilon$, $h = 2 + \varepsilon$,

$q(x,y) = 9 + 6x + 2y + \varepsilon(4 + 3x - 5y - 4xy)$, $u(0,y) = y + \varepsilon$,

$u_x(0,y) = 3 - 2\varepsilon y$.

22 $-\infty < x < \infty$, $y > 0$, $a = b = 0$, $c = 1$, $f = -3\varepsilon$, $g = 2 + \varepsilon$,
$h = 2$, $q(x,y) = 6x + 6xy + \varepsilon(x - 9y)$, $u(x,0) = -\varepsilon(x + 1)$,
$u_y(x,0) = 3x + \varepsilon$.

23 $-\infty < x < \infty$, $y > 0$, $a = b = 0$, $c = 1$, $f = 2\varepsilon$, $g = \varepsilon - 4$,
$h = 3 + \varepsilon$, $q(x,y) = 11 + 3x - 6y + \varepsilon(1 + 13x - 2y + 3x^2 - 9xy)$,
$u(x,0) = 1 + x + \varepsilon x^2$, $u_y(x,0) = -2 - 3\varepsilon x$.

24 $x > 0$, $-\infty < y < \infty$, $a = 1$, $b = c = 0$, $f = \varepsilon - 2$, $g = 2\varepsilon$,
$h = 3\varepsilon$, $q(x,y) = 2 + \varepsilon\left[-3x - 2y - 1 + (4 + y)e^{2x}\right]$, $u(0,y) = 2y - 3\varepsilon y$,
$u_x(0,y) = 4y - 1 + \varepsilon y$.

25 $x > 0$, $-\infty < y < \infty$, $a = 1$, $b = c = 0$, $f = 2\varepsilon - 4$, $g = -\varepsilon$, $h = 4 - \varepsilon$,
$q(x,y) = 4xy + 8x - 4y - 20 + \varepsilon(4x^2 - xy - 7x - 2y + 5)$,
$u(0,y) = -3 - \varepsilon y$, $u_x(0,y) = 2 + y + \varepsilon$.

26 $-\infty < x < \infty$, $y > 0$, $a = b = 0$, $c = 1$, $f = -\varepsilon$, $g = -2\varepsilon$, $h = 1 + \varepsilon$,
$q(x,y) = x - 2y + \varepsilon(4 + x - 5y)$, $u(x,0) = x + \varepsilon(x + 1)$, $u_y(x,0) = -2 - 3\varepsilon$.

27 $x > 0$, $-\infty < y < \infty$, $a = 1$, $b = 0$, $c = \varepsilon - 4$, $f = 2\varepsilon$, $g = 0$, $h = \varepsilon$,
$q(x,y) = 8x + \varepsilon(4 - 2x - 2y^2 - xy^2)$, $u(0,y) = 2 - \varepsilon$, $u_x(0,y) = -y^2 + 4\varepsilon y$.

28 $-\infty < x < \infty$, $y > 0$, $a = 1$, $b = 0$, $c = -1 - \varepsilon$, $f = \varepsilon$, $g = 0$, $h = 2\varepsilon$,
$q(x,y) = 4y + 4\varepsilon(1 + xy + x^2 y)$, $u(x,0) = 1$, $u_y(x,0) = 2x^2 + 3\varepsilon x$.

29 $-\infty < x < \infty$, $y > 0$, $a = 9 + \varepsilon$, $b = 0$, $c = -1$, $f = -2\varepsilon$,
$g = 0$, $h = \varepsilon$, $q(x,y) = 18 + \varepsilon(2 - 4x + 4y + x^2 - 2xy)$,
$u(x,0) = x^2 + 2\varepsilon$, $u_y(x,0) = -2x - \varepsilon(x + 1)$.

30 $x > 0$, $-\infty < y < \infty$, $a = 1$, $b = 0$, $c = 2\varepsilon - 1$, $f = 0$, $g = 3\varepsilon$,
$h = 2\varepsilon$, $q(x,y) = 14x + \varepsilon(6 - 4x - 6xy + 4x^3 - 2xy^2)$, $u(0,y) = \varepsilon$,
$u_x(0,y) = -y^2 - 2\varepsilon y$.

31 $x > 0$, $y > 0$, $a = 0$, $b = 1$, $c = 0$, $f = 2 + \varepsilon$, $g = 0$, $h = -3\varepsilon$,
$q(x,y) = 3 + 8x + 6y + \varepsilon(2 + 7y - 6x^2 - 9xy + 3y^2)$, $u(0,y) = -y^2 + \varepsilon$,
$u(x,0) = 2x^2 + \varepsilon(1 - x^2)$.

32 $x > 0$, $y > 0$, $a = 0$, $b = 1$, $c = f = 0$, $g = 2\varepsilon - 1$, $h = -2\varepsilon$,
$q(x,y) = 3 - 3x + 2y + \varepsilon(3 + 5x - 4y - 4x^2 - 6xy + 2y^2)$, $u(0,y) = -y^2 - 2\varepsilon y$,
$u(x,0) = 2x^2 + 4\varepsilon x$.

33 $x > 0$, $y > 0$, $a = 0$, $b = 1$, $c = 0$, $f = \varepsilon$, $g = 3\varepsilon - 2$, $h = \varepsilon$,
$q(x,y) = \varepsilon\left[1 - 2x + 4y + (4 + 3x + 3y)e^{2x}\right]$, $u(0,y) = y + \varepsilon(1 - y^2)$,
$u(x,0) = xe^{2x} + \varepsilon$.

34 $x > 0$, $y > 0$, $a = 0$, $b = 1$, $c = 0$, $f = 3 - \varepsilon$, $g = \varepsilon$, $h = 0$,

$\quad q(x, y) = \varepsilon[6x - 6y - 2 - (3x + 1)e^{-3y}]$, $u(0, y) = -2\varepsilon$,

$\quad u(x, 0) = x + \varepsilon(x^2 - 2)$.

35 $x > 0$, $y > 0$, $a = \varepsilon$, $b = 1$, $c = -\varepsilon$, $f = 2\varepsilon + 1$, $g = 0$, $h = -2\varepsilon$,

$\quad q(x, y) = 2x + \varepsilon(8x + 5y - 3 - 2x^2)$, $u(0, y) = 2 - 3y + 3\varepsilon$,

$\quad u(x, 0) = x^2 + 2 + \varepsilon(2x^2 + 3)$.

36 $x > 0$, $y > 0$, $a = \varepsilon$, $b = 1$, $c = -\varepsilon$, $f = 0$, $g = 4 - 2\varepsilon$, $h = 2\varepsilon$,

$\quad q(x, y) = -2 - 8x + \varepsilon(2x^3 - 4xy + 14x - 8y + 3)$, $u(0, y) = 1 - \varepsilon y^2$,

$\quad u(x, 0) = x^3 + 1 + \varepsilon x$.

Answers to Odd-Numbered Exercises

1 $u(r) = (1/4)(3 + r^2) + (\varepsilon/16)(29 - 12r^2 - r^4) + O(\varepsilon^2)$.

3 $u(r) = -3 - (13\varepsilon/2)(r^2 - 1) + O(\varepsilon^2)$.

5 $u(x, t) = 2xt + \varepsilon x e^{-t} + O(\varepsilon^2)$.

7 $u(x, t) = x^2 + t + \varepsilon(x - t)e^{-3t} + O(\varepsilon^2)$.

9 $u(x, t) = x - 3t + 2\varepsilon t(x - 3t) + O(\varepsilon^2)$.

11 $u(x, t) = (2t - x)e^t - 2 + \varepsilon x t e^t + O(\varepsilon^2)$.

13 $u(x, t) = e^{x-2t} + t^2 - \varepsilon t e^{x-2t} + O(\varepsilon^2)$.

15 $u(x, y) = xy^2 + \varepsilon y(e^x - 2x) + O(\varepsilon^2)$.

17 $u(x, y) = y^2 e^{-x} + \varepsilon(x^2 - y) + O(\varepsilon^2)$.

19 $u(x, y) = y - 2x^2 + \varepsilon y \cos x + O(\varepsilon^2)$.

21 $u(x, y) = 3x + y + \varepsilon(1 - 2xy) + O(\varepsilon^2)$.

23 $u(x, y) = 1 + x - 2y + \varepsilon(x^2 - 3xy) + O(\varepsilon^2)$.

25 $u(x, y) = xy + 2x - 3 + \varepsilon(x^2 + x - y) + O(\varepsilon^2)$.

27 $u(x, y) = 2 - xy^2 + \varepsilon(x^2 + 4xy - 1) + O(\varepsilon^2)$.

29 $u(x, y) = x^2 - 2xy + \varepsilon(2 - y - xy) + O(\varepsilon^2)$.

31 $u(x, y) = 2x^2 + 3xy - y^2 + \varepsilon(1 - x^2 + 2xy) + O(\varepsilon^2)$.

33 $u(x, y) = (x + y)e^{2x} + \varepsilon(xy - y^2 + 1) + O(\varepsilon^2)$.

35 $u(x, y) = x^2 - 3y + 2 + \varepsilon(2x^2 - xy + 3) + O(\varepsilon^2)$.

13.2.2 Secular Terms

In certain situations, we need to make sure that the terms of the asymptotic approximation of a solution are correctly ordered by magnitude on the entire range of the independent variables.

13.18 Example. Consider the regular perturbation problem

$$u_t(x,t) + 3\varepsilon u_x(x,t) = 0, \quad -\infty < x < \infty, \ t > 0,$$
$$u(x,0) = \sin x, \quad -\infty < x < \infty,$$

where $0 < \varepsilon \ll 1$. Replacing $u(x,t) = u_0(x,t) + \varepsilon u_1(x,t) + O(\varepsilon^2)$ in the PDE and IC and following normal procedure, we find that u_0 and u_1 are, respectively, the solutions of the IVPs

$$(u_0)_t(x,t) = 0, \quad -\infty < x < \infty, \ t > 0,$$
$$u_0(x,0) = \sin x, \quad -\infty < x < \infty$$

and

$$(u_1)_t(x,t) + 3(u_0)_x(x,t) = 0, \quad -\infty < x < \infty, \ t > 0,$$
$$u_1(x,0) = 0, \quad -\infty < x < \infty.$$

From the first IVP, it follows that $u_0(x,t) = \sin x$, which, replaced in the second, leads to $u_1(x,t) = -3t \cos x$, so

$$u(x,t) = \sin x - 3\varepsilon t \cos x + O(\varepsilon^2).$$

This series is a good asymptotic approximation for the solution of the given IVP as $\varepsilon \to 0$ if t is restricted to any fixed finite interval $[0, t_0]$. But in the absence of such a restriction, we see that when $t = O(\varepsilon^{-1})$, the term $-3\varepsilon t \cos x$ is of the same order of magnitude as the leading $O(1)$-term. (The situation becomes worse as t increases further.) Such a term is called *secular*, and its presence means that the perturbation series is not valid for very large values of t. To extend the validity of the series for all $t > 0$, we use the power series expansions of $\sin \alpha$ and $\cos \alpha$ and write

$$\sin(x - 3\varepsilon t) = \sin x \cos(3\varepsilon t) - \cos x \sin(3\varepsilon t)$$
$$= \sin x [1 + O(\varepsilon^2)] - \cos x [3\varepsilon t + O(\varepsilon^3)]$$
$$= \sin x - 3\varepsilon t \cos x + O(\varepsilon^2).$$

Then the above asymptotic solution series can be rearranged in the form

$$u(x,t) = \sin(x - 3\varepsilon t) + O(\varepsilon^2),$$

which does not contain the secular term identified earlier. Higher-order terms in ε may, however, contain additional secular terms.

VERIFICATION WITH MATHEMATICA®. The input

```
u = Sin[x - 3 * ε * t];
f = Sin[x];
Simplify[Series[{D[u,t] + 3 * ε * D[u,x], (u /. t -> 0) - f}, {ε,0,1}]
    // Normal]
```

generates the output $\{0,0\}$.

13.19 Remark. According to the discussion at the beginning of Section 12.1, the exact solution of the IVP in Example 13.18 is, in fact,

$$u(x,t) = \sin(x - 3\varepsilon t).$$

The additional $O(\varepsilon^2)$-term represents 'noise' generated by the approximating nature of the asymptotic expansion technique.

13.20 Example. The first-order IVP

$$u_t(x,t) + u_x(x,t) - \varepsilon u(x,t) = 0, \quad -\infty < x < \infty, \ t > 0,$$
$$u(x,0) = \cos x, \quad -\infty < x < \infty,$$

where $0 < \varepsilon \ll 1$, is a regular perturbation problem because the reduced problem is also of first order and both problems have unique solutions. Writing $u(x,t) = u_0(x,t) + \varepsilon u_1(x,t) + O(\varepsilon^2)$, we arrive at the IVPs

$$(u_0)_t(x,t) + (u_0)_x(x,t) = 0, \quad -\infty < x < \infty, \ t > 0,$$
$$u_0(x,0) = \cos x, \quad -\infty < x < \infty$$

and

$$(u_1)_t(x,t) + (u_1)_x(x,t) = u_0(x,t), \quad -\infty < x < \infty, \ t > 0,$$
$$u_1(x,0) = 0, \quad -\infty < x < \infty.$$

Using the method of characteristics (see Section 12.1), we find that the solutions of these problems are, respectively,

$$u_0(x,t) = \cos(x - t), \quad u_1(x,t) = t\cos(x - t);$$

hence,

$$u(x,t) = \cos(x - t) + \varepsilon t\cos(x - t) + O(\varepsilon^2).$$

As explained in Example 13.18, $\varepsilon t\cos(x - t)$ is a secular term. Since the procedure that removed such a term in the preceding example does not work here, we try another technique, called the *method of multiple scales*, which consists in introducing an additional variable

$$\tau = \varepsilon t.$$

Denoting by $v(x, t, \tau)$ the new unknown function and remarking that v depends on t both directly and through τ, we arrive at the problem

$$v_t(x, t, \tau) + \varepsilon v_\tau(x, t, \tau) + v_x(x, t, \tau) - \varepsilon v(x, t, \tau) = 0,$$
$$- \infty < x < \infty, \quad t, \tau > 0,$$
$$v(x, 0, 0) = \cos x, \quad -\infty < x < \infty.$$

We now set $v(x, t, \tau) = v_0(x, t, \tau) + \varepsilon v_1(x, t, \tau) + O(\varepsilon^2)$ and deduce in the usual way that v_0 and v_1 are, respectively, the solutions of the IVPs

$$(v_0)_t(x, t, \tau) + (v_0)_x(x, t, \tau) = 0, \quad -\infty < x < \infty, \quad t, \tau > 0,$$
$$v_0(x, 0, 0) = \cos x$$

and

$$(v_1)_t(x, t, \tau) + (v_1)_x(x, t, \tau) = -(v_0)_\tau(x, t, \tau) + v_0(x, t, \tau),$$
$$- \infty < x < \infty, \quad t, \tau > 0,$$
$$v_1(x, 0, 0) = 0, \quad -\infty < x < \infty.$$

We seek the solution of the former as

$$v_0(x, t, \tau) = f(x, t)\varphi(\tau),$$

where φ is required to satisfy the condition

$$\varphi(0) = 1$$

but is otherwise arbitrary. Then the problem for v_0 reduces to the IVP

$$f_t(x, t) + f_x(x, t) = 0, \quad -\infty < x < \infty, \quad t > 0,$$
$$f(x, 0) = \cos x, \quad -\infty < x < \infty,$$

with solution $f(x, t) = \cos(x - t)$, so

$$v_0(x, t, \tau) = \varphi(\tau) \cos(x - t).$$

Replacing this on the right-hand side in the PDE for v_1, we arrive at the equation

$$(v_1)_t(x, t, \tau) + (v_1)_x(x, t, \tau) = -\big[\varphi'(\tau) - \varphi(\tau)\big] \cos(x - t),$$

whose solution satisfying the condition $v_1(x, 0, 0) = 0$, computed by the method of characteristics, is

$$v_1(x, t, \tau) = -\big[\varphi'(\tau) - \varphi(\tau)\big] t \cos(x - t).$$

Consequently, the solution of the modified problem is

$$v(x, t, \tau) = \varphi(\tau) \cos(x - t) - \varepsilon \big[\varphi'(\tau) - \varphi(\tau)\big] t \cos(x - t) + O(\varepsilon^2).$$

To eliminate the secular term on the right-hand side above, we now choose φ so that $\varphi' - \varphi = 0$. In view of the earlier condition $\varphi(0) = 1$, we find that $\varphi(\tau) = e^{\tau}$. Hence,

$$v(x, t, \tau) = e^{\tau} \cos(x - t) + O(\varepsilon^2),$$

which, since $\tau = \varepsilon t$, means that

$$u(x, t) = e^{\varepsilon t} \cos(x - t) + O(\varepsilon^2).$$

The terms of the asymptotic solution written in this form are now well ordered with respect to magnitude for all $t > 0$.

As in Remark 13.19, we easily convince ourselves that the exact solution of the original IVP is $u(x, t) = e^{\varepsilon t} \cos(x - t)$, which, in the above approximation, is accompanied by computational 'noise'.

VERIFICATION WITH MATHEMATICA®. The input

```
u = E^(ε * t) * Cos[x - t];
f = Cos[x];
Simplify[Series[{D[u,t] + D[u,x] - ε * u, (u /. t -> 0) - f}, {ε, 0, 1}]
    // Normal]
```

generates the output $\{0, 0\}$.

13.3 Singular Perturbation Problems

In such problems, we often need to construct different solutions that are valid in different regions. These solutions are then matched in an appropriate manner for overall uniform validity.

13.21 Example. Consider the IVP

$$\varepsilon\big[u_t(x, t) + 2u_x(x, t)\big] + u(x, t) = \sin t, \quad -\infty < x < \infty, \ t > 0,$$
$$u(x, 0) = x, \quad -\infty < x < \infty,$$

where $0 < \varepsilon \ll 1$. This is a singular perturbation problem since the reduced (unperturbed) problem contains no derivatives of u.

Following the standard procedure, we try to seek a solution of the form $u(x, t) \approx \sum\limits_{n=0}^{\infty} u_n(x, t)\varepsilon^n$. Replacing in the PDE, we obtain

$$u_0(x, t) = \sin t,$$
$$u_n(x, t) = -(u_{n-1})_t(x, t) - 2(u_{n-1})_x(x, t), \quad n = 1, 2, \ldots.$$

From the above recurrence relation, it is easily seen that

$$u_{2n} = (-1)^n \sin t, \quad u_{2n+1} = (-1)^{n+1} \cos t, \quad n \geq 0;$$

therefore,

$$u(x,t) = \left[\sum_{n=0}^{\infty}(-1)^n \varepsilon^{2n}\right] \sin t - \varepsilon\left[\sum_{n=0}^{\infty}(-1)^n \varepsilon^{2n}\right] \cos t$$

$$= \frac{1}{1+\varepsilon^2}(\sin t - \varepsilon \cos t),$$

where we have used the formula for the sum of an infinite geometric progression with ratio $-\varepsilon^2$. However, it is immediately obvious that this function does not satisfy the initial condition. Also, when $t \approx \varepsilon$, using the power series for $\cos \varepsilon$ and $\sin \varepsilon$, we have

$$\sin t - \varepsilon \cos t \approx \sin \varepsilon - \varepsilon \cos \varepsilon = \left(\varepsilon + O(\varepsilon^3)\right) - \varepsilon\left(1 + O(\varepsilon^2)\right) = O(\varepsilon^2);$$

in other words, for $t = O(\varepsilon)$ the two terms in $u(x,t)$ are of the same order, which is not allowed in an asymptotic series. Consequently, the series is not well ordered in the region where $t = O(\varepsilon)$, so it is not a valid representation of the solution in that region. This means that we must seek a different series in a *boundary layer* of width $O(\varepsilon)$ near the x-axis ($t = 0$). To construct the new series, we make the change of variables

$$\tau = \frac{t}{\varepsilon}, \quad u(x,t) = u(x,\varepsilon\tau) = u^i(x,\tau),$$

which amounts to a stretching of the time argument near $t = 0$. Under this transformation, the IVP becomes

$$u^i_\tau(x,\tau) + 2\varepsilon u^i_x(x,\tau) + u^i(x,\tau) = \sin \varepsilon\tau = \varepsilon\tau + O(\varepsilon^3),$$

$$\qquad\qquad\qquad -\infty < x < \infty, \ \tau > 0,$$

$$u^i(x,0) = x, \quad -\infty < x < \infty.$$

For the *inner solution* u^i of this boundary layer IVP, we assume an asymptotic expansion of the form

$$u^i(x,\tau) \approx \sum_{n=0}^{\infty} u^i_n(x,\tau)\varepsilon^n.$$

Then the new PDE and IC yield, in the usual way, the sequence of IVPs

$$(u^i_0)_\tau(x,\tau) + u^i_0(x,\tau) = 0, \quad -\infty < x < \infty, \ \tau > 0,$$

$$u^i_0(x,0) = x, \quad -\infty < x < \infty,$$

$$(u^i_1)_\tau(x,\tau) + u^i_1(x,\tau) = \tau - 2(u^i_0)_x(x,\tau),$$

$$\qquad\qquad\qquad -\infty < x < \infty, \ \tau > 0,$$

$$u^i_1(x,0) = 0, \quad -\infty < x < \infty,$$

and so on. Restricting our attention to the first two terms, from the first problem we easily obtain

$$u_0^i(x, \tau) = xe^{-\tau},$$

and from the second problem (seeking a particular integral of the form $a\tau e^{-\tau}$, $a = \text{const}$),

$$u_1^i(x, \tau) = \tau - 1 + (1 - 2\tau)e^{-\tau};$$

hence,

$$u^i(x, \tau) = xe^{-\tau} + \varepsilon\left[\tau - 1 + (1 - 2\tau)e^{-\tau}\right] + O(\varepsilon^2)$$
$$= t - \varepsilon + (x - 2t + \varepsilon)e^{-t/\varepsilon} + O(\varepsilon^2).$$

We rename the first solution the *outer solution* and denote it by u^o:

$$u^o(x, t) = \frac{1}{1 + \varepsilon^2}(\sin t - \varepsilon \cos t) = \sin t - \varepsilon \cos t + O(\varepsilon^2).$$

Now we need to match u^o (valid for $t = O(1)$) and u^i (valid for $t = O(\varepsilon)$) up to the order of ε considered (here, it is $O(\varepsilon)$), in some common region of validity. To this end, first we write the outer solution in terms of the inner variable $\tau = t/\varepsilon$ and expand it for τ fixed and ε small, listing the terms up to $O(\varepsilon)$, after which we revert to t:

$$(u^o)^i \approx \sin t - \varepsilon \cos t + \cdots = \sin(\varepsilon\tau) - \varepsilon \cos(\varepsilon\tau) + \cdots$$
$$= \varepsilon(\tau - 1) + \cdots = t - \varepsilon + \cdots.$$

Next, we write the inner solution in terms of the outer variable $t = \varepsilon\tau$ and expand it for t fixed and ε small to the same order:

$$(u^i)^o \approx xe^{-t/\varepsilon} + \varepsilon\left[\frac{t}{\varepsilon} - 1 + \left(1 - 2\frac{t}{\varepsilon}\right)e^{-t/\varepsilon}\right] + \cdots = t - \varepsilon + \cdots$$

(the rest of the terms are $o(\varepsilon^n)$ for any positive integer n—see Example 13.2(iii)). Finally, we impose the condition

$$(u^o)^i = (u^i)^o$$

up to $O(\varepsilon)$–terms. In our case this is already satisfied, so the two solutions match.

Since the common region of validity is not clear, it is useful to consider a *composite solution* of the form

$$u^c = u^o + u^i - (u^o)^i = u^o + u^i - (u^i)^o.$$

This is valid uniformly for $t > 0$ since

$$(u^c)^o = (u^o)^o + (u^i)^o - \left((u^i)^o\right)^o = u^o + (u^i)^o - (u^i)^o = u^o,$$
$$(u^c)^i = (u^o)^i + (u^i)^i - \left((u^o)^i\right)^i = (u^o)^i + u^i - (u^o)^i = u^i.$$

Here, we have

$$u(x, t) \approx u^c(x, t) = \sin t - \varepsilon \cos t + (x - 2t + \varepsilon)e^{-t/\varepsilon} + \cdots.$$

13.22 Example. Consider the elliptic problem in a semi-infinite strip

$$\varepsilon(\Delta u)(x, y) + u_x(x, y) + u_y(x, y) = 0, \quad x > 0, \ 0 < y < 1,$$

$$u(x, 0) = e^{-x}, \quad u(x, 1) = p(x), \quad x > 0,$$

$$u(0, y) = y, \quad u(x, y) \text{ bounded as } x \to \infty, \quad 0 < y < 1,$$

where $0 < \varepsilon \ll 1$ and

$$p(x) = \begin{cases} 1 - x, & x < 1, \\ 0, & x \geq 1. \end{cases}$$

This is a singular perturbation problem because the perturbed BVP is of second order, whereas the reduced BVP is of first order.

Let

$$u(x, y) \approx \sum_{n=0}^{\infty} u_n(x, y)\varepsilon^n.$$

Then it is readily seen that u_0 is the solution of the BVP

$$(u_0)_x(x, y) + (u_0)_y(x, y) = 0, \quad x > 0, \ 0 < y < 1,$$

$$u_0(x, 0) = e^{-x}, \quad u_0(x, 1) = p(x), \quad x > 0,$$

$$u_0(0, y) = y, \quad u_0(x, y) \text{ bounded as } x \to \infty, \quad 0 < y < 1,$$

u_1 is the solution of the BVP

$$(u_1)_x(x, y) + (u_1)_y(x, y) = -(\Delta u_0)(x, y), \quad x > 0, \ 0 < y < 1,$$

$$u_1(x, 0) = 0, \quad u_1(x, 1) = 0, \quad x > 0,$$

$$u_1(0, y) = 0, \quad u_1(x, y) \text{ bounded as } x \to \infty, \quad 0 < y < 1,$$

and so on.

We find u_0 by the method of characteristics (see Chapter 12), using $y = 1$ as the data line. If $x = x(y)$ is a characteristic curve, then on it

$$\frac{d}{dy} u_0(x(y), y) = (u_0)_x(x(y), y)x'(y) + (u_0)_y(x(y), y).$$

Hence, $x'(y) = 1$ implies that $du_0/dy = 0$, so

$$x = y + c, \quad u_0(x, y) = c', \quad c, c' = \text{const}.$$

The equation of the characteristic through the points (x, y) and $(x_0, 1)$ is

$$x = y + x_0 - 1,$$

and on this line,

$$u_0(x, y) = c' = u_0(x_0, 1) = p(x_0) = p(x - y + 1);$$

therefore, the solution of the given BVP is

$$u(x,y) = u_0(x,y) + O(\varepsilon) = p(x - y + 1) + O(\varepsilon)$$

$$= \begin{cases} y - x + O(\varepsilon), & x < y, \\ O(\varepsilon), & x \geq y. \end{cases} \tag{13.3}$$

Since it is clear that this solution does not satisfy the other BCs, we need to introduce two boundary layers.

In the boundary layer near $y = 0$, we make the substitution $\eta = y/\delta(\varepsilon)$, write $u(x,y) = u(x, \delta(\varepsilon)\eta) = v(x, \eta)$, and arrive at the new BVP

$$\varepsilon v_{xx} + \frac{\varepsilon}{\delta^2(\varepsilon)} v_{\eta\eta} + v_x + \frac{1}{\delta(\varepsilon)} v_\eta = 0, \quad v(x,0) = e^{-x}.$$

The unspecified boundary layer width $\delta(\varepsilon)$ must be chosen so that the $v_{\eta\eta}$ term is one of the dominant ones in the above equation. Comparing the coefficients of all the terms in the new PDE, we see that there are three possibilities:

$$\text{(i)} \quad \frac{\varepsilon}{\delta^2(\varepsilon)} \approx \varepsilon; \quad \text{(ii)} \quad \frac{\varepsilon}{\delta^2(\varepsilon)} \approx 1; \quad \text{(iii)} \quad \frac{\varepsilon}{\delta^2(\varepsilon)} \approx \frac{1}{\delta(\varepsilon)}.$$

It is not difficult to check that cases (i) and (ii) do not satisfy our magnitude requirement, whereas (iii) does. Hence, we can take $\delta(\varepsilon) = \varepsilon$, for simplicity, and the BVP becomes

$$\varepsilon^2 v_{xx} + v_{\eta\eta} + \varepsilon v_x + v_\eta = 0, \quad v(x,0) = e^{-x}.$$

Writing $v(x,\eta) = \sum_{n=0}^{\infty} v_n(x, \eta)\varepsilon^n$, from the PDE and BC satisfied by v we find that

$$(v_0)_{\eta\eta} + (v_0)_\eta = 0, \quad v_0(x,0) = e^{-x},$$

with solution

$$v_0(x,\eta) = \alpha(x) + \left[e^{-x} - \alpha(x)\right]e^{-\eta},$$

where α is an arbitrary function; therefore,

$$u(x,y) = v(x,\eta) = v_0(x,\eta) + O(\varepsilon)$$

$$= \alpha(x) + \left[e^{-x} - \alpha(x)\right]e^{-\eta} + O(\varepsilon). \tag{13.4}$$

We rename (13.3) the outer solution u^o and (13.4) the first inner solution u^{i_1}, and match them by the method used in Example 13.21. Thus, expanding p in powers of ε, we have

$$(u^o)^{i_1} = p(x - \varepsilon\eta + 1) + O(\varepsilon) = p(x + 1) + O(\varepsilon) = O(\varepsilon),$$

$$(u^{i_1})^o = \alpha(x) + \left[e^{-x} - \alpha(x)\right]e^{-y/\varepsilon} + O(\varepsilon) = \alpha(x) + O(\varepsilon), \tag{13.5}$$

so $(u^o)^{i_1} = (u^{i_1})^o$ to $O(1)$–terms if

$$\alpha(x) = 0, \quad x > 0. \tag{13.6}$$

The second boundary layer needs to be constructed near $x = 0$. Proceeding as above (this time with the transformed u_{xx} term as one of the dominant terms in the new PDE), we again conclude that the correct layer width is $\delta(\varepsilon) \approx \varepsilon$, so we set

$$\xi = \frac{x}{\varepsilon}.$$

Writing

$$u(x, y) = u(\varepsilon\xi, y) = w(\xi, y),$$

we arrive at the new BVP

$$w_{\xi\xi} + \varepsilon^2 w_{yy} + w_\xi + \varepsilon w_y = 0, \quad w(0, y) = y.$$

If we assume that $w(\xi, y) \approx \sum_{n=0}^{\infty} w_n(\xi, y)\varepsilon^n$, then

$$(w_0)_{\xi\xi} + (w_0)_\xi = 0, \quad w_0(0, y) = y,$$

with solution

$$w_0(\xi, y) = \beta(y) + \big[y - \beta(y)\big]e^{-\xi},$$

where β is another arbitrary function; hence,

$$u(x, y) = w(\xi, y) = w_0(\xi, y) + O(\varepsilon)$$
$$= \beta(y) + \big[y - \beta(y)\big]e^{-\xi} + O(\varepsilon). \tag{13.7}$$

We call (13.7) the second inner solution u^{i_2} and match it with u^o up to $O(1)$–terms. As above, we have

$$(u^o)^{i_2} = p(\varepsilon\xi - y + 1) + O(\varepsilon) = p(-y + 1) + O(\varepsilon) = y + O(\varepsilon),$$
$$(u^{i_2})^o = \beta(y) + \big[y - \beta(y)\big]e^{-x/\varepsilon} + O(\varepsilon) = \beta(y) + O(\varepsilon), \tag{13.8}$$

so

$$\beta(y) = y. \tag{13.9}$$

It can be verified that the composite solution in this case is

$$u^c = u^o + u^{i_1} + u^{i_2} - (u^{i_1})^o - (u^{i_2})^o = u^o + u^{i_1} + u^{i_2} - (u^o)^{i_1} - (u^o)^{i_2},$$

since $(u^{i_1})^{i_2} = (u^{i_1})^o$ and $(u^{i_2})^{i_1} = (u^{i_2})^o$. Thus, by (13.3)–(13.9), the asymptotic solution of the given BVP to $O(1)$–terms is

$$u(x, y) \approx u^c(x, y)$$
$$= p(x - y + 1) + \big[e^{-x} - p(x + 1)\big]e^{-y/\varepsilon}$$
$$\qquad + \big[y - p(1 - y)\big]e^{-x/\varepsilon} + O(\varepsilon)$$
$$= \begin{cases} y - x + e^{-x-y/\varepsilon} + O(\varepsilon), & x < y, \\ e^{-x-y/\varepsilon} + O(\varepsilon), & x \geq y. \end{cases}$$

13.23 Remarks.   (i) We did not consider effects caused by the incompatibility of the boundary values at the 'corner points' $(0,0)$ and $(0,1)$.

(ii) If we had tried to construct the first boundary layer solution near $y = 1$ instead of $y = 0$ (and, thus, use $y = 0$ as the data line for computing the first term in u°), we would have failed. In that region, we would need to substitute

$$\eta = (1 - y)/\varepsilon,$$

and setting

$$u(x, y) = u(x, 1 - \varepsilon\eta) = v(x, \eta),$$

we would obtain the BVP

$$\varepsilon^2 v_{xx} + v_{\eta\eta} + \varepsilon v_x - v_\eta = 0,$$

$$v(x, 0) = p(x),$$

from which

$$(v_0)_{\eta\eta} - (v_0)_\eta = 0, \quad v_0(x, 0) = p(x).$$

This would yield

$$v_0(x, \eta) = \alpha(x) + [p(x) - \alpha(x)]e^\eta,$$

which is not good for matching since, with y fixed,

$$e^\eta = e^{y/\varepsilon} \to \infty \quad \text{as } \varepsilon \to 0+.$$

Exercises

In **1–6**, use the method of matched asymptotic expansions to compute a formal composite solution (to $O(\varepsilon)$–terms) for the singular perturbation problem

$$\varepsilon u_{xx}(x, y) + a u_x(x, y) + b u_y(x, y) + c u(x, y) = 0,$$

$$0 < x < L, \quad -\infty < y < \infty,$$

$$u(0, y) = f(y), \quad u(L, y) = g(y), \quad -\infty < y < \infty,$$

where $0 < \varepsilon \ll 1$, with functions f, g, coefficients a, b, c, and number L as indicated. (The location of the boundary layer is specified in each case.)

1 $f(y) = y + 1$, $g(y) = 2y$, $a = 1 + 2\varepsilon$, $b = 2\varepsilon$, $c = 1$, $L = 1$;
    boundary layer at $x = 0$.

2 $f(y) = y$, $g(y) = 2y - 1$, $a = 1 - \varepsilon$, $b = 0$, $c = 1$, $L = 2$;
    boundary layer at $x = 0$.

3 $f(y) = 2y + 3$, $g(y) = -y$, $a = \varepsilon - 1$, $b = \varepsilon$, $c = 0$, $L = 2$;
    boundary layer at $x = 2$.

4 $f(y) = -3y$, $g(y) = y + 2$, $a = -1 - \varepsilon$, $b = 2\varepsilon$, $c = 1$, $L = 1$;
   boundary layer at $x = 1$.

5 $f(y) = y^2$, $g(y) = 1 - y$, $a = 1 - 2\varepsilon$, $b = -\varepsilon$, $c = -1$, $L = 1$;
   boundary layer at $x = 0$.

6 $f(y) = 2$, $g(y) = 2 - 3y$, $a = 1$, $b = -2\varepsilon$, $c = 1$, $L = 2$;
   boundary layer at $x = 0$.

In **7–12**, use the method of matched asymptotic expansions (in conjunction with the method of characteristics) to compute a formal composite solution (to $O(\varepsilon)$–terms) for the singular perturbation problem

$$\varepsilon\big[u_{xx}(x, y) + au_{yy}(x, y)\big] + bu_x(x, y) + cu_y(x, y) = 0,$$

$$x > 0, \ 0 < y < L,$$

$$u(x, 0) = f(x), \quad u(x, L) = g(x), \quad x > 0,$$
$$u(0, y) = h(y), \quad 0 < y < L,$$

where $0 < \varepsilon \ll 1$, with functions f, g, h, coefficients a, b, c, and number L as indicated. (The locations of the boundary layers are specified in each case.)

7 $f(x) = x^2 + 1$, $g(x) = 2x - 3$, $h(y) = y^2$,
   $a = 1$, $b = 1$, $c = 2$, $L = 1$; boundary layers at $x = 0$, $y = 0$.

8 $f(x) = x^2$, $g(x) = 1 - x$, $h(y) = y^2 - 1$,
   $a = 2$, $b = 1$, $c = 1$, $L = 2$; boundary layers at $x = 0$, $y = 0$.

9 $f(x) = 2x^2 - 1$, $g(x) = x + 2$, $h(y) = y^2 + 1$,
   $a = 2$, $b = 1$, $c = -1$, $L = 1$; boundary layers at $x = 0$, $y = 1$.

10 $f(x) = 1 - x^2$, $g(x) = 2x + 4$, $h(y) = 2y^2$,
   $a = 1$, $b = 2$, $c = -2$, $L = 2$; boundary layers at $x = 0$, $y = 2$.

11 $f(x) = x + 1$, $g(x) = x^2 - 2$, $h(y) = 1 - 2y$,
   $a = 4$, $b = 2$, $c = 1$, $L = 1$; boundary layers at $x = 0$, $y = 0$.

12 $f(x) = x^2 + x$, $g(x) = -3x$, $h(y) = y^2 - 2y$,
   $a = 3$, $b = 1$, $c = -1$, $L = 2$; boundary layers at $x = 0$, $y = 2$.

Answers to Odd-Numbered Exercises

1 $u^c(x, y) = 2ye^{1-x} + 2\varepsilon(xy - 2x - y + 2)e^{1-x}$

$$+ [1 - x + (1 - 2e)y - (1 - 2e)xy + 2\varepsilon e(y - 2)]e^{-x/\varepsilon} + O(\varepsilon^2).$$

3 $u^c(x, y) = 2y + 3 + 2\varepsilon x + (3xy + 3x - 9y - 9 - 4\varepsilon)e^{-(2-x)/\varepsilon} + O(\varepsilon^2)$.

5 $u^c(x, y) = (1 - y)e^{x-1} + \varepsilon y(1 - x)e^{x-1}$

$\qquad + (exy^2 + xy + ey^2 - x + y - 1 - \varepsilon y)e^{-1-x/\varepsilon} + O(\varepsilon^2)$.

7 $u^c(x, y) = 2x - y - 2 + (x^2 + xy - 2x - y + 3)e^{-2y/\varepsilon}$

$\qquad + (4xy + y^2 + 2x + y + 2)e^{-x/\varepsilon} + O(\varepsilon^2)$.

9 $u^c(x, y) = 2x^2 + 4xy + 2y^2 - 1 + 12\varepsilon y$

$\qquad + (-2x^2 + 4xy - 7x + 3y - 2 - 12\varepsilon)e^{-(1-y)/(2\varepsilon)}$

$\qquad + (2xy - y^2 + 2 - 12\varepsilon y)e^{-x/\varepsilon} + O(\varepsilon^2)$.

11 $u^c(x, y) = x^2 - 4xy + 4y^2 + 4x - 8y + 2 + 34\varepsilon(1 - y)$

$\qquad - (x^2 + 4xy + 3x + 6y + 1 + 34\varepsilon)e^{-y/(4\varepsilon)}$

$\qquad + [-4xy - 4y^2 + 3x + 6y - 1 + 34\varepsilon(y - 1)]e^{-2x/\varepsilon} + O(\varepsilon^2)$.

CHAPTER 14

Complex Variable Methods

Certain linear two-dimensional elliptic problems turn out to be difficult to solve in a Cartesian coordinate setup. In many such cases, it is advisable to go over to equivalent formulations in terms of complex variables, which may be able to help us find the solutions much more readily and elegantly. Although complex numbers have already been mentioned in Chapters 1, 3, 8, 9, and 11, we start by giving a brief presentation of their basic rules of manipulation and a few essential details about complex functions.

14.1 Elliptic Equations

A complex number is an expression of the form

$$z = x + iy, \quad x, y \text{ real}, \quad i^2 = -1,$$

where x and y are called, respectively, the *real part* and the *imaginary part* of z. The number

$$\bar{z} = x - iy$$

is the *complex conjugate* of z and

$$r = |z| = (z\bar{z})^{1/2} = (x^2 + y^2)^{1/2}$$

is the *modulus* of z.

By Euler's formula, a complex number can also be written in polar form as

$$z = re^{i\theta} = r(\cos\theta + i\sin\theta),$$

where θ, $-\pi < \theta \leq \pi$, called the *argument* of z, is determined from the equalities

$$\cos\theta = \frac{x}{r}, \quad \sin\theta = \frac{y}{r}.$$

Obviously,

$$\bar{z} = r(\cos\theta - i\sin\theta) = re^{-i\theta}.$$

Addition and multiplication of complex numbers are performed according to the usual algebraic rules for real numbers.

A complex function of a complex variable has the general form

$$f(z) = (\operatorname{Re} f)(x, y) + i(\operatorname{Im} f)(x, y),$$

where $\operatorname{Re} f$ and $\operatorname{Im} f$ are its real and imaginary parts. Such a function f is called *holomorphic* if its derivative $f'(z)$ exists at all points in its domain of definition. A holomorphic function is *analytic*—that is, it can be expanded in a convergent power series.

14.1 Example. If $f(z) = z^2$, then, since $(x + iy)^2 = x^2 + 2ixy - y^2$,

$$(\operatorname{Re} f)(x, y) = x^2 - y^2, \quad (\operatorname{Im} f)(x, y) = 2xy.$$

14.2 Theorem. *A function f is holomorphic if and only if it satisfies the Cauchy–Riemann relations*

$$(\operatorname{Re} f)_x = (\operatorname{Im} f)_y, \quad (\operatorname{Im} f)_x = -(\operatorname{Re} f)_y. \tag{14.1}$$

In this case, both $\operatorname{Re} f$ and $\operatorname{Im} f$ are solutions of the Laplace equation.

14.3 Remarks. (i) Suppose that a smooth function f of real variables x and y is expressed in terms of the complex variables z and $\bar{z}$; that is,

$$f(x, y) = g(z, \bar{z}).$$

From the chain rule of differentiation, it follows that

$$\begin{aligned}
f_x &= g_z + g_{\bar{z}}, \\
f_y &= i(g_z - g_{\bar{z}}), \\
f_{xx} &= g_{zz} + 2g_{z\bar{z}} + g_{\bar{z}\bar{z}}, \\
f_{yy} &= -g_{zz} + 2g_{z\bar{z}} - g_{\bar{z}\bar{z}}, \\
f_{xy} &= f_{yx} = i(g_{zz} - g_{\bar{z}\bar{z}}).
\end{aligned} \tag{14.2}$$

(ii) The Laplace equation, which is not easily integrated in terms of real variables, has a very simple solution in terms of complex variables. Thus, if $u(x, y) = v(z, \bar{z})$, then, by (14.2),

$$\Delta u = u_{xx} + u_{yy} = 4v_{z\bar{z}}, \tag{14.3}$$

so $\Delta u(x, y) = 0$ is equivalent to

$$v_{z\bar{z}}(z, \bar{z}) = 0.$$

It is now trivial to see that the latter has the general solution

$$v(z, \bar{z}) = \varphi(z) + \bar{\psi}(\bar{z}),$$

where φ and ψ are arbitrary analytic functions of z.

If we want the *real* general solution, then we must have $v(z, \bar{z}) = \bar{v}(\bar{z}, z)$; that is,

$$\varphi(z) + \bar{\psi}(\bar{z}) = \bar{\varphi}(\bar{z}) + \psi(z),$$

or

$$\varphi(z) - \bar{\varphi}(\bar{z}) = \psi(z) - \bar{\psi}(\bar{z}),$$

which means that $\operatorname{Im} \varphi = \operatorname{Im} \psi$. Using this equality and (14.1), we find that $\operatorname{Re} \varphi = \operatorname{Re} \psi$ as well, so $\psi = \varphi$; therefore, the real general solution of the two-dimensional Laplace equation is

$$v(z, \bar{z}) = \varphi(z) + \bar{\varphi}(\bar{z}), \tag{14.4}$$

where φ is an arbitrary analytic function.

(iii) A similar treatment can be applied to the biharmonic equation

$$\Delta \Delta u(x, y) = 0.$$

Since, as we have seen, $\Delta v = 4v_{z\bar{z}}$, where $v(z, \bar{z}) = u(x, y)$, we easily deduce that $\Delta \Delta v(x, y) = 16 v_{zz\bar{z}\bar{z}}(z, \bar{z})$; hence, the biharmonic equation is equivalent to

$$v_{zz\bar{z}\bar{z}}(z, \bar{z}) = 0.$$

Then, by (ii) above,

$$\Delta v(z, \bar{z}) = 4v_{z\bar{z}}(z, \bar{z}) = \Phi(z) + \bar{\Phi}(\bar{z}).$$

Integrating and applying the argument in (ii), we arrive at the real general solution

$$v(z, \bar{z}) = \bar{z}\Phi(z) + z\bar{\Phi}(\bar{z}) + \varphi(z) + \bar{\varphi}(\bar{z}),$$

where Φ and φ are arbitrary analytic functions of z.

14.4 Example. Consider the BVP

$$\Delta u = 8 \quad \text{in } D,$$

$$u = 1 - 2\cos\theta - \cos(2\theta) + 2\sin(2\theta) \quad \text{on } \partial D,$$

where D and ∂D are the unit circular disk and its boundary, respectively, and r, θ are polar coordinates with the pole at the center of the disk. On ∂D we have

$$z = e^{i\theta} = \sigma, \quad \bar{z} = e^{-i\theta} = \sigma^{-1},$$

$$\cos(n\theta) = \tfrac{1}{2}\left(e^{in\theta} + e^{-in\theta}\right) = \tfrac{1}{2}\left(\sigma^n + \sigma^{-n}\right), \quad n = 1, 2, \ldots, \tag{14.5}$$

$$\sin(n\theta) = -\tfrac{1}{2}i(e^{in\theta} - e^{-in\theta}) = -\tfrac{1}{2}i(\sigma^n - \sigma^{-n});$$

hence, using (14.3), we bring the given BVP to the equivalent form

$$v_{z\bar{z}} = 2 \quad \text{in } D,$$

$$v = -\left(\tfrac{1}{2} - i\right)\sigma^{-2} - \sigma^{-1} + 1 - \sigma - \left(\tfrac{1}{2} + i\right)\sigma^2 \quad \text{on } \partial D.$$

It is easily seen that $2z\bar{z}$ is a particular solution of the PDE, so, by (14.4), the general solution of the equation is

$$v(z, \bar{z}) = \varphi(z) + \bar{\varphi}(\bar{z}) + 2z\bar{z}. \tag{14.6}$$

Since the arbitrary function φ is analytic, it admits a series expansion of the form

$$\varphi(z) = \sum_{n=0}^{\infty} a_n z^n. \tag{14.7}$$

Replacing (14.7) in (14.6), and then v with $z = \sigma$ and $\bar{z} = \sigma^{-1}$ in the BC, and performing the usual comparison of coefficients, we find that

$$a_0 + \bar{a}_0 + 2 = 1, \quad a_1 = -1, \quad a_2 = -\left(\tfrac{1}{2} + i\right), \quad a_n = 0 \ (n \neq 0, 1, 2).$$

Therefore, $a_0 + \bar{a}_0 = -1$, and (14.6) and (14.7) yield the solution

$$v(z, \bar{z}) = a_0 + \bar{a}_0 + a_1 z + \bar{a}_1 \bar{z} + a_2 z^2 + \bar{a}_2 \bar{z}^2 + 2z\bar{z}$$
$$= -1 - z - \bar{z} - \left(\tfrac{1}{2} + i\right)z^2 - \left(\tfrac{1}{2} - i\right)\bar{z}^2 + 2z\bar{z}.$$

In Cartesian coordinates with the origin at the center of the disc, this becomes

$$u(x, y) = -1 - 2x + x^2 + 4xy + 3y^2.$$

VERIFICATION WITH MATHEMATICA®. The input

```
u = - 1 - 2 * x + x^2 + 4 * x * y + 3 * y^2;
Polaru = u /. {x -> r * Cos[θ] , y -> r * Sin[θ]};
f = 1 - 2 * Cos[θ] - Cos[2 * θ] + 2 * Sin[2 * θ];
Simplify[{D[u,x,x] + D[u,y,y] - 8, (Polaru /. r -> 1) - f}]
```

generates the output $\{0, 0\}$.

14.5 Example. To solve the elliptic BVP

$$u_{xx} - 2u_{xy} + 2u_{yy} = 4 \quad \text{in } D,$$
$$u = \tfrac{3}{2} - \tfrac{1}{2}\cos(2\theta) + \tfrac{3}{2}\sin(2\theta) \quad \text{on } \partial D,$$

where the notation is the same as in Example 14.4, we use a slightly altered procedure. First, by (14.2), we easily see that $v(z, \bar{z}) = u(x, y)$ is the solution of the problem

$$(1 + 2i)v_{zz} - 6v_{z\bar{z}} + (1 - 2i)v_{\bar{z}\bar{z}} = -4 \quad \text{in } D,$$
$$v = -\tfrac{1}{4}(1 - 3i)\sigma^{-2} + \tfrac{3}{2} - \tfrac{1}{4}(1 + 3i)\sigma^2 \quad \text{on } \partial D. \tag{14.8}$$

We now perform a simple transformation of the form

$$\zeta = z + \bar{\alpha}\bar{z}, \quad \bar{\zeta} = \bar{z} + \alpha z, \quad v(z, \bar{z}) = w(\zeta, \bar{\zeta}),$$

where α is a complex number to be chosen so that the left-hand side of the PDE for w consists only of the mixed second-order derivative. Thus, by the chain rule,

$$v_{zz} = w_{\zeta\zeta} + 2\alpha w_{\zeta\bar{\zeta}} + \alpha^2 w_{\bar{\zeta}\bar{\zeta}},$$
$$v_{z\bar{z}} = \bar{\alpha} w_{\zeta\zeta} + (1 + \alpha\bar{\alpha}) w_{\zeta\bar{\zeta}} + \alpha w_{\bar{\zeta}\bar{\zeta}},$$
$$v_{\bar{z}\bar{z}} = \bar{\alpha}^2 w_{\zeta\zeta} + 2\bar{\alpha} w_{\zeta\bar{\zeta}} + w_{\bar{\zeta}\bar{\zeta}}.$$

When we replace this in the PDE in (14.8), we see that the coefficients of both $w_{\zeta\zeta}$ and $w_{\bar{\zeta}\bar{\zeta}}$ vanish if α is a root of the quadratic equation

$$(1 + 2i)\alpha^2 - 6\alpha + 1 - 2i = 0;$$

that is, if

$$\alpha = 1 - 2i \quad \text{or} \quad \alpha = \tfrac{1}{5}(1 - 2i).$$

Choosing, say, the first root, we have the transformation

$$\zeta = z + (1 + 2i)\bar{z}, \quad \bar{\zeta} = \bar{z} + (1 - 2i)z, \tag{14.9}$$

which leads to the equation

$$w_{\zeta\bar{\zeta}} = \tfrac{1}{4}$$

with general solution

$$w(\zeta, \bar{\zeta}) = \varphi(\zeta) + \bar{\varphi}(\bar{\zeta}) + \tfrac{1}{4}\zeta\bar{\zeta}, \tag{14.10}$$

where $\varphi(\zeta)$ is an arbitrary analytic function. As in (14.7), and taking (14.9) into account, we now write

$$\varphi(\zeta) = \sum_{n=0}^{\infty} a_n \zeta^n = \sum_{n=0}^{\infty} a_n \big(z + (1 + 2i)\bar{z}\big)^n,$$

so, by (14.9) and (14.10),

$$v(z, \bar{z}) = \sum_{n=0}^{\infty} \Big[a_n \big(z + (1 + 2i)\bar{z}\big)^n + \bar{a}_n \big(\bar{z} + (1 - 2i)z\big)^n\Big]$$
$$+ \tfrac{1}{4}\big(z + (1 + 2i)\bar{z}\big)\big(\bar{z} + (1 - 2i)z\big). \tag{14.11}$$

On the boundary ∂D, (14.11) and the BC in (14.8) give rise to the equality

$$\sum_{n=0}^{\infty} \Big[a_n \big(\sigma + (1 + 2i)\sigma^{-1}\big)^n + \bar{a}_n \big(\sigma^{-1} + (1 - 2i)\sigma\big)^n\Big]$$
$$+ \tfrac{1}{4}\Big[(1 + 2i)\sigma^{-2} + 6 + (1 - 2i)\sigma^2\Big]$$
$$= -\tfrac{1}{4}(1 - 3i)\sigma^{-2} + \tfrac{3}{2} - \tfrac{1}{4}(1 + 3i)\sigma^2.$$

Expanding the binomials on the left-hand side and equating the coefficients of each power of σ on both sides, we immediately note that

$$a_n = 0, \quad n = 3, 4, \ldots,$$

and that, in this case,

$$a_0 + \bar{a}_0 + 2(1 + 2i)a_2 + 2(1 - 2i)\bar{a}_2 + \tfrac{3}{2} = \tfrac{3}{2},$$

$$a_1 + (1 - 2i)\bar{a}_1 = 0,$$

$$a_2 - (3 + 4i)\bar{a}_2 + \tfrac{1}{4}(1 - 2i) = -\tfrac{1}{4}(1 + 3i).$$

The second and third equalities, taken together with their conjugates, yield the systems

$$a_1 + (1 - 2i)\bar{a}_1 = 0,$$

$$(1 + 2i)a_1 + \bar{a}_1 = 0,$$

and

$$a_2 - (3 + 4i)\bar{a}_2 = -\tfrac{1}{4}(2 + i),$$

$$(-3 + 4i)a_2 + \bar{a}_2 = -\tfrac{1}{4}(2 - i),$$

from which

$$a_1 = 0, \quad a_2 = \tfrac{1}{16}(2 + i).$$

Replacing a_2 in the first equality, we find that

$$a_0 + \bar{a}_0 = 0.$$

Hence, by (14.11),

$$v(z, \bar{z}) = a_0 + \bar{a}_0 + a_1\big(z + (1 + 2i)\bar{z}\big) + \bar{a}_1\big(\bar{z} + (1 - 2i)z\big)$$

$$+ a_2\big(z + (1 + 2i)\bar{z}\big)^2 + \bar{a}_2\big(\bar{z} + (1 - 2i)z\big)^2$$

$$= -\tfrac{1}{4}(1 + 3i)z^2 + \tfrac{3}{2}z\bar{z} - \tfrac{1}{4}(1 - 3i)\bar{z}^2,$$

or, in Cartesian coordinates,

$$u(x, y) = x^2 + 3xy + 2y^2.$$

VERIFICATION WITH MATHEMATICA®. The input

```
u = x^2 + 3 * x * y + 2 * y^2;
Polaru = u /. {x -> r * Cos[θ], y -> r * Sin[θ]};
f = 3/2 - (1/2) * Cos[2 * θ] + (3/2) * Sin[2 * θ];
Simplify[{D[u,x,x] - 2 * D[u,x,y] + 2 * D[u,y,y] - 4,
    (Polaru /. r -> 1) - f}]
```

generates the output $\{0, 0\}$.

Exercises

In **1–6**, find the solution $u(x, y)$ of the BVP

$$\Delta u = q \quad \text{in } D,$$
$$u = f \quad \text{on } \partial D,$$

where D and ∂D are the disc with the center at the origin and radius 1 and its circular boundary, respectively, θ is the polar angle mentioned earlier in this chapter, and the number q and function f are as indicated.

1 $q = -6$, $f(\theta) = (1/2)[1 + 5\cos(2\theta)]$.

2 $q = 8$, $f(\theta) = 2 + 2\cos\theta - 3\cos(2\theta)$.

3 $q = -2$, $f(\theta) = (1/2)[-5 + 2\sin\theta - 3\cos(2\theta) + 3\sin(2\theta)]$.

4 $q = 2$, $f(\theta) = (1/2)[1 + 6\sin\theta - 5\cos(2\theta) + \sin(2\theta)]$.

5 $q = 4$, $f(\theta) = 1 + 2\cos\theta - 3\sin\theta + 2\cos(2\theta) - 2\sin(2\theta)$.

6 $q = -4$, $f(\theta) = 2 + 2\cos\theta - \sin\theta + 2\cos(2\theta) - 4\sin(2\theta)$.

In **7–14**, find the solution $u(x, y)$ of the BVP

$$u_{xx} + a u_{xy} + b u_{yy} = q \quad \text{in } D,$$
$$u = f \quad \text{on } \partial D,$$

where D and ∂D are the disc with the center at the origin and radius 1 and its circular boundary, respectively, θ is the polar angle with the pole at the center of the disc, and the numbers a, b, q and function f are as indicated.

7 $a = -4$, $b = 5$, $q = 18$, $f(\theta) = (1/2)[3 + \cos(2\theta) - \sin(2\theta)]$.

8 $a = -2$, $b = 5$, $q = -8$, $f(\theta) = 2\cos\theta + 2\cos(2\theta) + \sin(2\theta)$.

9 $a = 2$, $b = 2$, $q = 2$, $f(\theta) = (1/2)[-1 - 2\sin\theta + 3\cos(2\theta) + 4\sin(2\theta)]$.

10 $a = 4$, $b = 5$, $q = -28$,
$f(\theta) = 2 + \cos\theta - 2\sin\theta + 3\cos(2\theta) + \sin(2\theta)$.

11 $a = -1$, $b = 5/4$, $q = 25/2$,
$f(\theta) = (1/2)[-4 + 4\cos\theta - 2\cos(2\theta) - 3\sin(2\theta)]$.

12 $a = -4$, $b = 17/4$, $q = -67/2$,
$f(\theta) = (1/2)[-5 + 2\cos\theta + 5\cos(2\theta) + 3\sin(2\theta)]$.

13 $a = -2$, $b = 10$, $q = 66$,
$f(\theta) = 3 + \cos\theta - 2\sin\theta - \cos(2\theta) - \sin(2\theta)$.

14 $a = -4$, $b = 13$, $q = -4$,
$f(\theta) = 1 + 2\cos\theta - 3\sin\theta + 2\cos(2\theta) - 2\sin(2\theta)$.

Answers to Odd-Numbered Exercises

1 $u(x,y) = x^2 - 4y^2 + 2$.

3 $u(x,y) = -2x^2 + 3xy + y^2 + y - 2$.

5 $u(x,y) = 3x^2 - 4xy - y^2 + 2x - 3y$.

7 $u(x,y) = 2x^2 - xy + y^2$.

9 $u(x,y) = x^2 + 4xy - 2y^2 - y$.

11 $u(x,y) = x^2 - 3xy + 3y^2 + 2x - 4$.

13 $u(x,y) = x^2 - 2xy + 3y^2 + x - 2y + 1$.

14.2 Systems of Equations

As an illustration of the efficiency of the complex variable method in solving linear two-dimensional systems of partial differential equations, we consider the mathematical model of plane deformation of an elastic body. This state is characterized by a two-component displacement vector $u = (u_1, u_2)$ defined in the two-dimensional domain D occupied by the body. In the absence of body forces, u satisfies the system of PDEs

$$(\lambda + \mu)\big[(u_1)_{xx} + (u_2)_{xy}\big] + \mu \Delta u_1 = 0,$$
$$(\lambda + \mu)\big[(u_1)_{xy} + (u_2)_{yy}\big] + \mu \Delta u_2 = 0,$$

(14.12)

where λ and μ are physical constants and Δ is the Laplacian. We assume that D is finite, simply connected (roughly, this means that D has no 'holes'), and bounded by a simple, smooth, closed contour ∂D, and consider the Dirichlet problem for (14.12); that is, the BVP with the displacement components prescribed on the boundary:

$$u_1\big|_{\partial D} = f_1, \quad u_2\big|_{\partial D} = f_2,$$

where f_1 and f_2 are known functions. Our aim is to find u at every point (x, y) in D.

14.6 Example. Using the same notation as in Example 14.4, consider the system

$$2\big[(u_1)_{xx} + (u_2)_{xy}\big] + \Delta u_1 = 0, \qquad \text{in } D,$$
$$2\big[(u_1)_{xy} + (u_2)_{yy}\big] + \Delta u_2 = 0$$

(14.13)

with the BCs

$$u_1\big|_{\partial D} = \tfrac{1}{2}\sin(2\theta), \quad u_2\big|_{\partial D} = \cos\theta.$$

(14.14)

To treat this BVP problem in terms of complex variables, we define the complex displacement

$$U = u_1 + iu_2$$

and see that

$$(u_1)_x + (u_2)_y = u_{1,z} + u_{1,\bar{z}} + iu_{2,z} - iu_{2,\bar{z}}$$
$$= (u_1 + iu_2)_z + (u_1 - iu_2)_{\bar{z}} = U_z + \bar{U}_{\bar{z}}. \tag{14.15}$$

Next, we rewrite (14.13) in the form

$$2[(u_1)_x + (u_2)_y]_x + \Delta u_1 = 0,$$
$$2[(u_1)_x + (u_2)_y]_y + \Delta u_2 = 0.$$

Multiplying the second equation above by i, adding it to the first one, using (14.15), (14.3), and the fact that, by (14.2),

$$\partial_x + i\partial_y = 2\partial_{\bar{z}},$$

we have

$$0 = 2(\partial_x + i\partial_y)[(u_1)_x + (u_2)_y] + \Delta(u_1 + iu_2)$$
$$= 4\partial_{\bar{z}}(U_z + \bar{U}_{\bar{z}}) + \Delta U = 4[(U_z + \bar{U}_{\bar{z}})_{\bar{z}} + U_{z\bar{z}}],$$

or

$$(2U_z + \bar{U}_{\bar{z}})_{\bar{z}} = 0,$$

with general solution

$$2U_z + \bar{U}_{\bar{z}} = \tfrac{1}{2}\alpha'(z),$$

where α is an arbitrary analytic function of z and $\alpha' = d\alpha/dz$. The algebraic system formed by this equation and its conjugate now yields

$$U_z = \tfrac{1}{3}\alpha'(z) - \tfrac{1}{6}\bar{\alpha}'(\bar{z}); \tag{14.16}$$

therefore, by integration,

$$U(z,\bar{z}) = \tfrac{1}{3}\alpha(z) - \tfrac{1}{6}z\bar{\alpha}'(\bar{z}) + \bar{\beta}(\bar{z}), \tag{14.17}$$

where β is another arbitrary analytic function of z.

To investigate the arbitrariness of α and β, let p and q be functions of z such that $\alpha+p$ and $\beta+q$ generate the same displacement U as α and β. Then, by (14.16),

$$\tfrac{1}{3}[\alpha'(z) + p'(z)] - \tfrac{1}{6}[\bar{\alpha}'(\bar{z}) + \bar{p}'(\bar{z})] = \tfrac{1}{3}\alpha'(z) - \tfrac{1}{6}\bar{\alpha}'(\bar{z}),$$

from which $2p'(z) - \bar{p}'(\bar{z}) = 0$. This equation together with its conjugate yield $p'(z) = 0$, so $p(z) = c$, where c is a complex number. Using (14.17), we now see that

$$\tfrac{1}{3}\alpha(z) + \tfrac{1}{3}c - \tfrac{1}{6}z\bar{\alpha}'(\bar{z}) + \bar{\beta}(\bar{z}) + \bar{q}(\bar{z}) = \tfrac{1}{3}\alpha(z) - \tfrac{1}{6}z\bar{\alpha}'(\bar{z}) + \bar{\beta}(\bar{z}),$$

so $q(z) = -\bar{c}/3$. The arbitrariness produced by the complex constant c in the functions α and β can be eliminated by imposing an additional condition. For example, since the origin is in D, we may ask that

$$\alpha(0) = 0. \tag{14.18}$$

As in the preceding section, let σ be a generic point on the circle ∂D. Then, by (14.5) and the fact that $\bar{\sigma} = \sigma^{-1}$, the BC (14.14) can be written as

$$U\big|_{\partial D} = (u_1 + iu_2)\big|_{\partial D} = \tfrac{1}{4} i(\sigma^{-2} + 2\sigma^{-1} + 2\sigma - \sigma^2);$$

hence, in view of (14.17), we have

$$\tfrac{1}{3}\alpha(\sigma) - \tfrac{1}{6}\sigma\bar{\alpha}'(\sigma^{-1}) + \bar{\beta}(\sigma^{-1}) = \tfrac{1}{4} i(\sigma^{-2} + 2\sigma^{-1} + 2\sigma - \sigma^2). \tag{14.19}$$

Since D is finite and simply connected, we can consider series expansions of the analytic functions α and β, of the form

$$\alpha(z) = \sum_{n=0}^{\infty} a_n z^n, \quad \beta(z) = \sum_{n=0}^{\infty} b_n z^n.$$

Substituting these series in (14.19), we arrive at

$$\sum_{n=0}^{\infty} \left(\tfrac{1}{3} a_n \sigma^n - \tfrac{1}{6} n\bar{a}_n \sigma^{-n+2} + \bar{b}_n \sigma^{-n}\right) = \tfrac{1}{4} i(\sigma^{-2} + 2\sigma^{-1} + 2\sigma - \sigma^2).$$

The next step consists in equating the coefficients of each power of σ on both sides, and it is clear that $a_n = b_n = 0$ for all $n = 3, 4, \ldots$. Thus, since (14.18) implies that $a_0 = 0$, the only nonzero coefficients are given by the equalities

$$\bar{b}_2 = \tfrac{1}{4} i, \quad \bar{b}_1 = \tfrac{1}{2} i, \quad \bar{b}_0 - \tfrac{1}{3}\bar{a}_2 = 0, \quad \tfrac{1}{3} a_1 - \tfrac{1}{6}\bar{a}_1 = \tfrac{1}{2} i, \quad \tfrac{1}{3} a_2 = -\tfrac{1}{4} i.$$

The coefficient a_1 is computed by combining the equation that it satisfies with its complex conjugate form. In the end, we obtain

$$a_1 = i, \quad a_2 = -\tfrac{3}{4} i, \quad b_0 = -\tfrac{1}{4} i, \quad b_1 = -\tfrac{1}{2} i, \quad b_2 = -\tfrac{1}{4} i,$$

which generate the functions

$$\alpha(z) = iz - \tfrac{3}{4} iz^2, \quad \beta(z) = -\tfrac{1}{4} i - \tfrac{1}{2} iz - \tfrac{1}{4} iz^2$$

and, by (14.17), the complex displacement

$$U(z, \bar{z}) = \tfrac{1}{4} i(1 + 2z + 2\bar{z} - z^2 - z\bar{z} + \bar{z}^2).$$

Hence, in terms of Cartesian coordinates, the solution of our BVP is

$$u_1(x, y) = \mathrm{Re}\left(U(z, \bar{z})\right) = xy,$$
$$u_2(x, y) = \mathrm{Im}\left(U(z, \bar{z})\right) = \tfrac{1}{4}\left(1 + 4x - x^2 - y^2\right).$$

VERIFICATION WITH MATHEMATICA®. The input

```
{u1, u2} = {x*y, (1/4)*(1+4*x-x^2-y^2)};
{Polaru1, Polaru2} = {u1, u2} /. {x->r*Cos[θ], y->r*Sin[θ]};
{f1, f2} = {(1/2)*Sin[2*θ], Cos[θ]};
Simplify[{{2*(D[u1,x,x]+D[u2,x,y])+D[u1,x,x]+D[u1,y,y],
    2*(D[u1,x,y]+D[u2,y,y])+D[u2,x,x]+D[u2,y,y]},
    ({Polaru1, Polaru2}/.r->1)-{f1, f2}}]
```

generates the output $\{\{0,0\}, \{0,0\}\}$.

14.7 Remark. If the functions f_1 and f_2 prescribed on ∂D are not finite sums of integral powers of σ, they need to be expanded in full Fourier series. Using an argument similar to that in Section 8.1, we can write such a series in the form (see Chapter 8)

$$f(\theta) = \sum_{n=-\infty}^{\infty} c_n e^{-in\theta} = \sum_{n=-\infty}^{\infty} c_n \sigma^{-n},$$

where

$$c_n = \frac{1}{2\pi} \int_{-\pi}^{\pi} f(\theta) e^{in\theta}\, d\theta.$$

Exercises

In **1–10**, find the solution $(u_1(x,y), u_2(x,y))$ of the BVP

$$(\lambda + \mu)[(u_1)_{xx} + (u_2)_{xy}] + \mu \Delta u_1 = 0,$$
$$(\lambda + \mu)[(u_1)_{xy} + (u_2)_{yy}] + \mu \Delta u_2 = 0 \quad \text{in } D,$$
$$u_1 = f_1, \quad u_2 = f_2 \quad \text{on } \partial D,$$

where D and ∂D are the disc with the center at the origin and radius 1 and its circular boundary, respectively, Δ is the Laplacian, θ is the polar angle with the pole at the center of the disc, and the numbers λ, μ and functions f_1, f_2 are as indicated.

1 $\lambda = 1$, $\mu = 1/2$, $f_1(\theta) = 2\cos\theta - \cos(2\theta)$, $f_2(\theta) = \cos\theta + \sin(2\theta)$.

2 $\lambda = 1$, $\mu = 1$, $f_1(\theta) = \sin\theta + \cos(2\theta)$, $f_2(\theta) = 1 + 2\sin(2\theta)$.

3 $\lambda = 2$, $\mu = 1$, $f_1(\theta) = 2 - \cos\theta + 2\sin(2\theta)$, $f_2(\theta) = 1 - \cos(2\theta)$.

4 $\lambda = -1$, $\mu = 2$,

$f_1(\theta) = \cos\theta + 2\sin(2\theta)$, $f_2(\theta) = 1 + 2\cos(2\theta) + \sin(2\theta)$.

5 $\lambda = -2, \ \mu = 3,$

$f_1(\theta) = 1 + 2\cos(2\theta) - \sin(2\theta), \quad f_2(\theta) = \cos\theta - 2\sin(2\theta).$

6 $\lambda = 1, \ \mu = 2,$

$f_1(\theta) = 1 - \sin\theta + \cos(2\theta), \quad f_2(\theta) = \cos(2\theta) - 2\sin(2\theta).$

7 $\lambda = 1/2, \ \mu = 1,$

$f_1(\theta) = 2 + \sin\theta + \sin(2\theta), \quad f_2(\theta) = 3\cos(2\theta) - \sin(2\theta).$

8 $\lambda = -1/2, \ \mu = 1,$

$f_1(\theta) = -\cos\theta + 2\cos(2\theta), \quad f_2(\theta) = 2 - \cos(2\theta) - \sin(2\theta).$

9 $\lambda = 3, \ \mu = 1,$

$f_1(\theta) = \cos(2\theta) - \sin(2\theta), \quad f_2(\theta) = 1 + \sin\theta - 2\cos(2\theta).$

10 $\lambda = -1/2, \ \mu = 2,$

$f_1(\theta) = \cos\theta - 2\sin(2\theta), \quad f_2(\theta) = \cos(2\theta) + 3\sin(2\theta).$

Answers to Odd-Numbered Exercises

1 $u_1(x, y) = -x^2 + y^2 + 2x, \quad u_2(x, y) = 2xy + x.$

3 $u_1(x, y) = 4xy - x + 2, \quad u_2(x, y) = (1/5)(-14x^2 - 4y^2 + 14).$

5 $u_1(x, y) = 2x^2 - 2xy - 2y^2 + 1, \quad u_2(x, y) = (1/7)(x^2 - 28xy + y^2 + 7x - 1).$

7 $u_1(x, y) = (1/7)(3x^2 + 14xy + 3y^2 + 7y + 11),$

$u_2(x, y) = (1/7)(27x^2 - 14xy - 15y^2 - 6).$

9 $u_1(x, y) = (1/3)(x^2 - 6xy - 5y^2 + 2), \quad u_2(x, y) = (1/3)(-8x^2 + 4y^2 + 3y + 5).$

Appendix

A.1   Useful Integrals

For all $m, n = 1, 2, \ldots$,

$$\int_0^L \cos \frac{n\pi x}{L}\, dx = 0; \quad \int_{-L}^L \sin \frac{n\pi x}{L}\, dx = 0;$$

$$\int_0^L \sin \frac{n\pi x}{L} \sin \frac{m\pi x}{L}\, dx = \begin{cases} 0, & n \neq m, \\ L/2, & n = m; \end{cases}$$

$$\int_0^L \cos \frac{n\pi x}{L} \cos \frac{m\pi x}{L}\, dx = \begin{cases} 0, & n \neq m, \\ L/2, & n = m; \end{cases}$$

$$\int_{-L}^L \sin \frac{n\pi x}{L} \cos \frac{m\pi x}{L}\, dx = 0;$$

$$\int_0^L \sin \frac{(2n-1)\pi x}{2L} \sin \frac{(2m-1)\pi x}{2L}\, dx = \begin{cases} 0, & n \neq m, \\ L/2, & n = m; \end{cases}$$

$$\int_0^L \cos \frac{(2n-1)\pi x}{2L} \cos \frac{(2m-1)\pi x}{2L}\, dx = \begin{cases} 0, & n \neq m \\ L/2, & n = m. \end{cases}$$

For any real numbers a, b, c, and p,

$$\int (bx + c) \cos(px)\, dx = \frac{b}{p^2} \cos(px) + \frac{1}{p} (bx + c) \sin(px) + \text{const};$$

$$\int (bx + c) \sin(px)\, dx = -\frac{1}{p} (bx + c) \cos(px) + \frac{b}{p^2} \sin(px) + \text{const};$$

$$\int e^{ax} \cos(px)\, dx = \frac{e^{ax}}{a^2 + p^2} \left[a \cos(px) + p \sin(px) \right] + \text{const};$$

$$\int e^{ax} \sin(px)\, dx = \frac{e^{ax}}{a^2 + p^2} \left[-p \cos(px) + a \sin(px) \right] + \text{const}.$$

A.2   Eigenvalue–Eigenfunction Pairs

1. For the PDE

$$u_t(x,t) = ku_{xx}(x,t), \quad 0 < x < L, \; t > 0, \quad k = \text{const:}$$

BCs	Associated eigenvalues and eigenfunctions
$u(0,t) = 0,$ $u(L,t) = 0$	$\lambda_n = \dfrac{n^2\pi^2}{L^2}, \quad X_n(x) = \sin\dfrac{n\pi x}{L}, \quad n = 1, 2, \ldots$
$u_x(0,t) = 0,$ $u_x(L,t) = 0$	$\lambda_n = \dfrac{n^2\pi^2}{L^2}, \quad X_n(x) = \cos\dfrac{n\pi x}{L}, \quad n = 0, 1, 2, \ldots$
$u(0,t) = 0,$ $u_x(L,t) = 0$	$\lambda_n = \dfrac{(2n-1)^2\pi^2}{4L^2}, \quad X_n(x) = \sin\dfrac{(2n-1)\pi x}{2L}, \quad n = 1, 2, \ldots$
$u_x(0,t) = 0,$ $u(L,t) = 0$	$\lambda_n = \dfrac{(2n-1)^2\pi^2}{4L^2}, \quad X_n(x) = \cos\dfrac{(2n-1)\pi x}{2L}, \quad n = 1, 2, \ldots$

2. For the PDE

$$u_{tt}(x,t) = c^2 u_{xx}(x,t), \quad 0 < x < L, \; t > 0, \quad c = \text{const}$$

with any of the BCs in the table above, the pairs are the same as in **1**.

3. For the PDE

$$u_{xx}(x,y) + u_{yy}(x,y) = 0, \quad 0 < x < L, \; 0 < y < K$$

with any of the BCs in the table above given at $x = 0$ and $x = L$, the pairs are the same as in **1**. If the BCs are given at $y = 0$ and $y = K$, the pairs are the same as in **1** with x replaced by y and L replaced by K.

4. The pairs for the BVP

$$\alpha u_{tt}(x,t) + \beta u_t(x,t) = u_{xx}(x,t) + au_x(x,t) + bu(x,t),$$
$$0 < x < L, \; t > 0, \; \alpha, \beta, a, b = \text{const},$$
$$u(0,t) = 0, \quad u(L,t) = 0$$

with α and β not both zero, are

$$\lambda_n = \frac{n^2\pi^2}{L^2} + \tfrac{1}{4}a^2 - b, \quad X_n(x) = e^{-(a/2)x}\sin\frac{n\pi x}{L}, \quad n = 1, 2, \ldots.$$

5. The pairs for the PDE

$$u_{xx}(x,y) + u_{yy}(x,y) + Au_x(x,y) + Bu_y(x,y) + Cu(x,y) = 0,$$
$$0 < x < L, \; 0 < y < K, \; A, B, C = \text{const}$$

with $u(0,y) = 0$ and $u(L,y) = 0$ or $u(x,0) = 0$ and $u(x,K) = 0$, are determined by analogy to **4** in terms of the appropriate variable.

A.3   Table of Fourier Transforms

$f(x) = \mathcal{F}^{-1}[F](x)$	$F(\omega) = \mathcal{F}[f](\omega)$
1 $f'(x)$	$-i\omega F(\omega)$
2 $f''(x)$	$-\omega^2 F(\omega)$
3 $f(ax + b)$ $(a > 0)$	$\dfrac{1}{a} e^{-i(b/a)\omega} F(\omega/a)$
4 $(f * g)(x)$	$F(\omega)G(\omega)$
5 $\delta(x)$	$\dfrac{1}{\sqrt{2\pi}}$
6 $e^{iax} f(x)$	$F(\omega + a)$
7 $e^{-a^2 x^2}$ $(a > 0)$	$\dfrac{1}{\sqrt{2}\,a} e^{-\omega^2/(4a^2)}$
8 $xe^{-a^2 x^2}$ $(a > 0)$	$\dfrac{i}{2\sqrt{2}\,a^3} \omega e^{-\omega^2/(4a^2)}$
9 $x^2 e^{-a^2 x^2}$ $(a > 0)$	$\dfrac{1}{4\sqrt{2}\,a^5} (2a^2 - \omega^2) e^{-\omega^2/(4a^2)}$
10 $\dfrac{1}{x^2 + a^2}$ $(a > 0)$	$\sqrt{\dfrac{\pi}{2}}\dfrac{1}{a} e^{-a|\omega|}$
11 $\dfrac{x}{x^2 + a^2}$ $(a > 0)$	$-i\sqrt{\dfrac{\pi}{2}}\dfrac{1}{2a} \omega e^{-a|\omega|}$
12 $H(a - |x|) = \begin{cases} 1, & |x| \le a \\ 0, & |x| > a \end{cases}$	$\sqrt{\dfrac{2}{\pi}}\dfrac{\sin(a\omega)}{\omega}$
13 $xH(a - |x|) = \begin{cases} x, & |x| \le a \\ 0, & |x| > a \end{cases}$	$i\sqrt{\dfrac{2}{\pi}}\dfrac{1}{\omega^2}\left[\sin(a\omega) - a\omega\cos(a\omega)\right]$
14 $e^{-a|x|}$ $(a > 0)$	$\sqrt{\dfrac{2}{\pi}}\dfrac{a}{a^2 + \omega^2}$
15 $e^{-(x+b)^2/(4a)} + e^{-(x-b)^2/(4a)}$	$2\sqrt{2a}\,e^{-a\omega^2}\cos(b\omega)$
16 $\mathrm{erf}(ax)$	$i\sqrt{\dfrac{2}{\pi}}\dfrac{1}{\omega} e^{-\omega^2/(4a^2)}$

A.4 Table of Fourier Sine Transforms

$f(x) = \mathcal{F}_S^{-1}[F](x)$	$F(\omega) = \mathcal{F}_S[f](\omega)$
1 $f'(x)$	$-\omega \mathcal{F}_C[f](\omega)$
2 $f''(x)$	$\sqrt{\dfrac{2}{\pi}}\,\omega f(0) - \omega^2 F(\omega)$
3 $f(ax)\quad(a>0)$	$\dfrac{1}{a} F(\omega/a)$
4 $f(ax)\cos(bx)\quad(a,\,b>0)$	$\dfrac{1}{2a}\left[F\left(\dfrac{\omega+b}{a}\right) + F\left(\dfrac{\omega-b}{a}\right)\right]$
5 1	$\sqrt{\dfrac{2}{\pi}}\dfrac{1}{\omega}$
6 $e^{-ax}\quad(a>0)$	$\sqrt{\dfrac{2}{\pi}}\dfrac{\omega}{a^2+\omega^2}$
7 $xe^{-ax}\quad(a>0)$	$\sqrt{\dfrac{2}{\pi}}\dfrac{2a\omega}{(a^2+\omega^2)^2}$
8 $x^2 e^{-ax}\quad(a>0)$	$2\sqrt{\dfrac{2}{\pi}}\dfrac{3a^2\omega - \omega^3}{(a^2+\omega^2)^3}$
9 $\dfrac{x}{x^2+a^2}$	$\sqrt{\dfrac{\pi}{2}}\,e^{-a\omega}$
10 $H(a-x) = \begin{cases}1, & 0\le x\le a\\ 0, & x>a\end{cases}$	$\sqrt{\dfrac{2}{\pi}}\dfrac{1}{\omega}\left[1-\cos(a\omega)\right]$
11 $xH(a-|x|) = \begin{cases}x, & |x|\le a\\ 0, & |x|>a\end{cases}$	$\sqrt{\dfrac{2}{\pi}}\dfrac{1}{\omega^2}\left[\sin(a\omega) - a\omega\cos(a\omega)\right]$
12 $\operatorname{erfc}(ax)\quad(a>0)$	$\sqrt{\dfrac{2}{\pi}}\dfrac{1}{\omega}\left[1-e^{-\omega^2/(4a^2)}\right]$
13 $xe^{-a^2x^2}$	$\dfrac{1}{2\sqrt{2}}\dfrac{1}{a^3}\,\omega e^{-\omega^2/(4a^2)}$
14 $\tan^{-1}(x/a)\quad(a>0)$	$\sqrt{\dfrac{\pi}{2}}\dfrac{1}{\omega}e^{-a\omega}$

A.5 Table of Fourier Cosine Transforms

$f(x) = \mathcal{F}_C^{-1}[F](x)$	$F(\omega) = \mathcal{F}_C[f](\omega)$
1 $f'(x)$	$-\sqrt{\dfrac{2}{\pi}}\, f(0) + \omega \mathcal{F}_S[f](\omega)$
2 $f''(x)$	$-\sqrt{\dfrac{2}{\pi}}\, f'(0) - \omega^2 F(\omega)$
3 $f(ax) \quad (a > 0)$	$\dfrac{1}{a} F(\omega/a)$
4 $f(ax)\cos(bx) \quad (a,\, b > 0)$	$\dfrac{1}{2a}\left[F\left(\dfrac{\omega+b}{a}\right) + F\left(\dfrac{\omega-b}{a}\right)\right]$
5 $e^{-ax} \quad (a > 0)$	$\sqrt{\dfrac{2}{\pi}} \dfrac{a}{a^2 + \omega^2}$
6 $xe^{-ax} \quad (a > 0)$	$\sqrt{\dfrac{2}{\pi}} \dfrac{a^2 - \omega^2}{(a^2 + \omega^2)^2}$
7 $x^2 e^{-ax} \quad (a > 0)$	$2\sqrt{\dfrac{2}{\pi}} \dfrac{a^3 - 3a\omega^2}{(a^2 + \omega^2)^3}$
8 $e^{-a^2 x^2}$	$\dfrac{1}{\sqrt{2}} \dfrac{1}{|a|} e^{-\omega^2/(4a^2)}$
9 $\dfrac{1}{x^2 + a^2} \quad (a > 0)$	$\sqrt{\dfrac{\pi}{2}} \dfrac{1}{a} e^{-a\omega}$
10 $\dfrac{1}{(a^2 + x^2)^3} \quad (a > 0)$	$\sqrt{\dfrac{\pi}{2}} \dfrac{1}{8a^5} (a^2\omega^2 + 3a\omega + 3)e^{-a\omega}$
11 $\dfrac{x^2}{(a^2 + x^2)^3} \quad (a > 0)$	$\sqrt{\dfrac{\pi}{2}} \dfrac{1}{8a^3} (-a^2\omega^2 + a\omega + 1)e^{-a\omega}$
12 $\dfrac{x^4}{(a^2 + x^2)^3} \quad (a > 0)$	$\sqrt{\dfrac{\pi}{2}} \dfrac{1}{8a} (a^2\omega^2 - 5a\omega + 3)e^{-a\omega}$
13 $H(a - x) = \begin{cases} 1, & 0 \le x \le a \\ 0, & x > a \end{cases}$	$\sqrt{\dfrac{2}{\pi}} \dfrac{1}{\omega} \sin(a\omega)$
14 $\begin{cases} (1/b)e^{-bx}\cosh(ab), & x \ge a \\ (1/b)e^{-ab}\cosh(bx), & x < a \end{cases}$ $\qquad\qquad (a,\, b > 0)$	$\sqrt{\dfrac{2}{\pi}} \dfrac{\cos(a\omega)}{b^2 + \omega^2}$

A.6   Table of Laplace Transforms

$f(t) = \mathcal{L}^{-1}[F](t)$	$F(s) = \mathcal{L}[f](s)$
1 $f^{(n)}(t)$ (nth derivative)	$s^n F(s) - s^{n-1} f(0) - \cdots$ $\quad\quad\quad - f^{(n-1)}(0)$
2 $f(t-a)H(t-a)$	$e^{-as} F(s)$
3 $e^{at} f(t)$	$F(s-a)$
4 $(f * g)(t)$	$F(s)G(s)$
5 1	$\dfrac{1}{s}$ $(s > 0)$
6 t^n (n positive integer)	$\dfrac{n!}{s^{n+1}}$ $(s > 0)$
7 e^{at}	$\dfrac{1}{s-a}$ $(s > a)$
8 $\sin(at)$	$\dfrac{a}{s^2 + a^2}$ $(s > 0)$
9 $\cos(at)$	$\dfrac{s}{s^2 + a^2}$ $(s > 0)$
10 $\sinh(at)$	$\dfrac{a}{s^2 - a^2}$ $(s > |a|)$
11 $\cosh(at)$	$\dfrac{s}{s^2 - a^2}$ $(s > |a|)$
12 $\delta(t-a)$ $(a \geq 0)$	e^{-as}
13 $e^{a^2 t} \operatorname{erfc}\left(a\sqrt{t}\right)$ $(a > 0)$	$\dfrac{1}{s + a\sqrt{s}}$
14 $\dfrac{a}{2\sqrt{\pi}} t^{-3/2} e^{-a^2/(4t)}$ $(a > 0)$	$e^{-a\sqrt{s}}$
15 $\operatorname{erfc}\left(\dfrac{a}{2\sqrt{t}}\right)$ $(a > 0)$	$\dfrac{1}{s} e^{-a\sqrt{s}}$
16 $-a\sqrt{\dfrac{t}{\pi}} e^{-a^2/(4t)} + \left(\tfrac{1}{2} a^2 + t\right) \operatorname{erfc} \dfrac{a}{2\sqrt{t}}$ $\quad\quad (a > 0)$	$\dfrac{1}{s^2} e^{-a\sqrt{s}}$
17 $\dfrac{1}{\sqrt{t}}$	$\sqrt{\dfrac{\pi}{s}}$

A.7 Second-Order Linear Equations

Let

$$Au_{xx} + Bu_{xy} + Cu_{yy} + Du_x + Eu_y + Fu = G.$$

If new variables

$$r = r(x, y), \quad s = s(x, y)$$

are defined by means of the characteristic equations

$$\frac{dy}{dx} = \frac{B - \sqrt{B^2 - 4AC}}{2A},$$

$$\frac{dy}{dx} = \frac{B + \sqrt{B^2 - 4AC}}{2A},$$

and if

$$u(x, y) = u\big(x(r, s), y(r, s)\big) = v(r, s),$$

then

$$\bar{A}v_{rr} + \bar{B}v_{rs} + \bar{C}v_{ss} + \bar{D}v_r + \bar{E}v_s + \bar{F}v = \bar{G},$$

where

$$\bar{A} = A(r_x)^2 + Br_x r_y + C(r_y)^2,$$

$$\bar{B} = 2Ar_x s_x + B(r_x s_y + r_y s_x) + 2Cr_y s_y,$$

$$\bar{C} = A(s_x)^2 + Bs_x s_y + C(s_y)^2,$$

$$\bar{D} = Ar_{xx} + Br_{xy} + Cr_{yy} + Dr_x + Er_y,$$

$$\bar{E} = As_{xx} + Bs_{xy} + Cs_{yy} + Ds_x + Es_y,$$

$$\bar{F} = F,$$

$$\bar{G} = G.$$

Further Reading

Andrews, L.C.: Elementary Partial Differential Equations with Boundary Value Problems. Academic Press, Orlando (1986)

Bleecker, D.D., Csordas, G.: Basic partial Differential Equations. International Press of Boston, Boston, MA (1997)

Brown, J.W., Churchill, R.V.: Fourier Series and Boundary Value Problems, 6th ed. McGraw-Hill, New York (2000)

Davis, J.M.: Introduction to Applied Partial Differential Equations. W.H. Freeman, New York, NY (2012)

DuChateau, P., Zachmann, D.: Applied Partial Differential Equations. Dover Publications, New York, NY (2002)

Duffy, D.G.: Solutions of Partial Differential Equations. TAB Books, Blue Ridger Summit, PA (1986)

Evans, L.C.: Partial Differential Equations, 2nd ed. American Mathematical Society, Providence, RI (2010)

Farlow, S.J.: Partial Differential Equations for Scientists and Engineers. Dover Publications, New York, NY (1993)

Garabedian, P.R.: Partial Differential Equations, 2nd ed. American Mathematical Society, Providence, RI (1998)

Gustafson, K.E.: Introduction to Partial Differential Equations and Hilbert Space Methods, 3rd ed. Dover Publications, New York, NY (1997)

Haberman, R.: Applied Partial Differential Equations with Fourier Series and Boundary Value Problems, 5th ed. Pearson (2012)

Keane, M.K.: A Very Applied First Course in Partial Differential Equations. Prentice Hall, Upper Saddle River, NJ (2001)

Kevorkian, J., Cole, J.D.: Perturbation Methods in Applied Mathematics. Springer, New York, NY (1985)

Kovach, L.D.: Boundary–Value Problems. Addison–Wesley, Reading, MA (1984)

Leach, J.A., Needham, D.J.: Matched Asymptotic Expansions in Reaction–Diffusion Theory. Springer, London (2003)

Myint-U, T., Debnath, L.: Linear Partial Differential Equations for Scientists and Engineers, 4th ed. Birkhäuser, Boston, MA (2006)

Nayfeh, A.H.: Introduction to Perturbation Techniques. Wiley, New York, NY (1993)

Shearer, M., Levy, R.: Partial Differential Equations: An Introduction to Theory and Applications. Princeton University Press, Princeton, NJ (2015)

Stakgold, I., Holst, M.J.: Green's Functions and Boundary Value Problems, 3rd ed. Wiley, New York, NY (2011)

Strauss, W.A.: Partial Differential Equations: An Introduction. Wiley, New York, NY (2007)

Taylor, M.E.: Partial Differential Equations, vols. I–III, 2nd ed. Springer, New York, NY (2013)

Vvedensky, D.: Partial Differential Equations with Mathematica. Addison–Wesley, Reading, MA (1993)

Wan, F.Y.M.: Mathematical Models and Their Analysis. Harper & Row, New York, NY (1989)

Weinberger, H.F.: A First Course in Partial Differential Equations: with Complex Variables and Transform Methods. Dover Publications, New York, NY (1995)

Zachmanoglou, E.C., Thoe, D.W.: Introduction to Partial Differential Equations with Applications. Dover Publications, New York, NY (1987)

Zauderer, E.: Partial Differential Equations of Applied Mathematics, 3rd ed. Wiley-Interscience, Hoboken, NJ (2006)

Index

Printed in the United States
by Baker & Taylor Publisher Services